DIMACS
Series in Discrete Mathematics
and Theoretical Computer Science

Volume 64

Unusual Applications of Number Theory

DIMACS Workshop
Unusual Applications of Number Theory
January 10–14, 2000
DIMACS Center

Melvyn B. Nathanson
Editor

Center for Discrete Mathematics
and Theoretical Computer Science
A consortium of Rutgers University, Princeton University,
AT&T Labs–Research, Bell Labs (Lucent Technologies),
NEC Laboratories America, and Telcordia Technologies
(with partners at Avaya Labs, IBM Research, and Microsoft Research)

American Mathematical Society

This DIMACS volume presents the proceedings from the workshop Unusual Applications of Number Theory held at the DIMACS Center, Rutgers University, January 10–14, 2000.

2000 *Mathematics Subject Classification*. Primary 11Z05.

Library of Congress Cataloging-in-Publication Data

DIMACS Workshop on Unusual Applications of Number Theory (2000 : Piscataway, N.J.)
Unusual applications of number theory : DIMACS workshop, unusual applications of number theory, January 10–14, 2000, DIMACS Center / Melvyn B. Nathanson, editor.
p. cm. — (DIMACS series is discrete mathematics and theoretical computer science ; v. 64)
ISBN 0-8218-2703-0 (alk. paper)
1. Number theory–Congresses. I. Nathanson, Melvyn B. (Melvyn Bernard), 1944– II. Title. III. Series.

QA241.D56 2000
512.7–dc22 2004043691

∞ The paper used in this book is acid-free and falls within the guidelines
established to ensure permanence and durability.
Visit the AMS home page at http://www.ams.org/

10 9 8 7 6 5 4 3 2 1 09 08 07 06 05 04

Contents

Foreword

A workshop on Unusual Applications of Number Theory was held at Rutgers University on January 10-14, 2000. We would like to express our appreciation to George Andrews, David Chudnovsky, Ron Graham, Jeff Lagarias, Victor Miller, Mel Nathanson, Andrew Odlyzko, and Carl Pomerance for their efforts to organize and plan this successful conference.

Number theory is an old and fascinating topic of great interest in and of itself, but one with a remarkably large number of applications in and connections to other areas of research, many of them not so well known. This workshop was centered around applications and interactions of number theory with other areas of science and of mathematics. Topics explored included physics, combinatorics, ergodic theory, spectral theory, geometry, and numerical analysis, as well as the more "standard" topics of computer science and cryptography.

DIMACS gratefully acknowledges the generous support that makes these programs possible. The National Science Foundation and the New Jersey Commission on Science and Technology, DIMACS' partners at Rutgers, Princeton, AT&T Labs-Research, Bell Labs, NEC Laboratories America, and Telcordia Technologies, and its affiliated partners at Avaya Labs, HP Labs, IBM Research, and Microsoft Research have generously supported the workshop program at DIMACS that makes this type of workshop possible.

Fred S. Roberts
Director

Robert Tarjan
Co-Director for Princeton

Preface

These are the proceedings of the DIMACS Workshop on Unusual Applications of Number Theory, which was held at the DIMACS Center of Rutgers University in Piscataway, New Jersey, on January 10–14, 2000. The usual applications of number theory are to computer science and cryptology. In this workshop, the organizers looked for other applications of number theory, as well as topics in number theory with potential applications in science and engineering.

The organizing committee of this workshop was Mel Nathanson (Chair), George Andrews, David Chudnovsky, Ron Graham, Jeff Lagarias, Victor Miller, Andrew Odlyzko, and Carl Pomerance. I thank the members of this committee for their help in suggesting and inviting speakers to this meeting.

It was a pleasure to work with DIMACS Director Fred Roberts and his staff on this workshop. I am especially grateful to Mel Janowitz of DIMACS, who provided crucial and efficient editorial assistance with these proceedings, and who made sure that the LaTex files of the papers in this volume satisfied AMS publication requirements.

Partial support for this conference was provided by the National Science Foundation and the New Jersey Commission on Science and Technology.

Melvyn B. Nathanson
Short Hills, New Jersey
January 16, 2004

DIMACS Series in Discrete Mathematics
and Theoretical Computer Science
Volume **64**, 2004

A four parameter generalization of Göllnitz's (big) partition theorem

K. Alladi, G. E. Andrews, and A. Berkovich

Dedicated to our friend Barry McCoy on his sixtieth birthday

ABSTRACT. We state a new four parameter partition theorem from which the (big) theorem of Göllnitz follows by setting any one of the parameters equal to 0. This settles a problem of Andrews who asked whether there exists a result that goes beyond the partition theorem of Göllnitz. We state a four parameter q-series identity (key identity) which is the generating function form of this theorem. In a subsequent paper, the proof of the new four parameter key identity will be given.

1. Introduction

Our purpose here is to announce the following new partition theorem:

THEOREM 1. *Let $P(n)$ denote the number of partitions of n into distinct parts $\equiv -2^3, -2^2, -2^1, -2^0$ (mod 15).*

Let $G(n)$ denote the number of partitions of n into parts $\not\equiv 2^0, 2^1, 2^2, 2^3$ (mod 15) such that the difference between the non-multiples of 15 is ≥ 15 with equality only if a part is relatively prime to 15, parts which are not relatively prime to 15 are > 15, the difference between the multiples of 15 is ≥ 60, and the smallest multiple of 15 is

$$\geq 30 + 30\tau, \text{ if 7 is a part, and}$$
$$\geq 45 + 30\tau, \text{ otherwise,}$$

where τ is the number of non-multiples of 15 in the partition. Then

$$G(n) = P(n).$$

While it is obvious that this result is a partition theorem of the Schur-Göllnitz type, it is not clear that it lies beyond the (big) theorem of Göllnitz [15]. It is possible to obtain a four parameter refinement of Theorem 1; this is stated as

2000 *Mathematics Subject Classification.* Primary 05A15,05A19,11P81, 11P83.
Key words and phrases. partitions, Göllnitz theorem, four parameter key identity, q-series.
Research of the first and third authors was supported in part by a grant from the Number Theory Foundation and of the first two authors by grants from the National Science Foundation.

Theorem 2 in §2. From this it follows that Theorem 1 (and Theorem 2) generalize the Göllnitz theorem in the same sense that Göllnitz's theorem extends Schur's 1926 partition theorem [16]. Thus the question raised by Andrews [12] nearly 30 years ago whether there exists a partition theorem that goes beyond the (big) theorem of Göllnitz, is now answered in the affirmative.

Our purpose here is only to state the new results and describe how they extend Göllnitz's theorem. The proof of Theorem 2 (and consequently of Theorem 1) is given in full in [6].

In 1995, Alladi, Andrews, and Gordon [5] obtained a three parameter refinement of Göllnitz's theorem by the use of colored partitions. Because colors were labelled by letters and the integers occurring in the colors were indicated by subscripts (weights), this approach was called *the method of weighted words.* In §2 we will describe an extension of this method that leads to the new four parameter Theorem 2. In doing so some essentially new ideas are required. In §2 we will also state the generating function form of Theorem 2 which we call a four parameter *key identity.* By setting any one of the parameters equal to 0 in Theorem 2, we get the three parameter refinement of Göllnitz's theorem due to Alladi-Andrews-Gordon (see §3). Similarly by setting any one of the parameters equal to 0 in the new key identity (2.6), the three parameter key identity for Göllnitz's theorem in [5] falls out. Finally in §4 we conclude with a brief description of some problems for future research opened up by Theorem 2.

The two parameter key identity for Schur's theorem due to Alladi and Gordon [9] is essentially equivalent to the $q-$Chu-Vandermonde summation as shown by Alladi and Berkovich [8]. The three parameter key identity for Göllnitz's theorem due to Alladi, Andrews, and Gordon [5] that extends the identity in [9] is substantially deeper, and its proof utilizes either the ${}_6\psi_6$ summation of Bailey as in [5], or Jackson's $q-$analog of Dougall's summation as in [4]. The proof of the new four parameter key identity (2.6) also relies on the ${}_6\psi_6$ summation, but requires several new ideas and is significantly deeper than the proofs in [4] and [5]. That is why the proof of the new four parameter identity is presented separately [6].

2. Colored reformulation and a four parameter refinement

We consider the integer 1 occurring in four primatry colors $\mathbb{A}, \mathbb{B}, \mathbb{C}$, and $\mathbb{D}$, and integers $n \geq 2$ occurring in these four primary colors as well as in the six secondary colors $\mathbb{AB}, \mathbb{AC}, \mathbb{AD}, \mathbb{BC}, \mathbb{BD}$, and $\mathbb{CD}$. We assume that integers $n \geq 4$ occur in the quaternary color $\mathbb{ABCD}$ in addition to the ten colors above. The crucial thing is that we discard all ternary colors $\mathbb{ABC}, \mathbb{ABD}, \mathbb{ACD}$, and $\mathbb{BCD}$.

The integer n in color $\mathbb{A}$ is denoted by the symbol $\mathbb{A}_n$ with similar interpretation for $\mathbb{B}_n, \ldots, \mathbb{CD}_n, \mathbb{ABCD}_n$. In order to discuss colored partitions, we need an ordering among the symbols, and so we assume that

$$(2.1)\quad \begin{cases} \text{if } m < n \text{ as ordinary (uncolored) integers,} \\ \text{then } m \text{ in any color} < n \text{ in any color, and} \\ \text{if two equal integers appear in different colors, the order is given by} \\ \mathbb{ABCD} < \mathbb{AB} < \mathbb{AC} < \mathbb{AD} < \mathbb{A} < \mathbb{BC} < \mathbb{BD} < \mathbb{B} < \mathbb{CD} < \mathbb{C} < \mathbb{D}. \end{cases}$$

Next, given any partition into parts occurring in the eleven colors above, we denote by a the number of parts in color $\mathbb{A}$ (the frequency of $\mathbb{A}$), with similar interpretation for b, c, and d. Pursuing the same notation, ab will denote the number of parts in color $\mathbb{AB}$, with similar interpretation for $ac, \ldots, cd$. Note that ab is not a times b! Finally, Q denotes the number of parts in color $\mathbb{ABCD}$.

We are now in a position to state our main result..

THEOREM 2. *Let i, j, k, l be given nonnegative integers.*

Let $P(n; i, j, k, l)$ denote the number of partitions of n into parts occurring in the four primary colors, parts in the same color being distinct, and with i parts in color $\mathbb{A}$, j parts in color $\mathbb{B}$, k parts in color $\mathbb{C}$, and l parts in color $\mathbb{D}$.

Let $G(n; a, b, c, d, ab, \ldots, cd, Q)$ denote the number of partitions of n into colored parts occurring in the frequencies as indicated, and such that the difference between the nonquaternary parts is ≥ 1, with equality only if parts are either of the same primary color, or if the larger part occurs in a color of higher order as given in (2.1), and the gap between the quaternary parts is ≥ 4, with the added condition that the least quaternary part is

$$\begin{aligned} &\geq 3 + 2\tau, \quad \text{if } \mathbb{A}_1 \text{ is a part,} \\ &\geq 4 + 2\tau, \quad \text{otherwise,} \end{aligned} \tag{2.2}$$

where τ is the number of nonquaternary parts. Then

$$P(n; i, j, k, l) = \sum_{constraints} G(n; a, b, c, d, ab, \ldots, cd, Q). \tag{2.3}$$

where the summation is over the variables $a, b, \ldots, cd, Q$ satisfying the constraints

$$\begin{aligned} i &= a + ab + ac + ad + Q \\ j &= b + ab + bc + bd + Q \\ k &= c + ac + bc + cd + Q \\ l &= d + ad + bd + cd + Q. \end{aligned} \tag{2.4}$$

A strong four parameter refinement of Theorem 1 follows from Theorem 2 upon replacing

$$\left\{ \begin{array}{l} \mathbb{A}_n \mapsto 15n - 8, \mathbb{B}_n \mapsto 15n - 4, \mathbb{C}_n \mapsto 15n - 2, \mathbb{D}_n \mapsto 15n - 1, \text{ for } n \geq 1, \\ \text{and consequently } \mathbb{AB}_n \mapsto 15n - 12, \mathbb{AC}_n \mapsto 15n - 10, \mathbb{AD}_n \mapsto 15n - 9, \\ \mathbb{BC}_n \mapsto 15n - 6, \mathbb{BD}_n \mapsto 15n - 5, \mathbb{CD}_n \mapsto 15n - 3, \text{ for } n \geq 2, \\ \text{and } \mathbb{ABCD}_n \mapsto 15n - 15, \text{ for } n \geq 4. \end{array} \right. \tag{2.5}$$

The nice thing about the substitutions (2.5) is that the ordering (2.1) becomes

$$7 < 11 < 13 < 14 < 18 < 20 < 21 < 22 < 24 < 25 < 26 < 27 < 28 < 29 < 33 <, \ldots$$

the natural ordering among the integers $\not\equiv 2^0, 2^1, 2^2, 2^3 \pmod{15}$. These substitutions imply that the primary colors correspond to the residue classes

$$-2^3, -2^2, -2^1, -2^0 (\text{mod} \quad 15).$$

Since

$$2^0 + 2^1 + 2^2 + 2^3 = 15,$$

the ternary colors correspond to the residue classses

$$2^0, 2^1, 2^2, 2^3 (\text{mod} \quad 15).$$

These are the four residue classes not considered in Theorem 1. Also, since the residue classes relatively prime to 15 are $2^0, 2^1, 2^2, 2^3, -2^3, -2^2, -2^1, -2^0 \pmod{15}$, it follows that the secondary colors correspond to the nonmultiples of 15 which are not relatively prime to 15. Finally, the quaternary color corresponds to the multiples of 15 which are ≥ 45. These features make Theorem 1 particularly appealing.

The generating function form of Theorem 2 is the following remarkable four parameter *key identity*: If $T_n = n(n+1)/2$, and $\tau = a+b+c+d+ab+\cdots+cd$, then

$$\sum_{\text{constraints}} \frac{q^{T_\tau + T_{ab}+\cdots+T_{cd}-bc-bd-cd+4T_{Q-1}+3Q+2Q\tau}}{(q)_a(q)_b(q)_c(q)_d(q)_{ab}\cdots(q)_{cd}(q)_Q} \times \tag{2.6}$$
$$\left\{(1-q^a)+q^{a+bc+bd+Q}(1-q^b)+q^{a+bc+bd+Q+b+cd}\right\}$$
$$= \frac{q^{T_i+T_j+T_k+T_l}}{(q)_i(q)_j(q)_k(q)_l},$$

where the constraints are as in (2.4) and the summation is over $a, b, \ldots, Q$. In (2.6) we have made use of the standard notation

$$(A)_n = (A;q)_n = \begin{cases} \prod_{j=0}^{n-1}(1-Aq^j), & if \quad n>0, \\ 1, & if \quad n=0, \\ \prod_{j=1}^{-n}(1-Aq^{-j})^{-1}, & if \quad n<0. \end{cases}$$

In [6] we do not prove Theorem 2 combinatorially. Instead we prove (2.6) using $q-$series techniques and show that (2.6) is equivalent to Theorem 2.

3. Reduction to Göllnitz and Schur

If any one of the parameters i, j, k, l, is set equal to 0, Theorem 2 reduces to Theorem 2 of [5], the three parameter refinement of the colored version of Göllnitz's theorem. Note that in this case, the quaternary color does not occur at all. We state the result with $l = 0$.

THEOREM A. *Let $P(n; i, j, k)$ denote the number of partitions of n into parts occuring in three primary colors $\mathbb{A}, \mathbb{B}, \mathbb{C}$, parts in the same color being distinct, with i parts in color $\mathbb{A}$, j parts in color $\mathbb{B}$, and k parts in color $\mathbb{C}$.*

Let $G(n; a, b, c, ab, ac, bc)$ denote the number of partitions of n into parts occuring in colors $\mathbb{A}, \mathbb{B}, \mathbb{C}, \mathbb{AB}, \mathbb{AC}, \mathbb{BC}$, with indicated frequencies $a, b, ..., bc$, such that the difference between the parts is ≥ 1 with equality only if parts are either of the same primary color, or if the larger part occurs in a color of higher order as indicated in (2.1). Then

$$\sum_{\substack{i=a+ab+ac \\ j=b+ab+bc \\ k=c+ac+bc}} G(n; a, b, c, ab, ac, bc) = P(n; i, j, k).$$

In Theorem A, if we use the substitutions

$$\begin{cases} \mathbb{A}_n \mapsto 6n-4, \mathbb{B}_n \mapsto 6n-2, \mathbb{C}_n \mapsto 6n-1, \text{ for } n \geq 1, \\ \text{yielding } \quad \mathbb{AB} \mapsto 6n-6, \mathbb{AC}_n \mapsto 6n-5, \mathbb{BC}_n \mapsto 6n-3, \text{ for } n \geq 2, \end{cases} \tag{3.1}$$

we get a three parameter refinement of the following theorem of Göllnitz [15].

THEOREM G. *Let $P(n)$ denote the number of partitions of n into distinct parts $\equiv -2^2, -2^1, -2^0 \pmod 6$.*

Let $G(n)$ denote the number of partitions of n into parts $\neq 1$ or 3, such that the difference between the parts ≥ 6, with equality only if a part is $\equiv -2^2, -2^1, -2^0 \pmod 6$. Then

$$G(n) = P(n).$$

If any one of the parameters i, j, k, l is set equal to 0, then (2.6) reduces to the three parameter key identity for Göllnitz's theorem in [5]. For example, with $l = 0$, (2.6) reduces to

$$\sum_{\substack{i=a+ab+ac\\ j=b+ab+bc\\ k=c+ac+bc}} \frac{q^{T_\tau+T_{ab}+T_{ac}+T_{bc-1}}\{1-q^a+q^{a+bc}\}}{(q)_a(q)_b(q)_c(q)_{ab}(q)_{ac}(q)_{bc}} = \frac{q^{T_i+T_j+T_k}}{(q)_i(q)_j(q)_k}. \tag{3.2}$$

Here $\tau = a + b + c + ab + ac + bc$. If we further set one of i, j, k equal to 0, say $k = 0$, in (3.2), then we get the two parameter key identity for the colored version of Schur's theorem due to Alladi and Gordon [9], namely,

$$\sum_{\substack{i=a+ab\\ j=b+ab}} \frac{q^{T_{a+b+ab}+T_{ab}}}{(q)_a(q)_b(q)_{ab}} = \frac{q^{T_i+T_j}}{(q)_i(q)_j} \tag{3.3}$$

In extending the refined Schur theorem in [9] to the refined Göllnitz theorem in [5], the statement of the extension was routine once the theorems were phrased in the language of primary and secondary colors. The principal reason for the increase in difficulty in going up from the Schur theorem to the Göllnitz theorem is because in the lexicographic ordering, one of the secondary colors $\mathbb{BC}$ is of higher order than the primary color $\mathbb{A}$. In addition, even though the refined Göllnitz theorem uses three primary colors, the ternary color $\mathbb{ABC}$ is dropped and so only a proper subset of the complete alphabet of colors is used. Thus the Göllnitz theorem is an extension of Schur's theorem in a direction different from the one taken by Andrews [10], [11], who in retrospect used a complete alphabet of colors. In going beyond the Göllnitz theorem to Theorem 2, there is a significant increase in depth and complexity for a variety of reasons. The quaternary color enters in a rather unusual way - the quaternaries do not directly interact with primaries and secondaries. The only interaction between quaternaries and the other colors is through the lower bound imposed on the quaternaries. This, combined with the uncertainity of selecting a proper subset of colors from the complete alphabet of four primaries, six secondaries, four ternaries, and one quaternary, was perhaps the reason that the solution to the problem of Andrews [12] remained elusive for so long.

4. Problems for investigation

If the expressions in (2.6) are multiplied of $A^iB^jC^kD^l$ and summed over i, j, k, l, we get on the right hand side the quadruple infinite product

$$\prod_{m=1}^{\infty}(1+Aq^m)(1+Bq^m)(1+Cq^m)(1+Dq^m), \tag{4.1}$$

with four free parameters A, B, C, D. This opens up several avenues of exploration a few of which we briefly indicate here.

Theorem 2 may be viewed as a base level undilated version of Theorem 1. More precisely, the generating function form of Theorem 1 may be viewed as emerging out of (2.6) and (4.1) under the transformation

$$\begin{cases} \text{(dilation) } q \mapsto q^{15}, \\ \text{(translations) } A \mapsto Aq^{-8}, B \mapsto Bq^{-4}, C \mapsto Cq^{-2}, D \mapsto Dq^{-1}. \end{cases} \tag{4.2}$$

The size of the modulus 15 and the choice of the translations involving powers of 2 ensures that the colors in Theorem 2 translate into distinct residue classes mod 15. If a dilation smaller than $q \mapsto q^{15}$ is used, then the residue classes would overlap, and so we would be counting parts with weights attached. Alladi [1] has studied weighted partition identities in general and discussed certain interesting reformulations of Göllnitz's theorem and their applications [2], [3], emerging out of *small* dilations of (3.2). In a similar spirit it would be worthwhile to study weighted partition theorems emerging out of (2.6) by the use of dilations $q \mapsto q^M$, with $M < 15$. We anticipate obtaining new and different versions of partition theorems that have arisen in the study of affine Lie algebras, and representations of symmetric groups, by such weighted reformulations of Theorem 1.

Recently, Alladi and Berkovich [7] have obtained the following double bounded version of (3.3):

$$\sum_{k\geq 0} q^{T_{i+j-k}+T_k} \begin{bmatrix} M-i-j+k \\ k \end{bmatrix} \begin{bmatrix} M-j \\ i-k \end{bmatrix} \begin{bmatrix} L-i \\ j-k \end{bmatrix} = \begin{bmatrix} L \\ j \end{bmatrix} \begin{bmatrix} M-j \\ i \end{bmatrix} q^{T_i+T_j}. \tag{4.3}$$

In (4.3), the symbols $\begin{bmatrix} n+m \\ n \end{bmatrix}$ are the $q-$binomial coefficients given by

$$\begin{bmatrix} n+m \\ n \end{bmatrix}_q = \begin{bmatrix} n+m \\ n \end{bmatrix} = \begin{cases} \dfrac{(q^{m+1})_n}{(q)_n}, & if \quad n \geq 0, \\ 0, & if \quad n < 0. \end{cases}$$

If we let $L, M \to \infty$ in (4.3) and make the identifications $a = i-k$, $b = j-k$, and $ab = k$, then we get (3.3).

In addition, Alladi and Berkovich have also obtained a double bounded version of (3.2), namely,

(4.4)

$$\begin{aligned} &\sum_{constraints} q^{T_\tau + T_{ab} + T_{ac} + T_{bc-1}} \\ &\quad \Bigg\{ q^{bc} \begin{bmatrix} L-\tau+a \\ a \end{bmatrix} \begin{bmatrix} L-\tau+b \\ b \end{bmatrix} \begin{bmatrix} M-\tau+c \\ c \end{bmatrix} \begin{bmatrix} L-\tau \\ ab \end{bmatrix} \begin{bmatrix} M-\tau \\ ac \end{bmatrix} \begin{bmatrix} M-\tau \\ bc \end{bmatrix} \\ &\quad + \begin{bmatrix} L-\tau+a-1 \\ a-1 \end{bmatrix} \begin{bmatrix} L-\tau+b \\ b \end{bmatrix} \begin{bmatrix} M-\tau+c \\ c \end{bmatrix} \begin{bmatrix} L-\tau \\ ab \end{bmatrix} \begin{bmatrix} M-\tau \\ ac \end{bmatrix} \begin{bmatrix} M-\tau \\ bc-1 \end{bmatrix} \Bigg\} \\ &= \sum_{t\geq 0} q^{t(M+2)-T_t+T_{i-t}+T_{j-t}+T_{k-t}} \begin{bmatrix} L-t \\ t \end{bmatrix} \begin{bmatrix} L-2t \\ i-t \end{bmatrix} \begin{bmatrix} L-i-t \\ j-t \end{bmatrix} \begin{bmatrix} M-i-j \\ k-t \end{bmatrix}, \end{aligned}$$

where the constraints are as in (3.2) and $\tau = a+b+c+ab+ac+bc$. The proof (4.4), which is quite intricate, is given in [8]. If we let $L, M \to \infty$, then (3.2) follows because only the term corresponding to $t = 0$ on the right hand side in (4.4) makes a contribution. Based on the discovery of (4.4), we now ask whether there exists a finite bounded version of (2.6) (which reduces to (2.6) when certain parameters tend to infinity).

In the last decade, many new generalizations of the Rogers-Ramanujan identities were discovered and proved by McCoy and collaborators (see [14] for a review and references), using the so called thermodynamic Bethe ansatz (TBA) techniques. It would be highly desirable to find a TBA interpretation of the new identity (2.6). Such an interpretation, besides being of substantial interest in physics, may provide insight into how to extend Theorem 2 to five or more primary colors.

Acknowledgements

K.A. and A.B. would like to thank Carl Pomerance for support and encouragement, and Mel Nathanson and the other organizers of the DIMACS conference for the invitation to present this work.

References

[1] K. Alladi, Partition identities involving gaps and weights, *Trans. Amer. Math. Soc.* **349** (1997), 5001-5019.

[2] K. Alladi, A combinatorial correspondence related to Göllnitz's big partition theorem and applications, *Trans. Amer. Math. Soc.* **349** (1997), 2721-2735.

[3] K. Alladi, On a partition theorem of Göllnitz and quartic transformations, (with an appendix by B. Gordon), *J. Num. Th.* **69** (1998), 153-180.

[4] K. Alladi and G.E. Andrews, A quartic key identity for a partition theorem of Göllnitz, *J. Num. Th.* **75** (1999), 220-236.

[5] K. Alladi, G. E. Andrews, and B. Gordon, Generalizations and refinements of a partition theorem of Göllnitz, *J. Reine Angew. Math.* **460** (1995), 165-188.

[6] K. Alladi, G. E. Andrews, and A. Berkovich, A new four parameter $q-$series identity and its partition implications, *Invent. Math.* **153** (2003), 231-260.

[7] K. Alladi and A. Berkovich, A double bounded version of Schur's partition theorem, *Combinatorica - Erdös memorial issue*, **22** (2002), 151-168.

[8] K. Alladi and A. Berkovich, A double bounded key identity for a partition theorem of Göllnitz, in *Symbolic Computation, Number Theory, Special Functions, Physics, and Combinatorics* (Frank Garvan and Mourad Ismail Eds.), *Developments in Math.*, Kluwer Academic Publishers, **5** (2001), 13-32.

[9] K. Alladi and B. Gordon, Generalizations of Schur's partition theorem, *Manus. Math.* **79** (1993), 113-126.

[10] G. E. Andrews, A new generalization of Schur's second partition theorem, *Acta Arithmetica* **14** (1968), 429-434.

[11] G. E. Andrews, A general partition theorem with difference conditions, *Amer. J. Math.* **91** (1969), 18-24.

[12] G. E. Andrews, The use of computers in the search of identities of Rogers-Ramanujan type, in (A.O.L. Atkin and B. J. Birch eds.), *Computers in Number Theory*, Academic Press, 1971, 377-387.

[13] G. E. Andrews, The theory of partitions, *Encyclopedia of Math. and its Appl, Vol. 2,*, Addison Wesley, Reading, 1976.

[14] A. Berkovich, B. M. McCoy, and A. Schilling, Rogers-Schur-Ramanujan type identities for the $M(P, P')$ minimal models of conformal field theory, *Comm. Math. Phys.,* **191** (1998), 325-395.

[15] H. Göllnitz, Partitionen mit Differenzenbedingungen, *J. Reine Angew. Math* **225** (1967), 154-190.

[16] I. Schur, Zur Additiven Zahlentheorie, *Gesammelte Abhandlungen, Vol. 2*, Springer, 1973, 43-50.

Department of Mathematics, University of Florida, Gainesville, FL 32611
E-mail address: `alladi@math.ufl.edu`

Department of Mathematics, The Pennsylvania State University, University Park, PA 16802
E-mail address: `andrews@math.psu.edu`

Department of Mathematics, University of Florida, Gainesville, FL 32611
E-mail address: `alexb@math.ufl.edu`

DIMACS Series in Discrete Mathematics
and Theoretical Computer Science
Volume **64**, 2004

The Friedman-Joichi-Stanton monotonicity conjecture at primes

George E. Andrews

ABSTRACT. The object is to prove that if n is an odd prime, then the number of partitions of N into parts taken from $(n, n+1, \ldots, 2n-1)$ is weakly increasing for $N \geq 1$.

1. Introduction.

In [2] (recently summarized in [6]) Friedman, Joichi and Stanton posed the following:

Conjecture. *Suppose that $n \geq 3$ is an odd integer. If*

$$\frac{1}{(1-q^n)(1-q^{n+1})\ldots(1-q^{2n-1})} = \sum_{m=0}^{\infty} a_m q^m,$$

then a_m is weakly increasing for $m \geq 1$.

From this conjecture, they are able to characterize all sets of positive integers P such that

$$\prod_{n \in P} \frac{1}{1-q^n}$$

has a power series expansion with weakly increasing coefficients past the constant term.

The object here is to prove the following:

Theorem 3. *Suppose that $n \geq 3$ is a prime. If*

$$\frac{1}{(1-q^n)(1-q^{n+1})\ldots(1-q^{2n-1})} = \sum_{m=0}^{\infty} a_m q^m, \tag{1.1}$$

then a_m is weakly increasing for $m \geq 1$.

The proof is divided into two major parts. In Section 2, we shall derive a new identity to represent the product in (1.1). In Section 3, we shall discuss a new set of polynomials that arise in Section 2. In Section 4, we will tie these two results

---

2000 *Mathematics Subject Classification.* 11P81, 05A17, 33D15.

Key words and phrases. partitions, primes, monotonicity.

Partially supported by National Science Foundation Grant DMS-020004.

together to prove Theorem 1. We conclude with comments and a look at open problems.

2. The identity.

To make clear the significance of the identity to be proved in this section, we list its first two instances

$$\frac{(1-q)}{(1-q^3)(1-q^4)(1-q^5)} = \left(1 - q + \frac{q^3}{1-q^5}\right)\frac{1}{(1-q^{12})}, \tag{2.1}$$

and

$$\begin{aligned} &\frac{(1-q)}{(1-q^5)(1-q^6)(1-q^7)(1-q^8)(1-q^9)} \\ &\quad= \left(1 - q + \frac{q^5}{1-q^9} + \frac{q^{12}}{(1-q^6)(1-q^8)(1-q^9)}\right)\frac{1}{(1-q^{35})}. \end{aligned} \tag{2.2}$$

The truth of Theorem 3 in the cases $n = 3$ or 5 is readily deduced from (2.1) and (2.2) respectively.

We shall use the following notation:

$$(A;q)_n = (1-A)(1-Aq)\dots(1-Aq^{n-1}), \tag{2.3}$$

$${}_{r+1}\Theta_r\left(\begin{matrix} a_0, a_1, \dots, a_r; q, t \\ b_1, \dots, b_r \end{matrix}\right) = \sum_{n=0}^{\infty} \frac{(a_0;q)_n(a_1;q)_n\dots(a_r;q)_n t^n}{(q;q)_n(b_1;q)_n\dots(b_r;q)_n}, \tag{2.4}$$

$$\begin{bmatrix} N \\ M \end{bmatrix} = \frac{(q;q)_N}{(q;q)_M(q;q)_{N-M}}, \tag{2.5}$$

$$A(n,j) = \frac{(q^{n-j+1};q)_{j-1}}{(q^2;q)_{j-1}} = \frac{(1-q)}{(1-q^n)}\begin{bmatrix} n \\ j \end{bmatrix}. \tag{2.6}$$

Theorem 1. *For $n \geq 3$ and odd,*

$$\begin{aligned} &\tfrac{(1-q)}{(1-q^n)(1-q^{n+1})\dots(1-q^{2n-1})} \\ &= \tfrac{1}{(1-q^{n(3n-1)/2})}\left(1 - q + \sum_{j=0}^{\frac{n-3}{2}} \tfrac{q^{(j+1)(n+j)}A(n-j-1,j+1)}{(q^{n+1};q)_j(q^{2n-j-1};q)_{j+1}}\right) \end{aligned} \tag{2.7}$$

Proof. Multiplying both sides of (2.7) by $(1-q^{n(3n-1)/2})(q^{n+1};q)_{n-1}/(1-q)$, we see that (2.7) is equivalent to the assertion that

$$\frac{1-q^{n(3n-1)/2}}{1-q^n} = \sum_{j=0}^{\frac{n-1}{2}} \frac{q^{j(n+j-1)}(q^{n+j};q)_{n-2j}(q^{n-2j};q)_j}{(q;q)_j(1-q^{n-2j})}. \tag{2.8}$$

Now the expression on the right equals

$$\begin{aligned}
&\frac{(q;q)_{2n-1}}{(q;q)_n}\ {}_4\Theta_3\left(\begin{matrix} q^{-\frac{n}{2}}, -q^{-\frac{n}{2}}, q^{-\frac{n}{2}+\frac{1}{2}}, -q^{-\frac{n}{2}+\frac{1}{2}}; q, q \\ q^n, q^{-2n+1}, q^{-n+1} \end{matrix}\right) \\
= &\frac{(q;q)_{2n-1}(q^{-\frac{n}{2}};q)_{(\frac{n-1}{2})}(-q^{n+1};q)_{(\frac{n-1}{q})}(-q^{n+\frac{1}{2}};q)_{(\frac{n-1}{2})}}{(q;q)_n(q^{1-n};q)_{(\frac{n-1}{q})}(q^n;q)_{(\frac{n-1}{2})}(q^{(3n+1)/2};q)_{(\frac{n-1}{2})}} \\
&\times {}_4\Theta_3\left(\begin{matrix} q^{-(\frac{n-1}{2})}, q^{1-\frac{n}{2}}, q^{\frac{3n}{2}}, q^{\frac{3n}{2}+\frac{1}{2}}; q, q \\ -q^{n+1}, -q^{n+\frac{1}{2}}, q^{\frac{3}{2}} \end{matrix}\right) \\
&\text{(by [3; eq. (III.16), p. 242])} \\
= &\frac{(q;q)_{2n-1}(q^{-\frac{n}{2}};q)_{(\frac{n-1}{2})}(-q^{n+1};q)_{(\frac{n-1}{2})}(-q^{n+\frac{1}{2}};q)_{(\frac{n-1}{2})}}{(q;q)_n(q^{1-n};q)_{(\frac{n-1}{2})}(q^n;q)_{(\frac{n-1}{2})}(q^{(3n+1)/2};q)_{(\frac{n-1}{2})}} \\
&\times \frac{(1+q^n)(1-q^{\frac{1}{2}})q^{-\frac{1}{2}}}{(1-q^{(3n-1)/2})(1-q^{-\frac{n}{2}})} \sum_{j\geq 0} \frac{(q^{-\frac{n}{2}};q^{\frac{1}{2}})_{2j+1}(q^{(3n-1)/2};q^{\frac{1}{2}})_{2j+1}q^{j+\frac{1}{2}}}{(q^{\frac{1}{2}};q^{\frac{1}{2}})_{2j+1}(-q^n;q^{\frac{1}{2}})_{2j+1}} \\
= &\frac{(q;q)_{2n-1}(q^{-\frac{n}{2}};q)_{(\frac{n-1}{2})}(-q^{n+1};q)_{(\frac{n-1}{2})}(-q^{n+\frac{1}{2}};q)_{(\frac{n-1}{2})}}{(q;q)_n(q^{1-n};q)_{(\frac{n-1}{2})}(q^n;q)_{(\frac{n-1}{2})}(q^{(3n+1)/2};q)_{(\frac{n-1}{2})}} \\
&\times \frac{(1+q^n)(1-q^{\frac{1}{2}})q^{-\frac{1}{2}}}{(1-q^{(3n-1)/2})(1-q^{-\frac{n}{2}})} \sum_{j\geq 0} \frac{(q^{-\frac{n}{2}};q^{\frac{1}{2}})_j(q^{(3n-1)/2};q^{\frac{1}{2}})_j q^{j/2}}{(q^{\frac{1}{2}};q^{\frac{1}{2}})_j(-q^n;q^{\frac{1}{2}})_j}\left(\frac{1-(-1)^j}{2}\right) \\
= &\frac{(q;q)_{2n-1}(q^{-\frac{n}{2}};q)_{(\frac{n-1}{2})}(-q^{n+1};q)_{(\frac{n-1}{2})}(-q^{n+\frac{1}{2}};q)_{(\frac{n-1}{2})}}{2(q;q)_n(q^{1-n};q)_{(\frac{n-1}{2})}(q^n;q)_{(\frac{n-1}{2})}(q^{(3n+1)/2};q)_{(\frac{n-1}{2})}} \\
&\times \frac{(1+q^n)(1-q^{\frac{1}{2}})q^{-\frac{1}{2}}}{(1-q^{(3n-1)/2})(1-q^{-\frac{1}{2}})}\left(\frac{q^{(3n-1)n/2}(-q^{(1-n)/2};q^{\frac{1}{2}})_n}{(-q^n;q)_n} - \frac{(-q^{(1-n)/2};q^{\frac{1}{2}})_n}{(-q^n;q)_n}\right) \\
&\text{(by [3; eq.(II.6), p.236] applied to the first} \\
&\text{sum and [3; eq. (II.7), p.236] applied to the second)} \\
= &\frac{(1-q^{n(3n-1)/2})}{(1-q^n)},
\end{aligned}$$

where the last line follows from a routine but lengthy cancellation of most of the products. Having obtained the desired expression for our sum, we have proved Theorem 1. □

3. The $A(n,j)$.

Some of the $A(n,j)$ are polynomials. For example

$$A(8,3) = 1 + q^2 + q^3 + q^4 + q^5 + q^6 + q^8. \tag{3.1}$$

Some are not; for example

$$A(8,6) = \frac{1 + q^2 + q^3 + q^4 + q^5 + q^6}{1 + q}. \tag{3.2}$$

Theorem 2. *If n and j are positive integers with $\gcd(n,j) = 1$, then $A(n,j)$ is a reciprocal polynomial of degree $(j-1)(n-j-1)$ with non-negative coefficients. If $\gcd(n,j) > 1$, then $A(n,j)$ is not a polynomial.*

Proof. We recall from (2.6) that

$$A(n,j) = \frac{(1-q^{n-j+1})(1-q^{n-j+2})\dots(1-q^{n-1})}{(1-q^2)(1-q^3)\dots(1-q^j)}. \tag{3.3}$$

Let us first determine whether $A(n,j)$ is a polynomial or not. We note that, as a rational function of q, the only possible poles of $A(n,j)$ occur for primitive k-th roots of unity, ρ, with $2 \leq k \leq j$. Thus ρ is a root of the denominator of order $\lfloor \frac{j}{k} \rfloor$ and a root of the numerator of order $\lfloor \frac{n-1}{k} \rfloor - \lfloor \frac{n-j}{k} \rfloor$.

We write $n = k\nu + r$, $0 \leq r < k$, and $j = k\mu + s$, $0 \leq s < k$; so

$$\begin{aligned} & \left(\left\lfloor \tfrac{n-1}{k} \right\rfloor - \left\lfloor \tfrac{n-j}{k} \right\rfloor\right) - \left\lfloor \tfrac{j}{k} \right\rfloor \\ = & \left\lfloor \tfrac{r-1}{k} \right\rfloor - \left\lfloor \tfrac{r-s}{k} \right\rfloor - \left\lfloor \tfrac{s}{k} \right\rfloor \\ = & \left\lfloor \tfrac{r-1}{k} \right\rfloor - \left\lfloor \tfrac{r-s}{k} \right\rfloor \end{aligned}$$

because $0 \leq s < k$.

Case 1. $\gcd(n, j) = 1$. Consequently it is impossible that both r and s are 0 because if they were this would mean $k(> 1)$ is a common divisor of n and j. So if $r > 0$, then

$$\left\lfloor \frac{r-1}{k} \right\rfloor - \left\lfloor \frac{r-s}{k} \right\rfloor \geq - \left\lfloor \frac{r-s}{k} \right\rfloor = 0 \text{ or } 1.$$

If $r = 0$, then $s > 0$; so

$$\left\lfloor \frac{r-1}{k} \right\rfloor - \left\lfloor \frac{r-s}{k} \right\rfloor = -1 - \left\lfloor \frac{-s}{k} \right\rfloor = -1 - (-1) = 0.$$

In any event, we see that ρ is at least as often a zero of the numerator as it is of the denominator. Hence in this case $A(n, j)$ has no poles, and so it is a polynomial.

Case 2. $\gcd(n, j) = k > 1$. Then $n = k\nu$ and $j = k\mu$ (with $r = s = 0$). In this case, if ρ is a primitive k^{th} root of unity, then ρ is a root of the denominator of order μ, and a root of the numerator of order

$$\left\lfloor \frac{k\nu - 1}{k} \right\rfloor - \left\lfloor \frac{k\nu - k\mu}{k} \right\rfloor = \mu - 1.$$

Thus $A(n, j)$ has a pole at ρ and so cannot be a polynomial.

Next we note that when $\gcd(n, j) = 1$ (i.e. when $A(n, j)$ is a polynomial), then degree of $A(n, j)$ is

$$\binom{n}{2} - \binom{j+1}{2} - \binom{n-j+1}{2} + 1 = (j-1)(n-j-1).$$

Now if in $A(n, j)$ we replace q by q^{-1} and multiply by $q^{(j-1)(n-j-1)}$, we obtain

$$\begin{aligned} & q^{(j-1)(n-j-1)} \frac{(1-q^{-n+j-1})(1-q^{-n+j-2})\dots(1-q^{-n+1})}{(1-q^{-2})(1-q^{-3})\dots(1-q^{-j})} \\ = & \ q^{(j-1)(n-j-1)-(j-1)n+\binom{j}{2}\binom{j+1}{2}-1} \frac{(q^{n-j+1};q)_{j-1}}{(q^2;q)_{j-1}} \\ = & \ A(n, j). \end{aligned}$$

There, if $\gcd(n, j) = 1$, then $A(n, j)$ is a reciprocal polynomial.

Finally, we recall the second line of (2.6)

$$A(n, j) = \frac{(1-q)}{(1-q^n)} \begin{bmatrix} n \\ j \end{bmatrix}. \tag{3.4}$$

The Gassian polynomial $\begin{bmatrix} n \\ j \end{bmatrix}$ has been proved to have weakly increasing coefficients up to $q^{\lfloor (j(n-j)+1)/2 \rfloor}$ (cf. [1; p. 48], [4], [5; p. 76], [7]). In other words, $(1-q) \begin{bmatrix} n \\ j \end{bmatrix}$ has non-negative coefficients up through $q^{\lfloor (j(n-j)+1)/2 \rfloor}$. Consequently as a rational function expanded in a power series around $q = 0$, $A(n, j)$ has non-negative coefficients up through $q^{\lfloor (j(n-j)+1)/2 \rfloor}$. But $A(n, j)$ is a reciprocal polynomial of degree $(j-1)(n-j-1)$. Hence if it has non-negative coefficients up through $q^{\lfloor (j-1)(n-j-1)/2 \rfloor}$, then all its coefficients are non-negative. Since

$$\lfloor (j-1)(n-j-1)/2 \rfloor \le \lfloor (j(n-j)+1)/2 \rfloor,$$

we know that all the coefficients of $A(n,j)$ are non-negative whenever $\gcd(n,j)=1$. □

4. Proof of Theorem 3.

We suppose n is an odd prime. We note that for any j with $0 \le j \le \frac{n-3}{2}$, $\gcd(n-j-1, j+1) = 1$ because if those two numbers had a common factor $d > 1$, then d would divide $(n-j-1)+(j+1) = n$, a prime which is impossible because $d \le j+1 \le \frac{n-1}{2}$. Hence when we examine the entries in (2.7) we see that every summand in the sum of j is a rational function with non-negative coefficients by Theorem 2. Of course, the coefficient of q on the right-hand side of (2.7) is -1. However if we can show that $q^{n(3n-1)/2+1}$ appears with a positive coefficient in the term $j = (n-3)/2$, then we will have proved that $-q$ is the only term with a negative coefficient on the right hand side of (2.7) and Theorem 3 will be proved.

Case 1. $n = 4\nu + 1$. In this case, the $j = (n-3)/2$ term is

$$\frac{q^{12\nu^2}}{(1-q^{4\nu+2})(1-q^{4\nu+3})\dots(1-q^{6\nu})(1-q^{6\nu+2})(1-q^{6\nu+3})\dots(1-q^{8\nu+1})} \tag{4.1}$$

Now

$$\begin{aligned}
& 1 + n(3n-1)/2 \\
= \; & 1 + (n + (n+1) + (n+2) + \dots + (2n-1)) \\
= \; & 1 + (n + (5\nu+2) + (5\nu-1) + \dots + 7\nu) + (n+1) + (n+2) + \dots + (5\nu+1) \\
& \qquad + (7\nu+1) + \dots + (8\nu+1) \\
= \; & 12\nu^2 + 2 \times (n+2) + (n+3) + \dots + (5\nu+1) + (7\nu+1) + \dots + (8\nu+1).
\end{aligned}$$

This last expression clearly arises as a partition generated by (4.1). Hence the coefficient of $q^{n(3n-1)/2+1}$ is positive in (4.1).

Case 2. $n = 4\nu + 3$ with $\nu > 2$. In this case, the $j = (n-3)/2$ term is

$$\frac{q^{3(2\nu+1)^2}}{(1-q^{4\nu+4})(1-q^{4\nu+5})\dots(1-q^{6\nu+3})(1-q^{6\nu+5})(1-q^{6\nu+6})\dots(1-q^{8\nu+5})} \tag{4.2}$$

Again

$$\begin{aligned}
& 1 + n(3n-1)/2 \\
= \; & 1 + (n + (n+1) + (n+2) + \dots + (2n-1) \\
= \; & 1 + ((5\nu+3) + (5\nu+4) + \dots + (7\nu+3)) \\
& + ((4\nu+3) + (4\nu+4) + \dots + (5\nu+2)) \\
& + ((7\nu+4) + (7\nu+5) + \dots + (8\nu+5)) \\
= \; & 3(2\nu+1)^2 + 2 \times (4\nu+4) + (4\nu+5) + \dots + (5\nu+2) \\
& + (7\nu+4) + (7\nu+5) + \dots + (8\nu+5).
\end{aligned}$$

This last expression clearly arises as a partition generated by (4.2). Hence the coefficient of $q^{n(3n-1)/2+1}$ is positive in (4.2).

Case 3. $n = 3$. In this case $j = 0$, and we see immediately that since $13 = 3+2$, the coefficient of $q^{n(3n-1)/2+1}$ in

$$\frac{q^3}{(1-q^5)}$$

is positive.

Case 4. $n = 7$. In this case $j = 1$, and we see directly that since $71 = 27 + 8 + 11 + 12 + 13$ the coefficient of $q^{n(3n-1)/2+1}$ in

$$\frac{27}{(1-q^8)(1-q^9)(1-q^11)(1-q^12)(1-q^13)}$$

is positive.

Therefore, we may rewrite the right hand side of (2.7) as

$$= -q + \frac{1}{(1-q^{n(3n-1)/2})}\left\{1 - q^{n(3n-1)/2+1} + \sum_{j=0}^{\frac{n-3}{2}} \frac{q^{(j+1)(n+j)}A(n-j-1,j+1)}{(q^{n+1};q)_j(q^{2n-j-1};q)_{j+1}}\right\},$$

and we have just proved that the expression inside the brackeets has non-negative coefficients. Therefore every power of q in the expansion of

$$\frac{(1-q)}{(1-q^n)(1-q^{n+1})\dots(1-q^{2n-1})}$$

beyond q^1 has non-negative coefficients. Thus Theorem 3 is proved. □

5. Conclusion.

One would hope that it might be possible to prove the full Friedman-Joichi-Stanton conjecture from Theorem 1. This is tempting but not immediate. When n is composite some of the terms in the sum in (2.7) have negative coefficients.

Conjecture. For each k with $0 \le k \le (n-1)/2$ and n odd, the coefficients in

$$\sum_{j=0}^{k} \frac{q^{(j+1)(n+j)}A(n-j-1,j+1)}{(q^{n+1};q)_j(q^{2n-j-1};q)_{j+1}}$$

are non-negative.

The conjecture is clearly true for $k = 0$ and 1, because $A(n-1,1)$ and $A(n-2,2)$ are both polynomials with nonnegative coefficients since n is odd. Obviously the Friedman-Joichi-Stanton conjecture should follow with minimal effort from this conjecture.

Dennis Stanton has observed that one may alternatively deduce Theorem 1 from [3; eq. (III.14), p.242] under the substitution $a = q^n$, $b = 1$, $z = q^{2n}$. Further study of this identity might lead to new positivity results. He has also noted that one can extend the results in this paper to strict montonicity of the coefficients past q^{3n+4} for primes ≥ 7; furthermore this observation is required for the application in [2].

Finally it is of interest to learn more about the $A(n,j)$ especially when n and j are relatively prime. Since $A(\infty,j)$ is the generating function for partitions with no 1's and all parts $\le j$, we would expect that $A(n,j)$ is the generating function for some finite subsets of the partitions with no 1's and all parts $\le j$. Furthermore something must appear in the constraints defining these partitions that makes clear the role of $\gcd(n,j) = 1$.

References

[1] G.E.Andrews, *The Theory of Partitions*, Encycl. of Math. and Its Applications, Vol. 2, G.-C. Rota Ed., Addison-Wesley, Reading, 1976. (Reissued: Cambridge University Press, Cambridge, 1998).

[2] J.Friedman, J.T.Joichi, and D.Stanton, *More monotonicity theorems for partitions*, Exp. Math. 4 (1994), 31-37.

[3] G.Gasper and M.Rahman, *Basic Hypergeometric Series*, Encycl. of Math. and Its Applications, Vol. 35, G.-C. Rota Ed., Cambridge University Press, Cambridge, 1990.

[4] R.Proctor, *Solutions of two difficult combinatorial problems with linear algebra*, Amer. Math. Monthly 89(1982), 721-734.

[5] I.Schur, *Vorlesungen über Invariantentheorie*, Springer, Berlin, 1968.

[6] D.Stanton, *Open positivity conjectures for integer partitioins*, preprint.

[7] J.J.Sylvester, *Proof of the hitherto undemonstrated fundamental theorem of invariants*, Philosophical Mag., 5 (1878), 178-188. (Reprinted: Coll. Papers of J.J.Sylvester, Vol. 3, 117-126, Chelsea, New York, 1973).

Department of Mathematics, The Pennsylvania State University, University Park, PA 16802

E-mail address: andrews@math.psu.edu

DIMACS Series in Discrete Mathematics
and Theoretical Computer Science
Volume **64**, 2004

A Group Theoretic Approach to Public-Key Cryptography

Iris Anshel, Michael Anshel, Benji Fisher, and Dorian Goldfeld

ABSTRACT. A public key cryptosystem is an algorithmic method for securely sending private information over an insecure channel in which the communicating parties have no common key. At the heart of a public-key cryptosystem is a two-party secure computation referred to as a protocol. The first generation of public-key cryptosystems employ protocols whose security is based on the difficulty of solving algorithmic problems associated with finite abelian groups. We discuss a new method for constructing public-key cryptosystems that employ protocols whose security is based on the difficulty of solving systems of equations over non-abelian groups. We illustrate our method using braid groups and their representations. A further application of our approach yields a new class of Diffie-Hellman type protocols.

1. INTRODUCTION

A cryptosystem consists of algorithms for key establishment, encryption, and decryption. Contemporary applications require that the algorithms for encryption and decryption be highly efficient. Although it is desirable that key establishment also be efficient, the keys must be chosen in such a manner that the information to be transmitted is not compromised. A public-key cryptosystem has the additional property that security is not compromised when an adversary obtains the encryption key.

The first generation of public-key cryptosystems employ protocols whose security is based on the difficulty of solving algorithmic problems associated with finite abelian groups. It is pointed out by Odlyzko [**Odl**] in a survey of the discrete logarithm problem and its relationship to cryptography:

> *...that the first practical public-key systems to be published, the Diffie-Hellman key exchange algorithm, was based on the assumption that discrete logarithms are hard to compute. This intractability hypothesis is also the foundation for the presumed security of a variety of other public key schemes. While there have been substantial advances in discrete log algorithms in the last two decades, in general the discrete log still appears to be hard, especially for*

2000 *Mathematics Subject Classification.* Primary 11T71; Secondary 20F36, 68P25, 94A60.
Key words and phrases. group-theoretic, non-abelian, key establishment protocol, colored Burau matrices.

some groups, such as those from elliptic curves. Unfortunately no proofs of hardness are available in this area, so it is necessary to rely on experience and intuition in judging what parameters to use for cryptosystems.

Thus the search for computationally secure public-key cryptosystems continues.

Below we discuss a new method for constructing public-key cryptosystems that employ protocols whose security is based on the difficulty of solving algorithmic problems associated with presentations of groups (e.g., the conjugacy problem). One source of difficult algorithmic problems associated with group presentations arises from the theory of braids groups [**Bir**]. Braid groups have a well established history dating back to the seminal work of Emil Artin [**Art**]. We employ the Artin braid groups in our discussion below.

2. AN ALGEBRAIC KEY ESTABLISHMENT PROTOCOL

Following [**AAG**] the algebraic key establishment protocol, in its most general form, consists of a five-tuple $(\mathbf{U}, \mathbf{V}, \beta, \gamma_1, \gamma_2)$ where $\mathbf{U}$ and $\mathbf{V}$ are feasibly computable monoids, and

$$\beta : \mathbf{U} \times \mathbf{U} \longrightarrow \mathbf{V}, \qquad \gamma_i : \mathbf{U} \times \mathbf{V} \longrightarrow \mathbf{V} \quad (i = 1, 2)$$

are feasibly computable functions satisfying the following properties.

(i) For all elements $x, y_1, y_2 \in \mathbf{U}$

$$\beta(x, y_1 \cdot y_2) = \beta(x, y_1) \cdot \beta(x, y_2)$$

(ii) For all elements $x, y \in \mathbf{U}$

$$\gamma_1(x, \beta(y, x)) = \gamma_2(y, \beta(x, y)).$$

(iii) Suppose $y_1, y_2, \dots y_k \in \mathbf{U}$ and $\beta(x, y_1), \beta(x, y_2), \dots, \beta(x, y_k)$ are publicly known for some secret element $x \in \mathbf{U}$. Then, in general, it is infeasible to determine the secret element x.

The users A and B are publicly assigned submonoids, $S_A, T_B \subseteq U$, respectively. Suppose that S_A is generated by the elements

$$\{s_1, \dots, s_m\},$$

and T_B is generated by $\{t_1, \dots, t_n\}$. The protocol begins with user A choosing a secret element a in S_A and transmitting the elements

$$\beta(a, t_i) \qquad i = 1, \dots, n.$$

Likewise, user B chooses a secret element b in T_B, transmits

$$\beta(b, s_i) \qquad i = 1, \dots, m.$$

It follows from property (iii) that even though the transmission is over a public channel, the secret elements a and b are secure. Property (i) above insures that user A can compute the element

$$\beta(b, a),$$

and

$$\gamma_1(a, \beta(b, a)).$$

Likewise user B can compute $\beta(a, b)$ and $\gamma_2(b, \beta(a, b))$. Recalling property (ii) above we see that

$$\kappa = \gamma_1(a, \beta(b, a)) = \gamma_2(b, \beta(a, b))$$

can serve as an established key.

3. A GROUP THEORETIC PROTOCOL

An important instance of the algebraic key establishment protocol of §2 is described below. Here the monoid $\mathbf{U} = \mathbf{V}$ is chosen to be a group, and is denoted $\mathbf{G}$. The users A and B are publicly assigned subgroups

$$S_A = \langle s_1, s_2, \ldots, s_m \rangle, \qquad T_B = \langle t_1, \ldots, t_n \rangle.$$

Here the function $\beta : \mathbf{G} \times \mathbf{G} \to \mathbf{G}$ is chosen to be conjugation,

$$\beta(x, y) = x^{-1} y\, x,$$

and the functions γ_1, γ_2 are simply

$$\gamma_1(u, v) = u^{-1} v, \qquad \gamma_2(u, v) = v^{-1} u.$$

Users A and B choose secret elements $a \in S_A$ and $b \in T_B$ respectively, and user A begins the protocol by computing, rewriting, and transmitting the collection of elements

$$a^{-1} t_1 a,\ a^{-1} t_2 a,\ \ldots, a^{-1} t_n a.$$

Similarly, user B computes, rewrites, and transmits

$$b^{-1} s_1 b,\ b^{-1} s_2 b,\ \ldots, b^{-1} s_m b.$$

An adversary observing these transmissions is unable to determine a or b unless (s)he can solve a set of simultaneous conjugacy equations over the base group.

Multiplying two elements in the group can be accomplished by simply concatenating the two expressions representing the elements. The process of rewriting, while not unique, must be chosen so that no adversary can determine the conjugating element from viewing the publicly transmitted conjugates.

Recalling that the conjugate of the product of two elements is the product of the conjugates of those elements (i.e., property (i) of β), users A and B are now in a position to compute, respectively, the elements

$$\beta(b, a) = b^{-1} a b, \qquad \beta(a, b) = a^{-1} b a.$$

In order to attain a common key, user A computes

$$\kappa = \gamma_1(a, \beta(b, a)) = a^{-1} b^{-1} a b = [a, b],$$

and user by B computes

$$\kappa = \gamma_2(b, \beta(a, b)) = [a, b].$$

At this point each user has obtained the key κ in a (usually different) rewritten form.

As was observed in [**AAG**] there are various possible methods for extracting a common (identical) element from the element κ. A feasible algorithm to put every word in the group in canonical form can be applied to the word κ by either user provided such an algorithm exists. However, there exist groups where any such canonical form algorithm is quite slow relative to an algorithm to solve the word problem (i.e., an algorithm to determine if a word is the identity element). In such a case it may be more efficient to obtain a common key by having user B either send a rewritten form of κ or some other random word (not equal to κ) in the group to user A. User A can then use the word problem algorithm to determine if κ was sent or not. If κ was sent then this determines the bit 1, otherwise the bit 0.

By iterating m times (with a different κ at each iteration), an m-bit common key is exchanged resulting in a probabilistic protocol. Another approach to extracting a common key (which we elaborate upon further in an illustration below) lies in defining an easily computable function from $\mathbf{G}$ to any set whose elements have an easily computable canonical form.

The construction of an algebraic key establishment protocol based on the braid group serves as an excellent illustration. This is due to the fact that, at the present time, the best known algorithm to solve the conjugacy problem requires at least exponential running time. Furthermore, there are two different algorithmic approaches to the word problem. Both the Birman, Ko, Lee (see [**BKL**]) canonical form algorithm for elements in the braid group and the Dehornoy handle reduction method (see [**Deh**]) of comparing braids can be employed to rewrite publicly known elements and to obtain a common element once an exchange has been completed. As a third method of extracting a common key we will define a homomorphism from the braid to an semidirect product of a matrix group and a symmetric group. Evaluation of this homomorphism will yield a method of extracting common keys.

Recall that the braid group on $N+1$ strands is generated by the symbols

$$x_1, x_2, \ldots, x_N$$

and has defining relations given by

$$x_i x_j = x_j x_i, \qquad \text{if } |j-i| \geq 2,$$

$$x_i x_j x_i = x_j x_i x_j, \quad \text{if } |i-j| = 1.$$

The colored Burau matrices (see [**Mor**]) allows us to define a homomorphism on the whole braid group whose image lies in the semidirect product of the symmetric group, S_N, and the matrix group

$$GL_{N-1}(R),$$

where R is the ring of Laurent polynomials

$$R = Z[t_1, t_1^{-1}, \ \ldots \ , t_N, t_N^{-1}].$$

The symmetric group acts on the matrix group $GL_{N-1}(R)$ in a coefficient-by-coefficient manner, by permuting the variables t_i. Multiplication in the semi-direct product is given by

$$(M_1, p) * (M_2, q) = (M_1 * p(M_2), pq),$$

where M_1, M_2 are matrices in $GL_{N-1}(R)$; p, q are permutations in S_N, and $p(M_2)$ denotes the matrix M_2 whose variables have been permuted by p.

Each Artin generator x_i is mapped to the element $(C_i(t_i), (i\; i+1))$ where

$$(i\; i+1)$$

denotes the transposition and $C_i(t_i)$ denotes the matrix below (ones on the diagonal, zeros elsewhere, except in the ith row, where we have $0\;0\;\cdots 0\;\; t_i\; -t_i\;\; 1\;\; 0\;\; \cdots\;\; 0\;0$

with $-t_i$ on the diagonal).

$$C_i(t_i) = \begin{pmatrix} 1 & & & & & & \\ & 1 & & & & & \\ & & \ddots & & & & \\ & & & t_i & -t_i & 1 & \\ & & & & & \ddots & \\ & & & & & & 1 \end{pmatrix}$$

One checks that these elements of the semi-direct product obey the braid relations. It should also be remarked that the semi-direct product does have a faithful linear representation of dimension $n!(n-1)$.

The above homomorphism can be computed more rapidly by reducing (mod p) and letting $\{t_1, \ldots, t_{N-1}\}$ denote distinct and invertible integers (mod p). Since the result of applying this function to the exchanged element κ is an invariant of κ any part of the resulting image can be used as a shared secret.

Several cryptoanalytic attacks have been developed regarding the methods above. A complete discussion of this topic will appear at a later date.

4. A DIFFIE-HELLMAN TYPE PROTOCOL BASED ON NON-ABELIAN GROUPS

In the Diffie-Hellman system, secret keys are viewed as elements of a finite field $GF(q)$ in the following manner. Suppose users A and B agree on the following public information: a large prime power q (which determines the finite field GF(q)), and a fixed element g of the finite field GF(q). Both A and B begin the process by secretly choosing random integers, denoted by a, b, between 1 and $q-1$. By computing and making public the elements

$$g^a, \qquad g^b,$$

respectively, the (easily computed) secret key that is exchanged is given by g^{ab}.

The Diffie-Hellman assumption, according to Koblitz [**Kob**], is that it is computationally infeasible for a third individual C to compute g^{ab}. The essence of this assumption lies in the infeasibility of the **discrete logarithm problem**, i.e., it is difficult to compute the value of x given only g and g^x. In fact, were C able to solve the discrete logarithm problem then b could be recovered from g and g^b (both of which are publicly known), and then the secret key is simply given by

$$g^{ab} = (g^a)^b.$$

Clearly in the Diffie-Hellman system the one-way functions $g \mapsto g^a$ and $g \mapsto g^b$ commute. The general algebraic key establishment protocol from §2 can be specialized to yield a generalization of the the Diffie-Hellman protocol with commuting one-way functions in the following manner. Let G denote a non-abelian group with intractable conjugacy problem. Assume G contains commuting subgroups S_A and T_B i.e., for all $h \in S_A$ and $k \in T_B$

$$h\,k = k\,h.$$

Fix a public element $p \in G$. The element p will be used to specify the monoid V and the functions $\beta, \gamma_1, \gamma_2$ in the protocol of §2. Define the set

$$V = \{g^{-1} p\, g \mid g \in G\}.$$

The set V can be viewed as a semi-group (with identity) in the following manner. Define a law of composition by

$$x \circ y = x \qquad (\text{for all} \quad x, y \in V,\ x \neq p, y \neq p),$$

and

$$p \circ x = x \circ p = x, \qquad (\text{ for all} \quad x \in V).$$

With this semigroup V in place we define the function $\beta : G \times G \to V$ by

$$\beta(x, y) = x^{-1} p x \in V.$$

Clearly β satisfies property (i) of §2. Suppose now that S_A is generated by elements $\{s_1, \ldots, s_m\}$, and T_B is generated by $\{t_1, \ldots, t_n\}$. Following the general protocol, A chooses a secret element $a \in S_A$ and publishes the list

$$\beta(a, t_i) \qquad i = 1, \ldots, n.$$

Remark that in this case, the above list is just a redundant list of one element, $a^{-1} p\, a$. Likewise B chooses a secret element $b \in T_B$ and publishes the list

$$\beta(b, s_i) = b^{-1} p\, b \qquad i = 1, \ldots, m.$$

We now define functions γ_1, γ_2 that allow us to complete the protocol. Let $C(p)$ denote the centralizer of p in the group G, i.e.,

$$C(p) = \{g \in G \mid gp = pg\}.$$

Let

$$\gamma_1(u, v) = \begin{cases} p, & \text{if } u \notin S_A \cdot C(p); \\ p, & \text{if } u \in S_A \cdot C(p) \text{ and } v \neq b^{-1} p b \text{ for all } b \in S_B; \\ a^{-1} v a, & \text{otherwise (where } u \in a \cdot C(p)). \end{cases}$$

One checks that this definition is well defined and independent of the choice of a.

Similarly, we define

$$\gamma_2(u, v) = \begin{cases} p, & \text{if } u \notin T_B \cdot C(p); \\ p, & \text{if } u \in T_B \cdot C(p) \text{ and } v \neq a^{-1} p a \text{ for all } a \in S_A; \\ b^{-1} v b, & \text{otherwise (where } u \in b \cdot C(p)). \end{cases}$$

One easily checks that condition (ii) of §2 is satisfied. The common established key κ that both users A and B obtain is

$$\kappa = a^{-1} b^{-1} p\, b\, a = b^{-1} a^{-1} p\, a\, b.$$

Implicit in the above protocol is the fact that if $a \in S(A)$ and $b \in S(B)$ then the one-way functions

$$I_a(g) = a^{-1} g\, a, \quad I_b(g) = b^{-1} g\, b$$

commute, i.e.,

$$I_b(I_a(g)) = I_a(I_b(g)) = a^{-1} b^{-1} g\, a\, b.$$

A special case of the Diffie-Hellman type protocol above, which is based on the braid group, is discussed in [**KLChHKP**]. The authors indicate that they have implemented and tested their system. A confirmation of their results would strengthen the case for considering our group theoretic methods as a basis for second generation public key cryptosystems.

It should be noted that there are other instances of commuting functions on groups. In [**A**] the author studies a class of HNN extensions that possess commuting endomorphisms. Following the general construction it is possible to specify a Diffie-Hellman type protocol by fixing a public element p in the HNN extension and letting users publish the image of p under their respective (commuting) private endomorphisms. A more complete discussion of other possible group theoretic Diffie-Hellman type protocols will appear in another venue.

The authors wish to gratefully acknowledge the financial support of Arithmetica Inc. in the preparation of this paper. All works and inventions described in this paper were for the benefit of, and were funded by, Arithmetica Inc.

References

[A] M. Anshel, *Decision Problems for HNN Groups and Commutative Semigroups*, Houston Journal of Mathematics vol. **4** no. 2 (1978) p. 137–142.

[AAG] I. Anshel, M. Anshel, and D. Goldfeld, *An algebraic method for public-key cryptography*, Mathematical Research Letters **6** (1999), 287–291.

[Art] E. Artin *Theory of Braids*, Annals of Math. **48** (1947), 101–126.

[Bir] J.S. Birman, *Braids, Links, and mapping class groups*, Ann. of Math Studies, **82** (1975).

[BKL] J.S. Birman, K.H. Ko, and S.J. Lee, *A new approach to the word and conjugacy problems in the braid groups*, Advances in Mathematics **139** (1998), 322–353.

[Deh] P. Dehornoy, *A fast method for comparing braids*, Advances in Mathematics, **125** (1997), 200–235.

[KLChHKP] K.H. Ko, S.J. Lee, J.H. Cheon, J.W. Han, J.S. Kang, and C. Park, *New public-key cryptosystem using braid groups*, preprint.

[Kob] N. Koblitz, *A Course in Number Theory and Cryptography: Second Edition*, Springer-Verlag, New York (1994).

[Mor] H. R. Morton, *The Multivariable Alexander Polynomial for a Closed Braid*, Contemporary Mathematics **233** AMS (1999), 167–172.

[Odl] A. M. Odlyzko, *Discrete logarithms: The past and the future*, Designs, Codes, and Cryptography **19** (2000), pp. 129–145.

ARITHMETICA INC., 31 PETER LYNAS COURT, TENAFLY, NJ 07670 USA
E-mail address: goldfeld@optonline.net

DEPARTMENT OF COMPUTER SCIENCES, CITY COLLEGE OF NEW YORK, NEW YORK, NY 10031 USA
E-mail address: MikeAt1140@aol.com

DEPARTMENT OF MATHEMATICS, BOSTON COLLEGE, CHESTNUT HILL, MA 02467 USA
E-mail address: benji@member.AMS.org

DEPARTMENT OF MATHEMATICS, COLUMBIA UNIVERSITY, NEW YORK, NY 10027 USA
E-mail address: goldfeld@columbia.edu

DIMACS Series in Discrete Mathematics
and Theoretical Computer Science
Volume **64**, 2004

Dedekind sums: a combinatorial-geometric viewpoint

Matthias Beck and Sinai Robins

ABSTRACT. The literature on Dedekind sums is vast. In this expository paper we show that there is a common thread to many generalizations of Dedekind sums, namely through the study of lattice point enumeration of rational polytopes. In particular, there are some natural finite Fourier series which we call **Fourier-Dedekind sums**, and which form the building blocks of the number of partitions of an integer from a finite set of positive integers. This problem also goes by the name of the 'coin exchange problem'. Dedekind sums have enjoyed a resurgence of interest recently, from such diverse fields as topology, number theory, and combinatorial geometry. The Fourier-Dedekind sums we study here include as special cases generalized Dedekind sums studied by Berndt, Carlitz, Grosswald, Knuth, Rademacher, and Zagier. Our interest in these sums stems from the appearance of Dedekind's and Zagier's sums in lattice point count formulas for polytopes. Using some simple generating functions, we show that generalized Dedekind sums are natural ingredients for such formulas. As immediate 'geometric' corollaries to our formulas, we obtain and generalize reciprocity laws of Dedekind, Zagier, and Gessel. Finally, we prove a polynomial-time complexity result for Zagier's higher-dimensional Dedekind sums.

1. Introduction

In recent years, Dedekind sums and their various siblings have enjoyed a new renaissance. Historically, they appeared in analytic number theory (Dedekind's η-function [**De**]), algebraic number theory (class number formulae [**Me**]), topology (signature defects of manifolds [**HZ**]), combinatorial geometry (lattice point enumeration [**Mo**]), and algorithmic complexity (pseudo random number generators [**K**]). In this expository paper, we define some broad generalizations of Dedekind sums, which are in fact finite Fourier series. We show that they appear naturally in the enumeration of lattice points in polytopes, and prove reciprocity laws for them.

In combinatorial number theory, one is interested in partitions of an integer n from a finite set. That is, one writes n as a nonnegative integer linear combination of a given finite set of positive integers. We showed in [**BDR**] that the number of such partitions of n from a finite set is a quasipolynomial in n, whose coefficients are built up from the following generalization of Dedekind sums.

2000 *Mathematics Subject Classification.* Primary 05A15, 11L03; Secondary 11P21, 52C07.
Key words and phrases. Dedekind sums, rational polytopes, lattice points, partition function.
The second author is supported by the NSA Young Investigator Grant MSPR-OOY-196.

DEFINITION 1.1. For $a_0, \ldots, a_d, n \in \mathbb{Z}$, we define the **Fourier-Dedekind sum** as

$$\sigma_n(a_1, \ldots, a_d; a_0) := \frac{1}{a_0} \sum_{\lambda^{a_0}=1} \frac{\lambda^n}{(1-\lambda^{a_1})\cdots(1-\lambda^{a_d})} .$$

Here the sum is taken over all a_0'th roots of unity for which the summand is not singular.

In [**G**], Gessel systematically studied sums of the form

$$\sum_{\lambda^a=1} R(\lambda) ,$$

where R is a rational function, and the sum is taken over all a'th roots of unity for which R is not singular. He called them 'generalized Dedekind sums', since his definition includes various generalizations of the Dedekind sum as special cases. Hence we study Gessel's sums where the poles of R are restricted to be roots of unity.

In Section 2, we give a brief history on those generalizations of the classical Dedekind sum (due to Rademacher [**R**], and Zagier [**Z**]) which can be written as Fourier-Dedekind sums. Our interest in these sums stems from the appearance of Dedekind's and Zagier's sums in lattice point enumeration formulas for polytopes [**Mo, P, BV, DR**]. Using generating functions, we show in Section 3 that generalized Dedekind sums are natural ingredients for such formulas, which also apply to the theory of partition functions. In Section 4 we obtain and generalize reciprocity laws of Dedekind [**De**], Zagier [**Z**], and Gessel [**G**] as 'geometric' corollaries to our formulas. Finally, in Section 5, we prove that Zagier's higher-dimensional Dedekind sums are in fact polynomial-time computable in fixed dimension. For Dedekind sums in 2 dimensions, this fact follows easily from their reciprocity law; but for higher dimensional Dedekind sums the polynomial-time complexity does not seem to follow so easily, and we therefore invoke some recent work of [**BP**] and [**DR**].

2. Classical Dedekind sums and generalizations

According to Riemann's will, it was his wish that Dedekind should get Riemann's unpublished notes and manuscripts [**RG**]. Among these was a discussion of the important function

$$\eta(z) = e^{\frac{\pi i z}{12}} \prod_{n \geq 1} \left(1 - e^{2\pi i n z}\right) ,$$

which Dedekind took up and eventually published in Riemann's collected works [**De**].

DEFINITION 2.1. Let $((x))$ be the sawtooth function defined by

$$((x)) := \begin{cases} \{x\} - \frac{1}{2} & \text{if } x \notin \mathbb{Z} \\ 0 & \text{if } x \in \mathbb{Z} . \end{cases}$$

Here $\{x\} = x - [x]$ denotes the fractional part of x. For two integers a and b, we define the **Dedekind sum** as

$$s(a,b) := \sum_{k \bmod b} \left(\left(\frac{ka}{b}\right)\right) \left(\left(\frac{k}{b}\right)\right) .$$

Here the sum is over a complete residue system modulo b.

Through the study of the transformation properties of η under $\mathrm{SL}_2(\mathbb{Z})$, Dedekind naturally arrived at $s(a,b)$. The classic introduction to the arithmetic properties of the Dedekind sum is [**RG**]. The most important of these, already proved by Dedekind [**De**], is the famous reciprocity law:

THEOREM 2.2 (Dedekind). *If a and b are relatively prime then*

$$s(a,b)+s(b,a)=-\frac{1}{4}+\frac{1}{12}\left(\frac{a}{b}+\frac{1}{ab}+\frac{b}{a}\right).$$

This reciprocity law is easily seen to be equivalent to the transformation law of the η-function [**De**]. Due to the periodicity of $((x))$, we can reduce a modulo b in the Dedekind sum: $s(a,b)=s(a \bmod b, b)$. Therefore, Theorem 2.2 allows us to compute $s(a,b)$ in polynomial time, similar in spirit to the Euclidean algorithm.

The Dedekind sum $s(a,b)$ has various generalizations, two of which we introduce here. The first one is due to Rademacher [**R**], who generalized sums introduced by Meyer [**Me**] and Dieter [**D**]:

DEFINITION 2.3. For $a,b\in\mathbb{Z}$, $x,y\in\mathbb{R}$, the **Dedekind-Rademacher sum** is defined by

$$s(a,b;x,y):=\sum_{k \bmod b}\left(\left(\frac{(k+y)a}{b}+x\right)\right)\left(\left(\frac{k+y}{b}\right)\right).$$

This sum posesses again a reciprocity law:

THEOREM 2.4 (Rademacher). *If a and b are relatively prime and x and y are not both integers, then*

$$s(a,b;x,y)+s(b,a;y,x)=((x))((y))+\frac{1}{2}\left(\frac{a}{b}B_2(y)+\frac{1}{ab}B_2(ay+bx)+\frac{b}{a}B_2(x)\right).$$

Here

$$B_2(x):=(x-[x])^2-(x-[x])+\frac{1}{6}$$

is the periodized second Bernoulli polynomial.

If x and y are both integers, the Dedekind-Rademacher sum is simply the classical Dedekind sum, whose reciprocity law we already stated. As with the reciprocity law for the classical Dedekind sum, Theorem 2.4 can be used to compute $s(a,b;x,y)$ in polynomial time.

The second generalization of the Dedekind sum we mention here is due to Zagier [**Z**]. From topological considerations, he arrived naturally at expressions of the following kind:

DEFINITION 2.5. Let $a_1,\dots,a_d$ be integers relatively prime to $a_0\in\mathbb{N}$. Define the **higher-dimensional Dedekind sum** as

$$s(a_0;a_1,\dots,a_d):=\frac{(-1)^{d/2}}{a_0}\sum_{k=1}^{a_0-1}\cot\frac{\pi k a_1}{a_0}\cdots\cot\frac{\pi k a_d}{a_0}.$$

This sum vanishes if d is odd. It is not hard to see that this indeed generalizes the classical Dedekind sum: the latter can be written in terms of cotangents [**RG**], which yields

$$s(a,b)=\frac{1}{4b}\sum_{k \bmod b}\cot\frac{\pi k a}{b}\cot\frac{\pi k}{b}=-\frac{1}{4}s(b;a,1).$$

Again, there exists a reciprocity law for Zagier's sums:

THEOREM 2.6 (Zagier). *If $a_0, \ldots, a_d \in \mathbb{N}$ are pairwise relatively prime then*

$$\sum_{j=0}^{d} s(a_j; a_0, \ldots, \hat{a_j}, \ldots, a_d) = \phi(a_0, \ldots, a_d) .$$

Here ϕ is a rational function in $a_0, \ldots, a_d$, which can be expressed in terms of Hirzebruch L-functions [**Z**].

It should be mentioned that a version of the higher-dimensional Dedekind sums had already been introduced by Carlitz [**C**]:

$$\sum_{k_1,\ldots,k_d \bmod a_0} \left(\left(\frac{a_1k_1 + \cdots + a_dk_d}{a_0}\right)\right)\left(\left(\frac{a_1}{a_0}\right)\right)\cdots\left(\left(\frac{a_d}{a_0}\right)\right) .$$

Berndt [**B**] noticed that these sums are, up to trivial factor, Zagier's higher-dimensional Dedekind sums.

If we write the higher-dimensional Dedekind sum as a sum over roots of unity,

$$s(a_0; a_1, \ldots, a_d) = \frac{1}{a_0} \sum_{\lambda^{a_0}=1\neq\lambda} \frac{\lambda^{a_1}+1}{\lambda^{a_1}-1} \cdots \frac{\lambda^{a_d}+1}{\lambda^{a_d}-1} ,$$

it becomes clear that it suffices to study sums of the form

$$\frac{1}{a_0} \sum_{\lambda^{a_0}=1\neq\lambda} \frac{1}{(\lambda^{a_1}-1)\cdots(\lambda^{a_d}-1)} .$$

Zagier's Dedekind sum can be expressed as a sum of expressions of this kind. On the other hand, we consider special cases of the Dedekind-Rademacher sum, namely, for $n \in \mathbb{Z}$,

$$s\left(a, b; \frac{n}{b}, 0\right) = \sum_{k \bmod b} \left(\left(\frac{ka+n}{b}\right)\right)\left(\left(\frac{k}{b}\right)\right) .$$

Knuth [**K**] discovered that these generalized Dedekind sums describe the statistics of pseudo random number generators. In [**BR**], we used the convolution theorem for finite Fourier series to show that, if a and b are relatively prime,

$$s\left(a, b; \frac{n}{b}, 0\right) = -\frac{1}{b} \sum_{\lambda^b=1\neq\lambda} \frac{\lambda^{-n}}{(1-\lambda^a)(1-\lambda)} - \frac{1}{2}\left\{\frac{n}{b}\right\} + \frac{1}{4} - \frac{1}{4b} . \tag{2.1}$$

Here $\{x\} = x - [x]$ denotes the fractional part of x. Comparing this with the representation we obtained for Zagier's Dedekind sums motivates the study of the Fourier-Dedekind sum

$$\sigma_n(a_1, \ldots, a_d; a_0) = \frac{1}{a_0} \sum_{\lambda^{a_0}=1} \frac{\lambda^n}{(1-\lambda^{a_1})\cdots(1-\lambda^{a_d})} ,$$

a finite Fourier series in n. Gessel [**G**] gave a new reciprocity law for a special case of Fourier-Dedekind sums:

THEOREM 2.7 (Gessel). *Let p and q be relatively prime and suppose that $1 \le n \le p+q$. Then*

$$\frac{1}{p} \sum_{\lambda^p=1\neq\lambda} \frac{\lambda^n}{(1-\lambda^q)(1-\lambda)} + \frac{1}{q} \sum_{\lambda^q=1\neq\lambda} \frac{\lambda^n}{(1-\lambda^p)(1-\lambda)}$$
$$= -\frac{n^2}{2pq} + \frac{n}{2}\left(\frac{1}{p}+\frac{1}{q}+\frac{1}{pq}\right) - \frac{1}{4}\left(\frac{1}{p}+\frac{1}{q}+1\right) - \frac{1}{12}\left(\frac{p}{q}+\frac{1}{pq}+\frac{q}{p}\right) .$$

It is easy to see that the reciprocity law for classical Dedekind sums (Theorem 2.2) is a special case of Gessel's theorem. We can rephrase the statement of Gessel's theorem in terms of Dedekind-Rademacher sums by means of (2.1): for p and q relatively prime, and $1 \le n \le p+q$,

$$\begin{aligned} & s\left(q,p;\frac{-n}{p},0\right) + s\left(p,q;\frac{-n}{q},0\right) \\ & \stackrel{\text{def}}{=} \sum_{k=0}^{p-1} \left(\left(\frac{qk-n}{p}\right)\right)\left(\left(\frac{k}{p}\right)\right) + \sum_{k=0}^{q-1} \left(\left(\frac{pk-n}{q}\right)\right)\left(\left(\frac{k}{q}\right)\right) \\ & = \frac{n^2}{2pq} - \frac{n}{2}\left(\frac{1}{p}+\frac{1}{q}+\frac{1}{pq}\right) + \frac{1}{4} + \frac{1}{12}\left(\frac{p}{q}+\frac{1}{pq}+\frac{q}{p}\right) - \frac{1}{2}\left\{\frac{-t}{p}\right\} - \frac{1}{2}\left\{\frac{-t}{q}\right\} . \end{aligned}$$

We will now view the Fourier-Dedekind sum from a generating-function point of view, which will allow us to obtain and extend geometric proofs of Dedekind's, Zagier's and Gessel's reciprocity laws.

3. A new combinatorial identity for partitions from a finite set

The form of the Fourier-Dedekind sum

$$\sigma_{-n}(a_1,\ldots,a_d;a_0) = \frac{1}{a_0} \sum_{\lambda^{a_0}=1\neq\lambda} \frac{\lambda^{-n}}{(1-\lambda^{a_1})\cdots(1-\lambda^{a_d})}$$

suggests the use of a generating function

$$f(z) := \frac{1}{1-z^{a_0}} \frac{z^{-n}}{(1-z^{a_1})\cdots(1-z^{a_d})} .$$

In fact, let's expand this generating function into partial fractions: suppose, for simplicity, that $n>0$, and $a_0,\ldots,a_d$ are pairwise relatively prime. Then we can write

$$f(z) = \sum_{\lambda^{a_0}=1\neq\lambda} \frac{A_\lambda}{z-\lambda} + \cdots + \sum_{\lambda^{a_d}=1\neq\lambda} \frac{A_\lambda}{z-\lambda} + \sum_{k=1}^{d+1} \frac{B_k}{(z-1)^k} + \sum_{k=1}^{n} \frac{C_k}{z^k} .$$

The coefficient A_λ for, say, a nontrivial a_0'th root of unity λ can be derived easily:

$$A_\lambda = \lim_{z\to\lambda}(z-\lambda)f(z) = -\frac{\lambda}{a_0} \frac{\lambda^{-n}}{(1-\lambda^{a_1})\cdots(1-\lambda^{a_n})} .$$

Hence we obtain the Fourier-Dedekind sums if we consider the constant coefficient of f (in the Laurent series about $z=0$):

$$\begin{aligned} \text{(3.1)}\quad \operatorname{const}(f) &= \sum_{\lambda^{a_0}=1\neq\lambda} \frac{A_\lambda}{-\lambda} + \cdots + \sum_{\lambda^{a_d}=1\neq\lambda} \frac{A_\lambda}{-\lambda} + \sum_{k=1}^{d+1} (-1)^k B_k \\ &= \sigma_{-n}(a_1,\ldots,a_d;a_0) + \cdots + \sigma_{-n}(a_0,\ldots,a_{d-1};a_n) + \sum_{k=1}^{d+1} (-1)^k B_k . \end{aligned}$$

The coefficients B_k are simply the coefficients of the Laurent series of f about $z=1$, and are easily computed, by hand or using mathematics software such as

Maple or Mathematica. It is not hard to see that they are polynomials in n whose coefficients are rational functions of the $a_0, \ldots, a_d$.[1] To simplify notation, define

$$q(a_0, \ldots, a_d, n) := \sum_{k=1}^{d+1} (-1)^k B_k \ . \tag{3.2}$$

On the other hand, we can compute the constant coefficient of f by brute force: By expanding

$$f(z) = \left(\sum_{k_0 \geq 0} z^{k_0 a_0} \right) \cdots \left(\sum_{k_d \geq 0} z^{k_d a_d} \right) z^{-n} \ ,$$

we can see that const(f) enumerates the ways of writing n as a linear combination of the $a_0, \ldots, a_d$ with nonnegative coefficients:

$$\begin{aligned} \text{const}(f) &= \# \left\{ (k_0, \ldots, k_d) \in \mathbb{Z}^{d+1} : \ k_j \geq 0, \ k_0 a_0 + \cdots + k_d a_d = n \right\} \\ &= p_{\{a_0, \ldots, a_d\}}(n) \ . \end{aligned} \tag{3.3}$$

This defines the partition function with parts in the finite set $A := \{a_0, \ldots, a_n\}$. Geometrically, $p_A(n)$ enumerates the integer points in n-dilates of the rational polytope

$$\mathcal{P} := \left\{ (x_0, \ldots, x_d) \in \mathbb{R}^{d+1} : \ x_j \geq 0, \ x_0 a_0 + \cdots + x_d a_d = 1 \right\} \ .$$

This geometric interpretation allows us to use the machinery of Ehrhart theory, which will be advantageous in the following section. We next give an explicit formula for the famous 'coin-exchange problem'—that is, the number of ways to form n cents from a finite set of coins with given denominations $a_0, \ldots, a_d$: comparing (3.1) with (3.3) yields our central result [**BDR**].

THEOREM 3.1. *Suppose $a_0, \ldots, a_d$ are pairwise relatively prime, positive integers. We recall that the number of partitions of an integer n from the finite set of a_i's is defined by*

$$p_{\{a_0, \ldots, a_d\}}(n) := \left\{ (k_0, \ldots, k_d) \in \mathbb{Z}^{d+1} : \ k_j \geq 0, \ k_0 a_0 + \cdots + k_d a_d = n \right\} .$$

Then

$$p_{\{a_0, \ldots, a_d\}}(n) = q(a_0, \ldots, a_d, n) + \sum_{j=0}^{d} \sigma_{-n} \left(a_0, \ldots, \hat{a_j}, \ldots, a_d; a_j \right) \ ,$$

where $q(a_0, \ldots, a_d, n)$ is given by (3.2).

---

[1] After this paper was submitted, general formulas for these polynomials were discovered in [**BGK**].

The first few expressions for $q(a_0, \ldots, a_d, n)$ are

$$
\begin{aligned}
q(a_0, n) &= \frac{1}{a_0} \\
q(a_0, a_1, n) &= \frac{n}{a_0 a_1} + \frac{1}{2}\left(\frac{1}{a_0} + \frac{1}{a_1}\right) \\
q(a_0, a_1, a_2, n) &= \frac{n^2}{2a_0 a_1 a_2} + \frac{n}{2}\left(\frac{1}{a_0 a_1} + \frac{1}{a_0 a_2} + \frac{1}{a_1 a_2}\right) \\
&+ \frac{1}{12}\left(\frac{3}{a_0} + \frac{3}{a_1} + \frac{3}{a_2} + \frac{a_0}{a_1 a_2} + \frac{a_1}{a_0 a_2} + \frac{a_2}{a_0 a_1}\right) \\
q(a_0, a_1, a_2, a_3, n) &= \frac{n^3}{6a_0 a_1 a_2 a_3} \\
&+ \frac{n^2}{4}\left(\frac{1}{a_0 a_1 a_2} + \frac{1}{a_0 a_1 a_3} + \frac{1}{a_0 a_2 a_3} + \frac{1}{a_1 a_2 a_3}\right) \\
&+ \frac{n}{4}\left(\frac{1}{a_0 a_1} + \frac{1}{a_0 a_2} + \frac{1}{a_0 a_3} + \frac{1}{a_1 a_2} + \frac{1}{a_1 a_3} + \frac{1}{a_2 a_3}\right) \\
&+ \frac{n}{12}\left(\frac{a_0}{a_1 a_2 a_3} + \frac{a_1}{a_0 a_2 a_3} + \frac{a_2}{a_0 a_1 a_3} + \frac{a_3}{a_0 a_1 a_2}\right) \\
&+ \frac{1}{24}\left(\frac{a_0}{a_1 a_2} + \frac{a_0}{a_1 a_3} + \frac{a_0}{a_2 a_3} + \frac{a_1}{a_0 a_2} + \frac{a_1}{a_0 a_3} + \frac{a_1}{a_2 a_3}\right. \\
&\left. + \frac{a_2}{a_0 a_1} + \frac{a_2}{a_0 a_3} + \frac{a_2}{a_1 a_3} + \frac{a_3}{a_0 a_1} + \frac{a_3}{a_0 a_2} + \frac{a_3}{a_1 a_2}\right) \\
&+ \frac{1}{8}\left(\frac{1}{a_0} + \frac{1}{a_1} + \frac{1}{a_2} + \frac{1}{a_3}\right) .
\end{aligned}
$$

4. Reciprocity laws

We will now use Theorem 3.1 to prove and extend some of the reciprocity theorems stated earlier. We will make use of two results due to Ehrhart for **rational polytopes**, that is, polytopes whose vertices are rational. Ehrhart [**E**] initiated the study of the number of integer points ("lattice points") in integer dilates of such polytopes:

DEFINITION 4.1. Let $\mathcal{P} \subset \mathbb{R}^d$ be a rational polytope, and n a positive integer. We denote the number of lattice points in the dilates of the closure of $\mathcal{P}$ and its interior by

$$L(\overline{\mathcal{P}}, n) := \#\left(n\mathcal{P} \cap \mathbb{Z}^d\right) \qquad \text{and} \qquad L(\mathcal{P}^\circ, n) := \#\left(n\mathcal{P}^\circ \cap \mathbb{Z}^d\right) ,$$

respectively.

Ehrhart proved that $L(\overline{\mathcal{P}}, n)$ and $L(\mathcal{P}^\circ, n)$ are **quasipolynomials** in the integer variable n, that is, expressions of the form

$$c_d(n)\, n^d + \cdots + c_1(n)\, n + c_0(n) ,$$

where $c_0, \ldots, c_d$ are periodic functions in n. Ehrhart conjectured the following fundamental theorem, which establishes an algebraic connection between our two lattice-point-count operators. Its original proof is due to Macdonald [**Ma**].

THEOREM 4.2 (Ehrhart-Macdonald reciprocity law). *Suppose the rational polytope $\mathcal{P}$ is homeomorphic to a d-manifold. Then*

$$L(\mathcal{P}^\circ, -n) = (-1)^d L(\overline{\mathcal{P}}, n) .$$

This enables us to rephrase Theorem 3.1 for the quantity

$$p^\circ_{\{a_0,\dots,a_d\}}(n) := \#\left\{(k_0,\dots,k_d) \in \mathbb{Z}^{d+1} :\ k_j > 0,\ k_0a_0 + \cdots + k_da_d = n\right\} .$$

By Theorem 4.2, we have the following result.

COROLLARY 4.3. *Suppose $a_0,\dots,a_d \in \mathbb{N}$ are pairwise relatively prime. Then*

$$p^\circ_{\{a_0,\dots,a_d\}}(n) = (-1)^d \left(q(a_0,\dots,a_d,-n) + \sum_{j=0}^{d} \sigma_n\left(a_0,\dots,\hat{a}_j,\dots,a_d;a_j\right) \right) ,$$

where $q(a_0,\dots,a_d,n)$ is given by (3.2).

We note that we could have derived this identity from scratch in a similar way as Theorem 3.1, without using Ehrhart-Macdonald reciprocity.

The reason for switching to $p^\circ_{\{a_0,\dots,a_d\}}(n)$ is that

$$p^\circ_{\{a_0,\dots,a_d\}}(n) = 0$$

for $0 < n < a_0 + \cdots + a_d$, by the very definition of $p^\circ_{\{a_0,\dots,a_d\}}(n)$. This yields a reciprocity law:

THEOREM 4.4. *Let $a_0,\dots,a_d$ be pairwise relatively prime integers and $0 < n < a_0 + \cdots + a_d$. Then*

$$\sum_{j=0}^{d} \sigma_n(a_0,\dots,\hat{a}_j,\dots,a_d;a_j) = -q(a_0,\dots,a_d,-n) ,$$

where $q(a_0,\dots,a_d,n)$ is given by (3.2).

For $d = 2, a_0 = p, a_1 = q, a_2 = 1$, this is the statement of Gessel's Theorem 2.7, which, in turn, implies Dedekind's reciprocity law Theorem 2.2.

To prove Zagier's Theorem 2.6 in the language of Fourier-Dedekind sums, we make use of another result of Ehrhart [**E**] on **lattice polytopes**, that is, polytopes whose vertices have integer coordinates. Recall that the **reduced Euler characteristic** of a polytope $\mathcal{P}$ can be defined as

$$\chi(\mathcal{P}) := \sum_{\sigma} (-1)^{\dim \sigma} ,$$

where the sum is over all sub-simplices of $\mathcal{P}$.

THEOREM 4.5 (Ehrhart). *Let $\mathcal{P}$ be a lattice polytope. Then $L(\overline{\mathcal{P}}, n)$ is a polynomial in n whose constant term is $\chi(\mathcal{P})$.*

We note that the polytope $\mathcal{P}$ corresponding to $p_{\{a_0,\dots,a_d\}}(n)$ is convex and hence has Euler characteristic 1. If we now dilate $\mathcal{P}$ only by multiples of $a_0 \cdots a_d$, say $n = a_0 \cdots a_d m$, we obtain the dilates of a lattice polytope. Theorem 3.1 simplifies for these n to

$$p_{\{a_0,\dots,a_d\}}(a_0 \cdots a_d m) = q(a_0,\dots,a_d,a_0 \cdots a_d m) + \sum_{j=0}^{d} \sigma_0\left(a_0,\dots,\hat{a}_j,\dots,a_d;a_j\right) ,$$

by the periodicity of the Fourier-Dedekind sums. However, $\chi(\mathcal{P}) = 1$, and Theorem 4.5 yields a result equivalent to Zagier's reciprocity law for his higher-dimensional Dedekind sums, Theorem 2.6:

THEOREM 4.6. *For pairwise relatively prime integers $a_1, \dots, a_d$,*

$$\sum_{j=0}^{d} \sigma_0(a_0, \dots, \hat{a}_j, \dots, a_d; a_j) = 1 - q(a_0, \dots, a_d, 0) ,$$

where $q(a_0, \dots, a_d, n)$ is given by (3.2).

5. The computational complexity of Zagier's higher-dimensional Dedekind sums

In this section we give a proof of the polynomial-time complexity of Zagier's higher-dimensional Dedekind sums, in fixed dimension d. In [**BP**], there is a nice theorem due to Barvinok which guarantees the polynomial-time computability of the generating function attached to a rational polyhedron. We will use his theorem for a cone. First, we mention that a common way to enumerate lattice points in a cone $\mathcal{K}$ (and in polytopes) is to use the generating function

$$f(\mathcal{K}, \mathbf{x}) := \sum_{\mathbf{m} \in \mathcal{K} \cap \mathbb{Z}^d} \mathbf{x}^{\mathbf{m}} ,$$

where we use the standard multivariate notation $\mathbf{x}^{\mathbf{m}} := x_1^{m_1} \dots x_d^{m_d}$. It is an elementary fact that for rational cones these generating functions are always rational functions of the variable $\mathbf{x}$. Barvinok's theorem reads as follows:

THEOREM 5.1 (Barvinok). *Let us fix the dimension d. There exists a polynomial-time algorithm, which for a given rational polyhedron $\mathcal{K} \subset \mathbb{R}^d$,*

$$\mathcal{K} = \{\mathbf{x} \in \mathbb{R}^d : \langle \mathbf{c}_i, \mathbf{x} \rangle \le \beta_i, i = 1...m\}, \text{ where } \mathbf{c}_i \in \mathbb{Z}^d \text{ and } \beta_i \in \mathbb{Q}$$

computes the generating function

$$f(\mathcal{K}, \mathbf{x}) := \sum_{\mathbf{m} \in \mathcal{K} \cap \mathbb{Z}^d} \mathbf{x}^{\mathbf{m}}$$

in the form (a virtual decomposition)

$$f(\mathcal{K}, \mathbf{x}) = \sum_{i \in I} \epsilon_i \frac{\mathbf{x}^{a_i}}{(1 - \mathbf{x}^{b_{i1}}) \cdots (1 - \mathbf{x}^{b_{id}})} ,$$

where $\epsilon_i \in \{-1, 1\}, a_i \in \mathbb{Z}$, and $b_{i1}, ..., b_{id}$ is a basis of $\mathbb{Z}^d$ for each i. The computational complexity of the algorithm for finding this virtual decomposition is $\mathcal{L}^{O(d)}$, where $\mathcal{L}$ is the input size of $\mathcal{K}$. In particular, the number I of terms in the summand is $\mathcal{L}^{O(d)}$.

Thus Barvinok's algorithm finds the coefficients of the rational function $f(\mathcal{K}, \mathbf{x})$ in polynomial time. In [**DR**], on the other hand, the generating function $f(\mathcal{K}, \mathbf{x})$ is given in terms of an average over a finite abelian group of a product of d cotangent functions, whose arguments are in terms of the coordinates of the vertices which generate the cone $\mathcal{K}$ (these are the extreme points of $\mathcal{K}$ whose convex hull is $\mathcal{K}$). This is the main theorem in [**DR**] and we apply it below to a special lattice cone which will give us the Zagier-Dedekind sums we want to study.

The following theorem is part of a bigger project on the computability of generalized Dedekind sums in all dimensions. A slightly different proof is sketched in [**BP**].

THEOREM 5.2. *For fixed dimension d, the higher-dimensional Dedekind sums*

$$s(a_0; a_1, \ldots, a_d) = \frac{(-1)^{d/2}}{a_0} \sum_{k=1}^{a_0-1} \cot \frac{\pi k a_1}{a_0} \cdots \cot \frac{\pi k a_d}{a_0} .$$

are polynomial-time computable.

PROOF. Let $\mathcal{K} \subset \mathbb{R}^{d+1}$ be the cone generated by the positive real span of the vectors

$$\begin{aligned} v_1 &= (1, 0, \ldots, 0, a_1) \\ v_2 &= (0, 1, 0, \ldots, 0, a_2) \\ &\vdots \\ v_d &= (0, \ldots, 0, 1, a_d) \\ v_{d+1} &= (0, \ldots, 0, a_0) \end{aligned}$$

Then the right-hand side of the main theorem of [**DR**] is in this case

$$\frac{1}{2^{d+1} a_0} \sum_{k=1}^{a_0-1} \prod_{j=0}^{d} \left(1 + \coth \frac{\pi}{a_0} (s + i a_j k) \right) .$$

When we compute the coefficient of s^{-1} in this meromorphic function of s, we arrive at the following higher-dimensional Dedekind sum:

$$\sum_{k=1}^{a_0-1} \cot \frac{\pi k a_1}{a_0} \cdots \cot \frac{\pi k a_d}{a_0}$$

plus other products of lower-dimensional Zagier-Dedekind sums. By induction on the dimension, all of the lower-dimensional Zagier-Dedekind sums are polynomial-time computable, and since the left-hand side of the main theorem is polynomial-time computable by Barvinok's theorem, the above Zagier-Dedekind sums in dimension d is now also polynomial-time computable. □

References

[BP] A. I. BARVINOK, J. E. POMMERSHEIM, An algorithmic theory of lattice points in polyhedra, in *New perspectives in algebraic combinatorics, Berkeley, CA, (1996-97), Math. Sci. Res. Inst. Publ.* **38**, Cambridge Univ. Press, Cambridge (1999), 91–147.

[BDR] M. BECK, R. DIAZ, S. ROBINS, The Frobenius problem, rational polytopes, and Fourier-Dedekind sums, to appear in *J. Number Th.* **96** (2002), 1–21.

[BGK] M. BECK, I. M. GESSEL, T. KOMATSU, The polynomial part of a restricted partition function related to the Frobenius problem, *Electronic J. Combin.* **8**, no. 1 (2001), N 7.

[BR] M. BECK, S. ROBINS, Explicit and efficient formulas for the lattice point count inside rational polygons, *Discr. Comp. Geom.* **27** (2002), 443–459.

[B] B. BERNDT Reciprocity theorems for Dedekind sums and generalizations, *Adv. in Math.* **23**, no. 3 (1977), 285–316.

[BV] M. BRION, M. VERGNE An equivariant Riemann-Roch theorem for simplicial toric varieties, *J. reine angew. Math.* **482** (1997), 67–92.

[C] L. CARLITZ, A note on generalized Dedekind sums, *Duke Math. J.* **21** (1954), 399–404.

[De] R. DEDEKIND, Erläuterungen zu den Fragmenten XXVIII, in *Collected works of Bernhard Riemann*, Dover Publ., New York (1953), 466–478.

[DR] R. DIAZ, S. ROBINS, The Erhart polynomial of a lattice polytope, *Ann. Math.* **145** (1997), 503–518.

[D] U. DIETER, Das Verhalten der Kleinschen Funktionen $\log \sigma_{g,h}(w_1, w_2)$ gegenüber Modultransformationen und verallgemeinerte Dedekindsche Summen, *J. reine angew. Math.* **201** (1959), 37–70.

[E] E. EHRHART, Sur un problème de géométrie diophantienne linéaire II, *J. reine angewandte Math.* **227** (1967), 25–49.

[HZ] F. HIRZEBRUCH, D. ZAGIER, *The Atiyah-Singer theorem and elementary number theory*, Publish or Perish, Boston (1974).

[G] I. GESSEL, Generating functions and generalized Dedekind sums, *Electronic J. Comb.* **4**, no. 2 (1997), R 11.

[K] D. KNUTH, *The art or computer programming, vol. 2*, Addison-Wesley, Reading, Mass., (1981).

[Ma] I. G. MACDONALD, Polynomials associated with finite cell complexes, *J. London Math. Soc.* **4** (1971), 181–192.

[Me] C. MEYER, Über einige Anwendungen Dedekindscher Summen, *J. reine angewandte Math.* **198** (1957), 143–203.

[Mo] L. J. MORDELL, Lattice points in a tetrahedron and generalized Dedekind sums, *J. Indian Math.* **15** (1951), 41–46.

[P] J. POMMERSHEIM, Toric varieties, lattice points, and Dedekind sums, *Math. Ann.* **295** (1993), 1–24.

[R] H. RADEMACHER, Some remarks on certain generalized Dedekind sums, *Acta Aritm.* **9** (1964), 97–105.

[RG] H. RADEMACHER, E. GROSSWALD, *Dedekind sums*, Carus Mathematical Monographs, The Mathematical Association of America (1972).

[Z] D. ZAGIER, Higher dimensional Dedekind sums, *Math. Ann.* **202** (1973), 149–172.

DEPARTMENT OF MATHEMATICS, SAN FRANCISCO STATE UNIVERSITY, SAN FRANCISCO, CA 94132, USA
E-mail address: `beck@math.sfsu.org`

DEPARTMENT OF MATHEMATICS, TEMPLE UNIVERSITY, PHILADELPHIA, PA 19122, USA
E-mail address: `srobins@math.temple.edu`

DIMACS Series in Discrete Mathematics
and Theoretical Computer Science
Volume **64**, 2004

Noncommutative Number Theory

Paula B. Cohen

ABSTRACT. This paper comments on recent ideas of Bost-Connes and of Connes on possible ways to approach the study of the Riemann zeta function using ideas inspired by noncommutative geometry.

1. Introduction

In 1859 Riemann published an important foundational paper on the Riemann zeta function. Recall that this function is given for $\mathrm{Re}(s) > 1$ by

$$\zeta(s) = \sum_{n=1}^{\infty} \frac{1}{n^s}$$

and that it has a continuation to all the complex plane which is analytic except for a simple pole at $s = 1$. It is straightforward to show that the Riemann zeta function has zeros at the negative even integers and these are called the trivial zeros of the Riemann zeta function. The Riemann hypothesis predicts that the remaining zeros lie on the line $\mathrm{Re}(s) = 1/2$. One knows that the non-trivial zeros of $\zeta(s)$ lie in the band $\mathrm{Re}(s) \in]0,1[$. The generalisation of this function for a number field is known as the Dedekind zeta function. It encodes much important arithmetical information about the field. One of the major motivations of number theory is to understand more fully the Dedekind zeta function, the most famous challenge being to understand the locus of the zeros of this function and in particular to settle the validity of the Riemann hypothesis. Much powerful work has been done in analytic number theory in the attempt to solve the Riemann hypothesis directly. As the study of the zeros of the Riemann zeta function and its generalisations is so difficult, one may ask how it is possible to recast the problem. For example, Polya and Hilbert proposed that if one can construct a Hilbert space $\mathcal{H}$ and an operator D in $\mathcal{H}$ whose spectrum comprises the zeros of the Riemann zeta function in the band $\mathrm{Re}(s) \in]0,1[$, then possibly one can settle whether or not $\sqrt{-1}(D - 1/2)$ is self-adjoint or whether $D(1-D)$ is positive, which would imply the Riemann hypothesis. The point here is that the properties of self-adjointness or of positivity are hopefully easier to check. It is important that the construction of the Hilbert space and the operator should not depend *a priori* on the zeta function, to avoid tautologies. An

2000 *Mathematics Subject Classification.* 11M06, 11R56, 46L89.
Key words and phrases. Riemann hypothesis, adeles, ideles, noncommutative spaces.

as yet non-rigorous approach to the Riemann hypothesis initiated by Connes in [**5**] and further developed in [**6**],[**7**], includes an interesting and rigorous interpretation of the zeros of the L-functions with Grossencharacter of a global field in terms of the action of the idele class group on the coset space of the adeles modulo the principal ideles. An excellent motivational article about Connes approach to the Riemann hypothesis can be found in [**7**].

This work of Connes derives in particular from his work with Bost [**2**] which was in turn inspired by ideas of Julia [**11**] and others. The aim is to enrich our knowledge of the Riemann zeta function by creating a dictionary between its properties and phenomena in statistical mechanics. The starting point of these approaches is the observation that, just as the zeta functions encode arithmetic information, the partition functions of quantum statistical mechanical systems encode their large-scale thermodynamical properties. The first step therefore is to construct a quantum dynamical system with partition function the Riemann zeta function, or the Dedekind zeta function in the general number field case. In order for the quantum dynamical system to reflect the arithmetic of the primes, it must capture also some sort of interaction between them. This last feature translates in the statistical mechanical language into the phenomenon of spontaneous symmetry breaking at a critical temperature with respect to a natural symmetry group. In the region of high temperature, there is a unique equilibrium state as the system is in disorder and symmetric with respect to the action of the symmetry group. In the region of low temperature, a phase transition occurs and the symmetry is broken. This symmetry group acts transitively on a family of possible extremal equilibrium states. The construction of a quantum dynamical system with partition function the Riemann zeta function $\zeta(\beta)$ and spontaneous symmetry breaking or phase transition at its pole $\beta = 1$ with respect to a natural symmetry group was achieved by Bost and Connes in [**2**]. A different construction of the basic algebra using crossed products was proposed by Laca and Raeburn and extended to the number field case by them with Arledge in [**1**]. An extension of the work of Bost and Connes to general global fields was done by Harari and Leichtnam in [**10**]. The generalisation proposed by Harari and Leichtnam in [**10**] fails to capture the Dedekind zeta function as partition function in the case of a number field with class number greater than 1. Their partition function in that case is the Dedekind zeta function with a finite number of non-canonically chosen Euler factors removed. This prompted the author's paper [**3**] where the full Dedekind zeta function is recovered as partition function. This is achieved by recasting the original construction of Bost and Connes more completely in terms of adeles and ideles.

The symmetry group of the system constructed by Bost and Connes is a Galois group, in fact the Galois group over the rational number field of its maximal abelian extension. Using the Artin isomorphism, which says that this symmetry group is also the unit group of the finite ideles, Bost and Connes recover the actual Galois action on the elements of the maximal abelian extension via its action on the equilibrium states of the system. In the general number field case, the symmetry group is again the unit group of the finite ideles, but this group does not in general have a Galois interpretation. See [**10**] for a discussion of this point.

In this paper we provide an introduction both to the work of Bost–Connes and to the work of Connes on the Riemann hypothesis. Our purpose is to assist the interested reader to understand these works. We restrict our attention to the rational numbers and to the Riemann zeta function. It should be noted that a solution of the Riemann hypothesis may well necessarily involve studying the zeta function as part of a family of zeta functions in the spirit of [**12**].

Acknowledgements: The author thanks the Ellentuck Fund and the School of Mathematics of the Institute for Advanced Study, Princeton, for their support during the preparation of this paper.

2. The problem studied by Bost–Connes

Before stating the problem solved by Bost and Connes in [**2**] and its analogue for number fields, we recall a few basic notions from the C^*-algebraic formulation of quantum statistical mechanics. For the background, see [**4**]. Recall that a C^*-algebra B is an algebra over the complex numbers $\mathbb{C}$ with an adjoint $x \mapsto x^*$, $x \in B$, that is, an anti-linear map with $x^{**} = x$, $(xy)^* = y^*x^*$, $x, y \in B$, and a norm $\|\,.\,\|$ with respect to which B is complete and addition and multiplication are continuous operations. One requires in addition that $\|xx^*\| = \|x\|^2$ for all $x \in B$. All our C^*-algebras will be assumed unital. The most basic example of a noncommutative C^*-algebra is $B = M_N(\mathbb{C})$ for $N \geq 2$ an integer. The C^*-algebra plays the role of the "space" on which the system evolves, the evolution itself being described by a 1-parameter group of C^*-automorphisms $\sigma : \mathbb{R} \mapsto \mathrm{Aut}(B)$. The quantum dynamical system is therefore the pair (B, σ_t). It is customary to use the inverse temperature $\beta = 1/kT$ rather than the temperature T, where k is Boltzmann's constant. Then, one has the definition of Kubo-Martin-Schwinger (KMS) of an equilibrium state at inverse temperature β. Recall that a state φ on a C^*-algebra B is a positive linear functional on B satisfying $\varphi(1) = 1$. It is the generalisation of a probability distribution.

DEFINITION 1. *Let (B, σ_t) be a dynamical system, and φ a state on B. Then φ is an equilibrium state at inverse temperature β, or* KMS$_\beta$*-state, if for each $x, y \in B$ there is a function $F_{x,y}(z)$, bounded and holomorphic in the band $0 < \mathrm{Im}(z) < \beta$ and continuous on its closure, such that for all $t \in \mathbb{R}$,*

$$F_{x,y}(t) = \varphi(x\sigma_t(y)), \qquad F_{x,y}(t + \sqrt{-1}\beta) = \varphi(\sigma_t(y)x). \tag{1}$$

In the case where $B = M_N(\mathbb{C})$, every 1-parameter group σ_t of automorphisms of B can be written in the form,

$$\sigma_t(x) = e^{itH}xe^{-itH}, \qquad x \in B, \qquad t \in \mathbb{R},$$

for a self-adjoint matrix $H = H^*$. Then for $H > 0$ and for all $\beta > 0$, there is a unique KMS$_\beta$ equilibrium state for (B, σ_t) given by

$$\phi_\beta(x) = \mathrm{Trace}(xe^{-\beta H})/\mathrm{Trace}(e^{-\beta H}), \qquad x \in M_N(\mathbb{C}). \tag{2}$$

This has the familiar form of a Gibbs state and is easily seen to satisfy the KMS$_\beta$ condition of 1. The KMS$_\beta$ states can therefore be seen as generalisations of Gibbs states. The normalisation constant $\mathrm{Trace}(e^{-\beta H})$ is known as the partition function of the system. A symmetry group G of the dynamical system (B, σ_t) is a subgroup of $\mathrm{Aut}(B)$ commuting with σ:

$$g \circ \sigma_t = \sigma_t \circ g, \qquad g \in G, t \in \mathbb{R}.$$

Consider now a system (B, σ_t) with interaction. Then, guided by quantum statistical mechanics, we expect to see the following features. When the temperature is high, so that β is small, the system is in disorder, there is no interaction between its constituents and the state of the system does not see the action of the symmetry group G: the KMS$_\beta$-state is unique. As the temperature is lowered, the constituents of the system begin to interact. At a critical temperature β_0 a phase transition occurs and the symmetry is broken. The symmetry group G then permutes transitively a family of extremal KMS$_\beta$- states generating the possible states of the system after phase transition: the KMS$_\beta$-state is no longer unique. This phase transition phenomenon is known as spontaneous symmetry breaking at the critical inverse temperature β_0. The partition function should have a pole at β_0. For a fuller explanation, see [**2**]. The problem solved by Bost and Connes was the following.

Problem 1: *Construct a dynamical system (B, σ_t) with partition function the zeta function $\zeta(\beta)$ of Riemann, where $\beta > 0$ is the inverse temperature, having spontaneous symmetry breaking at the pole $\beta = 1$ of the zeta function with respect to a natural symmetry group.*

As mentioned in the introduction, the symmetry group is the unit group of the ideles, given by $W = \prod_p \mathbb{Z}_p^*$ where the product is over the primes p and $\mathbb{Z}_p^* = \{u_p \in \mathbb{Q}_p : |u_p|_p = 1\}$. We use here the normalisation $|p|_p = p^{-1}$. This is the same as the Galois group $\mathrm{Gal}(\mathbb{Q}^{ab}/\mathbb{Q})$. Here $\mathbb{Q}^{ab}$ is the maximal abelian extension of the rational number field $\mathbb{Q}$, which in turn is isomorphic to its maximal cyclotomic extension, that is the extension obtained by adjoining to $\mathbb{Q}$ all the roots of unity. The interaction detected in the phase transition comes about from the interaction between the primes coming from considering at once all the embeddings of the non-zero rational numbers $\mathbb{Q}^*$ into the completions $\mathbb{Q}_p$ of $\mathbb{Q}$ with respect to the prime valuations $|.|_p$.

The natural generalisation of this problem to the number field case was solved in [**3**] and is the following.

Problem 2: *Given a number field K, construct a dynamical system (B, σ_t) with partition function the Dedekind zeta function $\zeta_K(\beta)$, where $\beta > 0$ is the inverse temperature, having spontaneous symmetry breaking at the pole $\beta = 1$ of the Dedekind function with respect to a natural symmetry group.*

Recall that the Dedekind zeta function is given by

$$\zeta_K(s) = \sum_{\mathcal{C} \subset \mathcal{O}} \frac{1}{N(\mathcal{C})^s}, \qquad \mathrm{Re}(s) > 1. \tag{3}$$

Here $\mathcal{O}$ is the ring of integers of K and the summation is over the ideals $\mathcal{C}$ of K contained in $\mathcal{O}$. The symmetry group is the unit group of the finite ideles of K.

For the natural generalisation to the function field case see [**10**]. For the sake of exposition, we restrict ourselves in the sequel to the case of the rational numbers, that is to a discussion of Problem 1.

3. Construction of the C^*-algebra

We give a different construction of the C^*-algebra of [**2**] to that found in their original paper. It is essentially equivalent to the construction of [**1**], except that we work with adeles and ideles. In the generalisation to the number field case, this makes quite a difference. Let $\mathcal{A}$ denote the finite adeles of $\mathbb{Q}$, that is the restricted product of $\mathbb{Q}_p$ with respect to $\mathbb{Z}_p$. Recall that this restricted product consists of the infinite vectors $(a_p)_p$, indexed by the primes p, such that $a_p \in \mathbb{Q}_p$ with $a_p \in \mathbb{Z}_p$ for almost all primes p. The (finite) adeles form a ring under componentwise addition and multiplication. The (finite) ideles $\mathcal{J}$ are the invertible elements of the adeles. They form a group under componentwise multiplication. Let $\mathbb{Z}_p^*$ be those elements of $u_p \in \mathbb{Z}_p$ with $|u_p|_p = 1$. Notice that an idele $(u_p)_p$ has $u_p \in \mathbb{Q}_p^*$ with $u_p \in \mathbb{Z}_p^*$ for almost all primes p. Let

$$\mathcal{R} = \prod_p \mathbb{Z}_p, \qquad I = \mathcal{J} \cap \mathcal{R}, \qquad W = \prod_p \mathbb{Z}_p^*.$$

Further, let I denoted the semigroup of integral ideals of $\mathbb{Z}$. It is the semigroup of $\mathbb{Z}$-modules of the form $m\mathbb{Z}$ where $m \in \mathbb{Z}$. Notice that I as above is also a semigroup. We have a natural short exact sequence,

$$1 \to W \to I \to \mathrm{I} \to 1. \tag{4}$$

The map $I \to \mathrm{I}$ in this short exact sequence is given as follows. To $(u_p)_p \in I$ associate the ideal $\prod_p p^{\mathrm{ord}_p(u_p)}$ where $\mathrm{ord}_p(u_p)$ is determined by the formula $|u_p|_p = p^{-\mathrm{ord}_p(u_p)}$. It is clear that this map is surjective with kernel W, that is that the above sequence is indeed short exact. By the Strong Approximation Theorem we have

$$\mathbb{Q}/\mathbb{Z} \simeq \mathcal{A}/\mathcal{R} \simeq \oplus_p \mathbb{Q}_p/\mathbb{Z}_p \tag{5}$$

and we have therefore a natural action of I on $\mathbb{Q}/\mathbb{Z}$ by multiplication in $\mathcal{A}/\mathcal{R}$ and transport of structure. We use here that $I\mathcal{R} \subset \mathcal{R}$. Mostly we shall work in $\mathcal{A}/\mathcal{R}$ rather than $\mathbb{Q}/\mathbb{Z}$. We have the following straightforward Lemma (see [**3**]).

LEMMA 1. *For $a = (a_p)_p \in I$ and $y \in \mathcal{A}/\mathcal{R}$, the equation*

$$ax = y$$

has $n(a) =: \prod_p p^{\mathrm{ord}_p(a_p)}$ solutions in $x \in \mathcal{A}/\mathcal{R}$. Denote these solutions by $[x : ax = y]$.

In the above lemma it is important to bear in mind that we are computing modulo $\mathcal{R}$. Now, let $\mathbb{C}[\mathcal{A}/\mathcal{R}] =: \mathrm{span}\{\delta_x : x \in \mathcal{A}/\mathcal{R}\}$ be the group algebra of $\mathcal{A}/\mathcal{R}$ over $\mathbb{C}$, so that $\delta_x\delta_{x'} = \delta_{x+x'}$ for $x, x' \in \mathcal{A}/\mathcal{R}$. We have (see for comparison [**1**]),

LEMMA 2. *The formula*

$$\alpha_a(\delta_y) = \frac{1}{n(a)} \sum_{[x:ax=y]} \delta_x$$

for $a \in I$ defines an action of I by endomorphisms of $C^*(\mathcal{A}/\mathcal{R})$.

The endomorphism α_a for $a \in I$ is a one-sided inverse of the map $\delta_x \mapsto \delta_{ax}$ for $x \in \mathcal{A}/\mathcal{R}$, so it is like a semigroup "division". The C^*-algebra can be thought of as the operator norm closure of $\mathbb{C}[\mathcal{A}/\mathcal{R}]$ in its natural left regular representation in $l^2(\mathcal{A}/\mathcal{R})$. We now appeal to the notion of semigroup crossed product developed by Laca and Raeburn and used in [**1**], applying it to our situation. A covariant representation of $(C^*(\mathcal{A}/\mathcal{R}), I, \alpha)$ is a pair (π, V) where

$$\pi : C^*(\mathcal{A}/\mathcal{R}) \to B(\mathcal{H})$$

is a unital representation and

$$V : I \to B(\mathcal{H})$$

is an isometric representation in the bounded operators in a Hilbert space $\mathcal{H}$. The pair (π, V) is required to satisfy,

$$\pi(\alpha_a(f)) = V_a \pi(f) V_a^*, \qquad a \in I, \quad f \in C^*(\mathcal{A}/\mathcal{R}).$$

Notice that the V_a are not in general unitary. Such a representation is given by (λ, L) on $l^2(\mathcal{A}/\mathcal{R})$ with orthonormal basis $\{e_x : x \in \mathcal{A}/\mathcal{R}\}$ where λ is the left regular representation of $C^*(\mathcal{A}/\mathcal{R})$ on $l^2(\mathcal{A}/\mathcal{R})$ and

$$L_a e_y = \frac{1}{\sqrt{n(a)}} \sum_{[x:ax=y]} e_x.$$

The universal covariant representation, through which all other covariant representations factor, is called the (semigroup) crossed product $C^*(\mathcal{A}/\mathcal{R}) \times_\alpha I$. This algebra is the universal C^*-algebra generated by the symbols $\{e(x) : x \in \mathcal{A}/\mathcal{R}\}$ and $\{\mu_a : a \in I\}$ subject to the relations

$$\mu_a^* \mu_a = 1, \quad \mu_a \mu_b = \mu_{ab}, \qquad a, b \in I, \tag{6}$$

$$e(0) = 1, \quad e(x)^* = e(-x), \quad e(x)e(y) = e(x+y), \qquad x, y \in \mathcal{A}/\mathcal{R}, \tag{7}$$

$$\frac{1}{n(a)} \sum_{[x:ax=y]} e(x) = \mu_a e(y) \mu_a^*, \qquad a \in I, y \in \mathcal{A}/\mathcal{R}. \tag{8}$$

The relations in (6) reflect a multiplicative structure, those in (7) an additive structure and those in (8) how these multiplicative and additive structures are related via the crossed product action. Julia [**11**] observed that by using only the multiplicative structure of the integers one cannot hope to capture an interaction between the different primes. When $u \in W$ then μ_u is a unitary, so that $\mu_u^* \mu_u = \mu_u \mu_u^* = 1$ and we have for all $x \in \mathcal{A}/\mathcal{R}$,

$$\mu_u e(x) \mu_u^* = e(u^{-1}x), \qquad \mu_u^* e(x) \mu_u = e(ux). \tag{9}$$

Therefore we have a natural action of W as inner automorphisms of $C^*(\mathcal{A}/\mathcal{R}) \times_\alpha I$ using (9).

To recover the C^*-algebra of [**2**] we must split the short exact sequence (4). The ideals in I are all of the form $m\mathbb{Z}$ for some $m \in \mathbb{Z}$. This generator m is determined up to sign. Consider the image of $|m|$ in I under the diagonal embedding $q \mapsto (q)_p$ of $\mathbb{Q}^*$ into I, where the p-th component of $(q)_p$ is the image of q in $\mathbb{Q}_p^*$ under the natural embedding of $\mathbb{Q}^*$ in $\mathbb{Q}_p^*$. The map

$$+ : m\mathbb{Z} \mapsto (|m|)_p \tag{10}$$

defines a splitting of (4). Let I_+ denote the image and define B to be the semigroup crossed product $C^*(\mathcal{A}/\mathcal{R}) \times_\alpha I_+$ with the restricted action α from I to I_+. By transport of structure using (5), this algebra is easily seen to be isomorphic to a semigroup crossed product of $C^*(\mathbb{Q}/\mathbb{Z})$ by $\mathbb{N}_+$, where $\mathbb{N}_+$ denotes the positive natural numbers. This is the algebra constructed in [**2**] (see also [**1**]). From now on, we use the symbols $\{e(x) : x \in \mathbb{Q}/\mathbb{Z}\}$ and $\{\mu_a : a \in \mathbb{N}_+\}$. It is essential to split the short exact sequence in this way in order to obtain the symmetry breaking phenomenon.

In particular, this replacement of I by I_+ now means that the group W acts by outer automorphisms. For $x \in B$, one has that $\mu_u^* x \mu_u$ is still in B (computing in the larger algebra $C^*(\mathcal{A}/\mathcal{R}) \times_\alpha I$), but now this defines an outer action of W. This coincides with the definition of W as the symmetry group as in [**2**].

4. The Theorem of Bost–Connes

Using the abstract description of the C^*-algebra B of §3, to define the time evolution σ of our dynamical system (B, σ) it suffices to define it on the symbols $\{e(x) : x \in \mathbb{Q}/\mathbb{Z}\}$ and $\{\mu_a : a \in \mathbb{N}_+\}$. For $t \in \mathbb{R}$, let σ_t be the automorphism of B defined by

$$\sigma_t(\mu_m) = m^{it}, \quad m \in \mathbb{N}_+, \qquad \sigma_t(e(x)) = e(x), \quad x \in \mathbb{Q}/\mathbb{Z}. \tag{11}$$

By (6) and (9) we clearly have that the action of W commutes with this 1-parameter group σ_t. Hence W will permute the extremal KMS$_\beta$-states of (B, σ_t). To describe the KMS$_\beta$-states for $\beta > 1$, we shall represent (B, σ_t) on a Hilbert space. Namely, following [**2**], let $\mathcal{H}$ be the Hilbert space $l^2(\mathbb{N}_+)$ with canonical orthonormal basis $\{\varepsilon_m, m \in \mathbb{N}_+\}$. For each $u \in W$, one has a representation π_u of B in $B(\mathcal{H})$ given by,

$$\pi_u(\mu_m)\varepsilon_n = \varepsilon_{mn}, \; m, n \in \mathbb{N}_+$$

$$\pi_u(e(x))\varepsilon_n = \exp(2i\pi n u \circ x)\varepsilon_n, \; n \in \mathbb{N}_+, \, x \in \mathbb{Q}/\mathbb{Z}. \tag{12}$$

Here $u \circ x$ for $u \in W$ and $x \in \mathbb{Q}/\mathbb{Z}$ is the multiplication induced by transport of structure using (5). One verifies easily that (12) does indeed give a C^*-algebra representation of B. Let H be the unbounded operator in $\mathcal{H}$ whose action on the canonical basis is given by

$$H\varepsilon_n = (\log n)\varepsilon_n, \quad n \in \mathbb{N}_+. \tag{13}$$

Then clearly, for each $u \in W$, we have

$$\pi_u(\sigma_t(x)) = e^{itH}\pi_u(x)e^{-itH}, \qquad t \in \mathbb{R}, \, x \in B.$$

Notice that, for $\beta > 1$,

$$\mathrm{Trace}(e^{-\beta H}) = \sum_{n=1}^{\infty}\langle e^{-\beta H}\varepsilon_n, \varepsilon_n\rangle = \sum_{n=1}^{\infty} n^{-\beta}\langle \varepsilon_n, \varepsilon_n\rangle = \sum_{n=1}^{\infty} n^{-\beta},$$

so that the Riemann zeta function appears as a partition function of Gibbs state type. We can now state the main result of [**2**].

THEOREM 1. *(Bost-Connes) The dynamical system (B, σ_t) has symmetry group W. The action of $u \in W$ is given by $[u] \in \mathrm{Aut}(B)$ where*

$$[u] : e(y) \mapsto e(u \circ y), \quad y \in \mathbb{Q}/\mathbb{Z}, \qquad [u] : \mu_a \mapsto \mu_a, \quad a \in \mathbb{N}.$$

This action commutes with σ,

$$[u] \circ \sigma_t = \sigma_t \circ [u], \qquad u \in W, \quad t \in \mathbb{R}.$$

Moreover,

(1) for $0 < \beta \le 1$, there is a unique KMS$_\beta$ *state. (It is a factor state of Type* III$_1$ *with associated factor the Araki-Woods factor R_∞.)*

(2) for $\beta > 1$ and $u \in W$, the state

$$\phi_{\beta,u}(x) = \zeta(\beta)^{-1}\mathrm{Trace}(\pi_u(x)e^{-\beta H}), \qquad x \in B$$

is a KMS$_\beta$ *state for (B, σ_t). (It is a factor state of Type* I$_\infty$*). The action of W on B induces an action on these* KMS$_\beta$ *states which permutes them transitively and the map $u \mapsto \phi_{\beta,u}$ is a homomorphism of the compact group W onto the space $\mathcal{E}_\beta$ of extremal points of the simplex of* KMS$_\beta$ *states for (B, σ_t).*

(3) the ζ function of Riemann is the partition function of (B, σ_t).

Part (1) of the above theorem is difficult and the reader is referred to [**2**] for complete details, as for a full proof of (2). That for $\beta > 1$ the KMS_β-states given in part (2) fulfil 1 of §2 is a straightforward exercise. Notice that they have the form of Gibbs equilibrium states.

5. The space and group action in Connes' approach to the Riemann hypothesis

Theorem 1 solves Problem 1 of §2. More information is contained in its proof however. As mentioned in the §1, given the existence of the Artin isomorphism in class field theory for the rationals, one can recover the Galois action of W explicitly. It is still an open problem to exhibit this Galois action in terms of an analogue of (B, σ_t) in a satisfactory way for general number fields.

Another feature occurs in the analysis of the proof of part (1) of Theorem 1. One can treat the infinite places in a similar way to that already described for the finite places, so working with the (full) adeles A and (full) ideles J. The ring of adeles A of $\mathbb{Q}$ consists of the infinite vectors $(a_\infty, a_p)_p$ indexed by the archimedean place and the primes p of $\mathbb{Q}$ with $a_p \in \mathbb{Z}_p$ for all but finitely many p. The group J of ideles consists of the infinite vectors $(u_\infty, u_p)_p$ with $u_\infty \in \mathbb{R}, u_\infty \neq 0$ and $u_p \in \mathbb{Q}_p, u_p \neq 0$ and $|u_p|_p = 1$ for all but finitely many primes p. There is a norm $|\cdot|$ defined on J given by $|u| = |u_\infty| \prod_p |u_p|_p$. We have natural diagonal embeddings of $\mathbb{Q}$ in A and $\mathbb{Q}^* = \mathbb{Q} \setminus \{0\}$ in J induced by the embeddings of $\mathbb{Q}$ into its completions. Notice that by the product formula $\mathbb{Q}^* \subset \mathrm{Ker}|\cdot|$. We define an equivalence relation on A by $a \equiv b$ if and only if there exists a $q \in \mathbb{Q}^*$ with $a = qb$. With respect to this equivalence, we form the coset space $X = A/\mathbb{Q}^*$. The ideles J act on A by componentwise multiplication, which induces an action of $C = J/\mathbb{Q}^*$ on X. Notice that this action has fixed points. For example, whenever an adele a has $a_p = 0$ it is a fixed point of the embedding of $\mathbb{Q}_p^*$ into J (to $q_p \in \mathbb{Q}_p^*$ one assigns the idele with 1 in every place except the pth place.) On the other hand, every Type III_1 factor has a continuous decomposition, that is it can be written as a cross-product of $\mathbb{R}$ with a Type II_∞ factor. Connes has observed that the von-Neumann algebra of Type III_1 in the region $0 < \beta \leq 1$ of Theorem 1 has in its continuous decomposition the Type II_∞ factor given by the crossed product of $L^\infty(A)$ by the action of $\mathbb{Q}^*$ by multiplication. The associated von Neumann algebra has orbit space $X = A/\mathbb{Q}^*$.

The pair (X, C) plays a fundamental role in Connes's proposed approach to the Riemann hypothesis in [**5**] and can be thought of as playing the role for number fields of the curve and Frobenius for the proof of the Riemann hypothesis in the case of curves over finite fields. We remark that Connes's approach does not as yet reprove the established analogue of the Riemann hypothesis in this case: if it did so this would provide an alternative proof. In order for his approach to work, Connes would need to be able to prove an asymptotic formula for the trace of the action of C on X. A "spectral" analysis of this action, related to older ideas of Weil [**17**] and Tate, shows it to be related to the non-trivial zeros of the L-functions with Grossencharacter, and a heurisitic "geometric" analysis relates this action to the Weil distribution. The "geometric" heuristics aim at suggesting a proof for the positivity of the Weil distribution which is known to imply the Riemann hypothesis. In the subsequent sections we give a leisurely introduction to these ideas.

6. A trace formula over the reals

There is a rigorous local version of the geometric side of the asymptotic trace formula of [**6**]. For simplicity we work at the infinite place, the discussion at the prime places being analogous, and justify the formula in terms of the simple computation of the distributional trace. Consider the action of $\mathbb{R}^* = \mathbb{R} \setminus \{0\}$ on $\mathbb{R}$ by multiplication. This induces the action on smooth functions on $\mathbb{R}$,

$$U(u)g = g \circ u^{-1}, \qquad u \in \mathbb{R}^*, \quad g \in C^\infty(\mathbb{R}). \tag{14}$$

We average this action over a function of rapid decay. Namely, for $h \in \mathcal{S}(\mathbb{R}^*)$, we let $U(h)$ be the operator in $L^2(\mathbb{R})$ given by,

$$U(h) = \int_{\mathbb{R}^*} h(u)U(u)d^*u, \tag{15}$$

where the multiplicative Haar measure d^*u is normalised by,

$$\int_{|u|\in[1,\Lambda]} d^*u \simeq \log \Lambda, \qquad \Lambda \to \infty. \tag{16}$$

The associated kernel of $U(u)$ is

$$k(x, u^{-1}, y) = \delta(y - u^{-1}x), \tag{17}$$

and for $u \neq 1$,

$$\int_{\mathbb{R}} k(x, u^{-1}, x)dx = |1 - u^{-1}|^{-1}. \tag{18}$$

If $h \in \mathcal{S}(\mathbb{R}^*)$ and $h(1) = 0$, we therefore see that the distributional trace of $U(h)$ is given by,

$$\text{“Trace”}(U(h)) = \int_{\mathbb{R}^*} \frac{h(u^{-1})}{|1-u|} d^*u. \tag{19}$$

To state an ordinary trace formula in the general case $h(1) \neq 0$, we introduce the principal part of the integral in (19), defined by

$$\int' \frac{h(u^{-1})}{|1-u|} d^*u =< L, \frac{h(u^{-1})}{|u|} >, \tag{20}$$

where L is the unique distribution on $\mathbb{R}$ which agrees with $\frac{du}{|1-u|}$ for $u \neq 1$ and whose Fourier transform vanishes at 1. If $g(u) = h((u+1)^{-1})/|u+1|^{-1}$, then from [**6**], §V, we have the formula

$$\int' g(a) \frac{da}{|a|} = -\int_{\mathbb{R}} \widehat{g}(u) \log |u| du. \tag{21}$$

Let $\Lambda > 0$ and let P_Λ be the projection given by multiplication by the characteristic function of the set

$$\{\xi \in L^2(\mathbb{R}) : \xi(x) = 0, \text{ for } |x| > \Lambda\}. \tag{22}$$

Let F denote the Fourier transform in $L^2(\mathbb{R})$ and $\widehat{P}_\Lambda = FP_\Lambda F^{-1}$. Let $R_\Lambda = \widehat{P}_\Lambda P_\Lambda$. In [**6**] the following is proved.

PROPOSITION 1. *If $h \in \mathcal{S}_c(\mathbb{R}^*)$, then $R_\Lambda U(h)$ is a trace class operator in $L^2(\mathbb{R})$ and as $\Lambda \to \infty$ we have the asymptotic formula*

$$\text{Trace}(R_\Lambda U(h)) = 2h(1)\log\Lambda + \int' \frac{h(u^{-1})}{|1-u|} d^*u + o(1). \tag{23}$$

An analogous result holds for the local fields $\mathbb{Q}_p$ and also if one considers a finite number of places at a time. The problem is that the error term then implied by the "$o(1)$" is not uniform in the set of primes chosen so that one cannot pass to the limit over all primes. This seems to represent a deep difficulty of an analytic number theoretic nature.

7. The Polya-Hilbert space

Connes proposes a Polya-Hilbert space for the problem in the following way. Let $\mathcal{S}(A)_0$ denote the subspace of $\mathcal{S}(A)$ given by,

$$\mathcal{S}(A)_0 = \{f \in \mathcal{S}(A) : f(0) = \int f dx = 0\}. \tag{24}$$

Let E be the averaging over $\mathbb{Q}^*$ operator which to $f \in \mathcal{S}(A)_0$ associates the element of $\mathcal{S}(C)$ given by

$$E(f)(u) = |u|^{1/2} \sum_{q \in \mathbb{Q}^*} f(qu). \tag{25}$$

For $\delta \geq 0$, let $L^2(X)_{0,\delta}$ be the completion of $\mathcal{S}(A)_0$ with respect to the norm given by,

(26)

$$||f||_\delta^2 = \int_C |E(f)(u)|^2(1+\log^2|u|)^{\delta/2} d^*u, \quad \text{for} \quad \int_{|u|\in[1,\Lambda]} d^*u \simeq \log\Lambda, \qquad \Lambda \to \infty.$$

If $g(x) = f(qx)$ for some fixed $q \in \mathbb{Q}^*$, then $||g|| = ||f||$ and so one sees that this norm respects, in this sense, the passage to the quotient $A/\mathbb{Q}^*$. We define $L^2(X)_\delta$ by the short exact sequence

$$0 \to L^2(X)_{0,\delta} \to L^2(X)_\delta \to \mathbb{C} \oplus \mathbb{C}(1) \to 0. \tag{27}$$

When $\delta = 0$ we write $L^2(X)_0$ and $L^2(X)$ for the first two terms. Here $\mathbb{C}$ is the trivial C-module and $\mathbb{C}(1)$ is the C-module for which $u \in C$ acts by $|u|$, where $|\cdot|$ is the norm on C. Multiplication of C on A induces a representation of C on $L^2(X)$ given by,

$$(U(\lambda)\xi)(x) = \xi(\lambda^{-1}x). \tag{28}$$

We introduce a Hilbert space H_δ via another short exact sequence,

$$0 \to L^2(X)_{0,\delta} \to L^2(C)_\delta \to H_\delta \to 0. \tag{29}$$

Here $L^2(C)_\delta$ is the completion with respect to the weighted Haar measure as in(26), where we write $L^2(C)$ when $\delta = 0$. The spectral interpretation on H_δ of the critical zeros of the L-functions in [**5**] relies on taking $\delta > 0$. The regular representation descends to H_δ (it commutes with E up to a phase as an easy calculation shows) and we denote it by W. Connes describes (H_δ, W) as the Polya-Hilbert space with group action for his approach to the Riemann hypothesis. Now, the action of C is free on $L^2(C)_\delta$ so that the short exact sequence (29) tells us that the trace of the action of C on H_δ should be, up to a correction due to a regularisation, the negative of the trace of the action of C on $L^2(X)_{0,\delta}$. From (27), we see that the regularised trace of the action of C on $L^2(X)_\delta$ should involve the sum of the corresponding trace on $L^2(X)_{0,\delta}$ and the trace on $\mathbb{C}\oplus\mathbb{C}(1)$. Therefore the regularised trace of the action of C on $L^2(X)_\delta$ should involve the trace of the action on $\mathbb{C}\oplus\mathbb{C}(1)$ *minus* the trace of this action on H_δ. This *minus* sign is crucial for the comparison with the Weil distribution. The following result going back to Weil [**17**] shows how the operation E in (25) brings in the zeros of the L functions in the critical strip: it indicates that the non-trivial zeros of the L-functions should "span" H_δ as they are "orthogonal" to the image of E

PROPOSITION 2. *Let χ be a character on C. For any $\rho \in \mathbb{C}$ with* $\mathrm{Re}(\rho) \in]-\frac{1}{2},\frac{1}{2}[$, *we have*

$$\int_C E(\xi)(u)\chi(u)|u|^\rho d^*u = 0, \qquad \text{for all } \xi \in \mathcal{S}(A)_0$$

precisely when $L(\chi, \frac{1}{2}+\rho) = 0$.

This accounts for the appearance of the non-trivial zeros of the $L(\chi, s)$ in Proposition 3.

8. Relation to the Weil distribution

In [**6**] Connes works with a cut-off in order to avoid the parameter δ of §7. He introduces a family of subspaces $B_{\Lambda,0}$ of $L^2(X)_0$, depending on a real parameter $\Lambda > 0$, such that $E(B_{\Lambda,0}) \subset \mathbf{S}_\Lambda$ where,

$$\mathbf{S}_\Lambda = \{\xi \in L^2(C) : \xi(u) = 0, \text{ for } |u| \notin [\Lambda^{-1}, \Lambda]\}. \tag{30}$$

In the case of global fields of positive characteristic it makes sense to introduce $B_{\Lambda,0}$ as the space spanned by the $f \in \mathcal{S}(A)_0$ that together with their Fourier transform vanish for $|x| > \Lambda$. In the number field case, all such functions (on the reals) are trivial, and $B_{\Lambda,0}$ is described in terms of prolate spheroidal functions. This aspect of [**6**] is very technical and we do not attempt to go into it here as it offers little additional insight into the underlying ideas. As the asymptotic trace formula involving them is conjectural, we *assume* for our present purpose the existence of an appropriate family of $B_{\Lambda,0}$. Let $Q_{\Lambda,0}$ be the orthogonal projection onto $B_{\Lambda,0}$ and $Q'_{\Lambda,0} = EQ_{\Lambda,0}E^{-1}$. By assumption,

$$Q'_{\Lambda,0} \leq S_\Lambda, \tag{31}$$

where S_Λ denotes the projection onto the set $\mathbf{S}_\Lambda$ in (30). Therefore, for all $\Lambda > 0$, the following distribution is positive,

$$\Delta_\Lambda(f) = \mathrm{Trace}((S_\Lambda - Q'_{\Lambda,0})V(f)), \qquad f \in \mathcal{S}(C), \tag{32}$$

where V is the regular representation of C on $L^2(C)$ and $V(f) = \int_C f(h)V(u)d^*u$. The positivity of Δ_Λ signifies that for $f \in \mathcal{S}(C)$ we have

$$\Delta_\Lambda(f * f^*) \geq 0, \tag{33}$$

where $f^*(u) = \overline{f(u^{-1})}$. Therefore the limiting distribution is also positive,

$$\Delta_\infty = \lim_{\Lambda\to\infty} \Delta_\Lambda \geq 0. \tag{34}$$

In [**6**] the following conjecture is formulated, together with a proposed construction of the family $B_{\Lambda,0}$.

Conjecture 1. *One can find subspaces $B_{\Lambda,0}$ of $L^2(X)_0$ such that the distributions Δ_Λ are of positive type and converge to the Weil distribution, so that for $h \in \mathcal{S}(C)$ of compact support,*

$$\Delta_\infty(h) = \int_C h(u)(|u|^{1/2} + |u|^{-1/2})d^*u - \sum_v \int'_{\mathbb{Q}_v^*} \frac{h(u)}{|1-u|}|u|^{1/2}d^*u, \tag{35}$$

where v runs over the places of $\mathbb{Q}$.

It is known that the positivity of the Weil distribution, which would follow from Conjecture 1 using (34), implies the Riemann hypothesis (see for example [**14**]). One can view Δ_Λ, via (32), as a distribution associated to the regularised trace of the action of C on the Hilbert space H_δ of §7 with the cut-off at $|u| = \Lambda$ enabling one to take $\delta = 0$. The factor $|u|^{1/2}$ in the integrands of (35) as opposed to those of (23) is due to the passage to the quotient by the image of E.

9. Relation of the limit distribution to the zeros

The distributions Δ_Λ can be regarded as a sequence of distributions attached to the Polya-Hilbert space and action (H_δ, W) of §7. We have the following interpretation of Δ_∞ in terms of the zeros of the L-functions with Grossencharacter (compare with Proposition 2).

Proposition 3. *Let h be a function of compact support on $\mathcal{S}(C)$. Then,*

$$\Delta_\infty(h) = \sum_{\chi,\rho} N(\chi, \frac{1}{2} + \rho) \int_{z\in i\mathbb{R}} \widehat{h}(\chi, z) d\mu_\rho(z) \tag{36}$$

where the sum is over the pairs (χ, ρ) of characters of C and zeros ρ of $L(\chi, \frac{1}{2} + \rho)$ with $\mathrm{Re}(\rho) \in]-\frac{1}{2}, \frac{1}{2}[$. The number $N(\chi, \frac{1}{2} + \rho)$ is the multiplicity of the zero, the measure $d\mu_\rho(z)$ is the harmonic measure of ρ with respect to the line $i\mathbb{R} \subset \mathbb{C}$ and the Fourier transform $\widehat{h}$ of h is defined by

$$\widehat{h}(\chi, \rho) = \int_C h(u)\chi(u)|u|^\rho d^*u. \tag{37}$$

In [**6**] Proposition 3 is proven in the function field case, and the proof for number fields is outlined. It allows one to derive under Conjecture 1, the following conjectural explicit formula.

Conjecture 2. *Let $h \in \mathcal{S}(C)$ be of compact support, then*

$$\int_C h(u)(|u|^{1/2} + |u|^{-1/2})d^*u - \sum_v \int'_{\mathbb{Q}_v^*} \frac{h(u)}{|1-u|}|u|^{1/2}d^*u$$
$$= \sum_{\chi,\rho} N(\chi, \frac{1}{2} + \rho) \int_{z\in i\mathbb{R}} \widehat{h}(\chi, z) d\mu_\rho(z),$$

where v runs over the places of $\mathbb{Q}$.

Proposition 3 gives an interpretation of the non-trivial zeros of the Riemann zeta function with the non-critical zeros appearing as resonances. From the comments preceding Proposition 2, one sees that for the action of C on $L^2(X)$, the critical zeros would enter the trace formula with a *minus* sign, which Connes interprets as their being an absorption spectrum for the action of C on $A/\mathbb{Q}^*$.

10. A global geometric trace formula

One of the most interesting heuristic aspects of Connes's ideas, which appears already in [**5**], is the interpretation of the global analogue of Proposition 1 as a non-commutative trace formula for a flow. This conjectured trace formula is as follows.

CONJECTURE 3. *There is a sequence of closed projections Q_Λ, $\Lambda > 0$, in $L^2(X)$ extending the $Q_{\Lambda,0}$ of §8 such that for $h \in \mathcal{S}(C)$ of compact support we have, as $\Lambda \to \infty$,*

$$\text{Trace}(Q_\Lambda U(h)) = 2h(1)\log'\Lambda + \sum_v \int'_{\mathbb{Q}_v^*} \frac{h(u^{-1})}{|1-u|} d^*u + o(1), \tag{38}$$

where the sum is over the places v of $\mathbb{Q}$. Here $2\log'\Lambda = \int_{|\lambda|\in[\Lambda^{-1},\Lambda]} d^*u$ which is asymptotic to $2\log\Lambda$ for the correct choice of Haar measure d^*u on C.

As in §8, for a global field of positive characterstic, it makes sense to introduce Q_Λ as the orthogonal projection onto the set of $f \in \mathcal{S}(A)$ that together with their Fourier transforms vanish for $|x| > \Lambda$. For number fields, Connes proposes in [**6**] a construction using prolate spheroidal functions.

In [**5**], the validity of Conjecture 3 was interpreted as the existence of a non-commutative generalisation of a distributional trace formula for flows on compact manifolds (see [**9**]). If M is a smooth compact manifold with an everywhere non-vanishing vector field ξ, then the associated flow is given by the 1-parameter group $\{F_t = \exp t\xi\}_{t\in\mathbb{R}}$ which induces an action of $\mathbb{R}$ on smooth functions on M by,

$$U(t)f = f \circ F_t, \qquad f \in C^\infty(M). \tag{39}$$

For $h \in \mathcal{S}(M)$, consider the operator,

$$U(h) = \int_{\mathbb{R}} U(t)h(t)dt, \tag{40}$$

where we suppose for simplicity that $h(0) = 0$. Then we know from [**9**] that the distributional trace is given by,

$$\text{“Trace”}(U(h)) = \sum_\pi \int_{H_\pi} h(t)\frac{1}{|1-P_t|} d_\pi t, \tag{41}$$

where the sum is over the primitive periodic orbits π and where H_π is the isotropy group of any $x \in \pi$ with measure $d_\pi t$ normalised so that H_π has covolume 1. We have denoted by P_t the restriction of the differential $d(\exp(t\xi))_x$ to the space transverse to the orbits (the Poincaré map) and by $|1-P_t|$ the absolute value of the determinant of $1-P_t$ which is assumed non-degenerate. Now, observe the following fact about the pair (X, C) proved in [**5**].

LEMMA 3. *For $x \in X, x \neq 0$, the isotropy group of x in C is compact if and only if there exists exactly one place v of $\mathbb{Q}$ with $x_v = 0$ where $(x_\infty, x_p)_p$ is any lift of x to A.*

If $x \in A$ with $x_v = 0$ at only one place v of $\mathbb{Q}$, then the isotropy group of x in C is isomorphic to $\mathbb{Q}_v^*$ and the transverse space is isomorphic to $\mathbb{Q}_v$. Hence the analogue of the Poincaré return map for this fixed point is multiplication of $\mathbb{Q}_v^*$ on $\mathbb{Q}_v$ whose trace formula we discussed in §6 for $v = \infty$. We therefore write suggestively $|1-P_v| = |1-u|$, $u \in \mathbb{Q}_v^*$. Let $h \in \mathcal{S}(C)$ with compact support and

$h(1) = 0$. The analogue of the trace formula (41) becomes, once we agree to single out the fixed points with compact isotropy group determined by Lemma 3,

$$\text{“Trace”}(U(h)) = \sum_v \int_{\mathbb{Q}_v^*} \frac{h(u^{-1})}{|1-u|} d^*u. \tag{42}$$

This is consistent with the local computation of §6 when $h(1) = 0$ and introduces the sum over local terms in (35) as coming from an analogue for $A/\mathbb{Q}^*$ of a geometric trace formula for flows.

11. Concluding remarks

The work of Bost–Connes and Connes leads naturally to investigating $(A/\mathbb{Q}^*, C)$ as the geometric site and (H_δ, W) as the spectral site for the study of the generalised Riemann Hypothesis. Although much of Connes's suggested approach to the Riemann hypothesis is conjectural, it seems likely that a deeper understanding of the space $A/\mathbb{Q}^*$ should have interesting consequences. The analogy between the trace formula for flows on smooth compact manifolds of [**9**] and the conjectured trace formula in (42), suggests the interpretation of Weil's explicit formula as a trace formula on $A/\mathbb{Q}^*$. It should be noted that Paul Cohen proposed in unpublished remarks in the 1970's that in order to understand the Riemann Hypothesis one should understand the nature of measurable $\mathbb{Q}^*$-invariant functions on A. The fact that the von-Neumann algebra associated to $A/\mathbb{Q}^*$ is of Type II_∞, as shown by the work of Bost–Connes, indicates that it is not possible to do classical measure (Type I) theory on $A/\mathbb{Q}^*$. Non-classical ideas are therefore needed and the work of Connes is a bold step in that direction.

References

[1] J. Arledge, M. Laca, I. Raeburn, *Semigroup crossed products and Hecke algebras arising from number fields*, Doc. Mathematica **2** (1997) 115–138.

[2] J-B. Bost, A. Connes, *Hecke Algebras, Type III factors and phase transitions with spontaneous symmetry breaking in number theory*, Selecta Math. (New Series), **1**, 411–457 (1995).

[3] P.B. Cohen, *A C^*-dynamical system with Dedekind zeta partition function and spontaneous symmetry breaking*, J. Théorie des Nombres de Bordeaux, **11** (1999), 15–30

[4] A. Connes, Noncommutative Geometry, Academic Press, 1994. (Version française: Géométrie non commutative, InterEditions, Paris, 1990.)

[5] A. Connes, *Formule de trace en géométrie non commutative et hypothese de Riemann*, C.R. Acad. Sci. Paris, **t.323**, Série 1 (Analyse),1231–1236 (1996).

[6] A. Connes, *Trace formula in Noncommutative Geometry and the zeros of the Riemann zeta function*, Selecta Mathematica, New Series **5**, n. 1 (1999), 29–106.

[7] A. Connes, *Noncommutative geometry and the Riemann zeta function*, in Mathematics: frontiers and perspectives, eds, Vladimir Arnold...[et al], IMU, AMS 1999, 35–54.

[8] J. Dixmier, Les C^*-algebres et leurs représentations, Gauthier-Villars, Paris, 1964.

[9] V. Guillemin, *Lectures on spectral theory of elliptic operators*, Duke Math. J., **44** no. 3, 485–517 (1977).

[10] D. Harari, E. Leichtnam, *Extension du phénomene de brisure spontanée de symétrie de Bost-Connes au cas des corps globaux quelconques*, Selecta Mathematica, New Series 3 (1997), 205-243.

[11] B. JULIA, *Statistical Theory of Numbers*, in Number Theory and Physics, Springer Proceedings in Physics, Vol. 47, 1990.

[12] N. KATZ, P. SARNAK, Random matrices, Frobenius eigenvalues and Monodromy, *AMS Colloquium Publications* **45** 1999.

[13] K.R. PARTHASARATHY, An Introduction to Quantum Stochastic Calculus, *Monographs in Mathematics* Vol. 85, Birkhaüser, 1992.

[14] S. PATTERSON, An Introduction to the Riemann zeta function, *Cambridge studies in advanced mathematics* Vol 14, Cambridge Uni. Press, 1988.

[15] B. RIEMANN, Ueber die Anzahl der Primzahlen unter einer gegebenen Grösse, *Monat der Königl. Preuss. Akad. der Wissen. zu Berlin aus der Jahre 1859* (1860), 671–680; also, *Gesammelte math. Werke und wissensch. Nachlass,* 2. Aufl. 1892, 145–155. English translation in H.M. EDWARDS, *Riemann's zeta function*, Academic Press New York–London 1974.

[16] E.C. TITCHMARSH, The Theory of the Riemann Zeta-function, Second Edition revised by D.R. Heath-Brown, Oxford University Press, New York, 1988.

[17] A. WEIL, *Fonctions zeta et distributions*, Séminaire Bourbaki **312** (1966).

DEPARTMENT OF MATHEMATICS, TAMU 3368, TEXAS A&M UNIVERSITY, COLLEGE STATION, TX 77843-3368

E-mail address: pcohen@math.tamu.edu

DIMACS Series in Discrete Mathematics
and Theoretical Computer Science
Volume **64**, 2004

An Application of Structure Theory of Set Addition to a Question in Ergodic Theory

Jean-Marc Deshouillers, Gregory A. Freiman, and Alexander A. Yudin

ABSTRACT. We consider a dynamical system with discrete time on a set Ω. Let $A(x, B)$ be a sequence of return times of a point x to a subset $B \subset \Omega$ of positive measure. The structure theory of set addition is applied in some cases to the study of such a sequence.

1. Let $(\Omega, \mathcal{A}, \mu)$ be a probability space and T be an ergodic transformation of Ω onto itself preserving the measure μ.

For each pair (x, B) in $\Omega \times \mathcal{A}$ with $\mu(B) > 0$ we denote as

$$A(x, B) = \{0 \leq n_1 < n_2 < \cdots < n_k < \cdots / T^{n_k}x \in B\}$$

the sequence of return times of a point x to a subset $B \subset \Omega$, and we let

$$A_N(x, B) = A(x, B) \cap [0, N-1].$$

The existence of an infinite sequence $A(x, B)$ for almost all x follows from Poincaré's Recurrence Theorem. Our aim is to provide, in some special cases, some information on the *structure* of the sets $A_N(x, B)$. For this, we need some basic notions of the structure theory of set addition (cf. [1], [2], [3] and [5]).

Let G be an abelian group and A and B two subsets of G; a map $\varphi : A \to B$ is said to be a 2-*homomorphism* if for any a_1, a_2, a_3, a_4 in A we have

$$a_1 + a_2 = a_3 + a_4 \Rightarrow \varphi(a_1) + \varphi(a_2) = \varphi(a_3) + \varphi(a_4).$$

The map φ is said to be a 2-*isomorphism* if it is a bijection and if φ^{-1} is also a 2-homomorphism. A subset P of G is said to be a *generalized arithmetic progression of rank at most* m if there exist $x_0, x_1, \ldots, x_m$ in A and positive integers $k_1, \ldots, k_m$ such that

$$P = \{x_0 + h_1x_1 + \cdots + h_mx_m / 0 \leq h_i < k_i \quad \text{for all} \quad i\}.$$

2000 *Mathematics Subject Classification.* Primary 11P70, 11J13, 11J20, 11J70, 37A45.

Key words and phrases. inverse problems in additive number theory, systems of linear diophantine inequalities, dynamical systems.

This work has been realized with the support of CNRS, Universities of Bordeaux 1 and 2, Tel Aviv University, Vladimir Pedagogical University and the grant RFFI (Russia) No. 96-01-00502.

Finally, we say that P is a 2-*progression of rank at most* m if the map defined by

$$\{0,1,\dots,k_1-1\}\times\cdots\times\{0,1,\dots,k_m-1\}\longrightarrow P$$
$$(h_1,\dots,h_m)\longmapsto x_0+h_1x_1+\cdots+h_mx_m$$

is a 2-isomorphism.

Theorem 1. *Let T be an ergodic measure-preserving transform on a probability space $(\Omega,\mathcal{A},\mu)$. Then for any pair $0<\mu,\mu'\le 1$ there exists $C>0$ such that for any triple $x\in\Omega$, $B,B'\in\mathcal{A}$ with $\mu(B)=\mu$, $\mu(B')=\mu'$ and the properties*

$$\lim_{N\to\infty}\frac{1}{N}|A_N(x,B)|=\mu(B) \tag{1}$$

$$\lim_{N\to\infty}\frac{1}{N}|A_N(x,B')|=\mu(B') \tag{2}$$

$$T^nx\in B \quad\text{and}\quad T^mx\in B \quad\text{imply}\quad T^{n+m}x\in B' \tag{3}$$

for each sufficiently large N the set $A_N(x,B)$ is a subset of a 2-progression of cardinality at most $C|A_n(x,B)|$ and rank at most the least integer strictly larger than $2\frac{\mu'}{\mu}-2$.

This result is indeed a corollary to the main result in the structure theory of set addition (cf. [1], [2], [5]), namely

Theorem F. *Let G be a torsion-free abelian group and σ a real number ≥ 2. There exist real numbers c_0 and c_1 (depending on σ only) such that any subset A of G satisfying*

$$|A+A|\le\sigma|A| \quad\text{and}\quad |A|\ge c_0$$

is a subset of a set which is 2-isomorphic to a 2-progression of rank at most $[\sigma-1]$ and cardinality at most $c_1|A|$.

Proof of Theorem 1. Relation (iii) implies that we have

$$A_N(x,B)+A_N(x,B)\subset A_{2N}(x,B').$$

This, (i) and (ii) imply that for any $\delta>0$ and N sufficiently large we have

$$|A_N(x,B)+A_N(x,B)|\le\left(2\frac{\mu(B')}{\mu(B)}+\delta\right)|A_N(x,B)|.$$

Thus, we are in a position to apply Theorem F and see that when N is sufficiently large, the set $A_N(x,B)$ is a subset of a 2-progression of rank at most $\left[2\frac{\mu(B')}{\mu(B)}-1+\delta\right]$ and cardinality at most $c_1|A_N(x,B)|$. □

2. We present some special situations connected with Theorem 1. The first one deals with the diophantine approximation of (αn): a complete self-contained proof is given. The second one concerns the simultaneous approximation of $(\alpha_1 n,\dots,\alpha_t n)$, and gives a new description of the set of solutions which seems to be of interest in itself. In the third subsection we consider the diophantine approximation of the linear form $\alpha_1n_1+\cdots+\alpha_tn_t$ which suggests natural extensions of Theorem 1.

2.1 The function αn.

Let α be an irrational number and ε be a given real number in $(0, 1/4)$. We consider the probability set $(\mathbb{R}/\mathbb{Z}, \mathcal{B}, \mu)$, where μ is the Haar measure, the measure-preserving and ergodic transform $T : x \mapsto x + \alpha$ and the sets $B = \{x/\|x\| < \varepsilon\}$ and $B' = \{x/\|x\| < 2\varepsilon\}$. It is well known (cf. [4]) that the conditions (i), (ii) and (iii) in Theorem 1 are fulfilled. As a conclusion, when N is large enough, the set $A_N(x, B)$ is a subset of a 2-progression of rank at most 3 and cardinality at most $C\varepsilon N$. We give a direct proof of the more precise following result.

Theorem 2. *Let α be an irrational number and ε be a real number in $(0, 1/4)$. When N is sufficiently large, the set*

$$A(\alpha, \varepsilon, N) := \{n \in [-N, N] : \|\alpha n\| < \varepsilon\}$$

is a subset of a 2-progression of rank at most 2 and cardinality at most $56\,\varepsilon N$.

The first step in the proof consists in showing that $A(\alpha, \varepsilon, N)$ is 2-isomorphic to the set of integral points in $\mathbb{R}^2$ lying in a parallelogram of area at most $4\varepsilon N$. This part uses the approximation of α by the convergents of its continued fraction.

Let us write P_r/Q_r the rth-convergent of α, and let us define $\gamma_r = \alpha Q_r - P_r$. We choose s such that

$$\frac{1}{2}Q_{s-2}Q_{s-1} \leq N < \frac{1}{2}Q_{s-1}Q_s.$$

Let us define the map f from $\mathbb{Z}^2$ to $\mathbb{Z}$ by

$$f(x, y) = Q_{s-1}x + Q_s y;$$

since Q_{s-1} and Q_s are coprime, f is surjective. The kernel of f is the set of integral multiples of $(Q_s, -Q_{s-1})$. For a real number u, we denote by $\mathcal{L}(u)$ the line defined by the equation

$$\gamma_{s-1}x + \gamma_s y = u.$$

By the well-known property of consecutive convergents to α, we have

$$(*) \qquad \gamma_{s-1}Q_s - \gamma_s Q_{s-1} = -P_{s-1}Q_s + P_s Q_{s-1} = (-1)^{s-1}\ .$$

This implies that the set $S_{\mathbb{Z}}$ of points with integral coordinates lying in the strip S limited by the lines $\mathcal{L}(-\frac{1}{2})$ and $\mathcal{L}(\frac{1}{2})$ is a set of representatives of $\mathbb{Z}^2/\operatorname{Ker} f$. (We may notice that the lines $\mathcal{L}(-\frac{1}{2})$ and $\mathcal{L}(\frac{1}{2})$ contain no point with integral coordinates, since α is irrational).

We now define

$$S_{\mathbb{Z},\varepsilon} = \{(x, y) \in S_{\mathbb{Z}}/\|\alpha f(x, y)\| < \varepsilon\},$$

so that $S_{\mathbb{Z},\varepsilon}$ consists of the integral points in $\mathbb{R}^2$ lying inside the strip defined by the two lines $\mathcal{L}(-\varepsilon)$ and $\mathcal{L}(\varepsilon)$. An integral point in $S_{\mathbb{Z}}$ which is mapped by f into $[-N, N]$ belongs to the strip R_N lying between the two lines $\mathcal{M}(-N)$ and $\mathcal{M}(N)$, where $\mathcal{M}(v)$ has the equation

$$Q_{s-1}x + Q_s y = v.$$

The map f induces a bijection between the two sets $S_{\mathbb{Z},\varepsilon} \cap R_N$ and $A(\alpha, \varepsilon, N)$. We shall show that this bijection is a 2-isomorphism. Since f is linear, it is clearly a 2-homomorphism. Let us consider 4 points (x_i, y_i), for $1 \leq i \leq 4$, such that $f(x_1, y_1) + f(x_2, y_2) = f(x_3, x_3) + f(x_4, y_4)$, and let us write $x = x_1 + x_2 - x_3 - x_4$ and $y = y_1 + y_2 - y_3 - y_4$. Since f is linear, (x, y) belongs to the kernel of f. On the other hand, since each (x_i, y_i) belongs to $S_{\mathbb{Z},\varepsilon}$ and ε is less than $1/4$, the point (x, y) belongs to $S_{\mathbb{Z}}$. Since (x, y) belongs to $S_{\mathbb{Z}}$ and $\operatorname{Ker} f$, it is the point $(0,0)$, whence

$$(x_1, y_1) + (x_2, y_2) = (x_3, y_3) + (x_4, y_4).$$

Using the relation $(*)$, it is a routine computation to check that the area of the parallelogram in $\mathbb{R}^2$ bounded by the lines $\mathcal{L}(-\varepsilon)$, $\mathcal{L}(\varepsilon)$, $\mathcal{M}(-N)$ and $\mathcal{M}(N)$ is equal to

$$4\varepsilon\mathcal{N}|Q_s\gamma_{s-1} - Q_{s-1}\gamma_s| = 4\varepsilon N.$$

So far, we have proved that the set $A(\alpha, \varepsilon, N)$ is 2-isomorphic to the set of integral points lying in a parallelogram of volume $4\varepsilon N$. This description has its own interest; e.g. it can be used to show that the sequence (αn) is uniformly distributed modulo 1. However, although the parallelogram we have constructed is not too far from being a rectangle, we have not yet described $A(\alpha, \varepsilon, N)$ as the 2-isomorphic image of a subset of a 2-progression of rank at most 2. In [1] and [5], a general construction is given, showing that the set of integral points lying in a convex subset of $\mathbb{R}^d$ can be viewed as the 2-isomorphic image of a part of some d-progression. Here we give a direct proof for our special case (Lemma 2); we start with an elementary technical result.

Lemma 1. *Let (p, q) and (a, b) be two noncolinear vectors in $\mathbb{R}^2$ such that (p, q) has integral coprime coordinates. There exists a pair of integers (p', q') such that $pq' - qp' = 1$ and*

$$|g(a, b) \cdot (1, 0)| \leq \frac{1}{\sqrt{5}} \|g(a, b)\|,$$

where g is the linear map from $\mathbb{R}^2$ to $\mathbb{R}^2$ defined by $g(p, q) = (1, 0)$, $g(p', q') = (0, 1)$ and $\mathbb{R}^2$ is equipped with its natural Euclidean structure.

Proof. Let (p_0, q_0) be a pair of integers such that $pq_0 - qp_0 = 1$ (Bezout's theorem). There exist real numbers u and v, with $v \neq 0$ such that $(a, b) = u(p, q) + v(p_0, q_0)$. We choose λ an integer such that $|\frac{u}{v} - \lambda| \leq \frac{1}{2}$, and let $p' = p_0 + \lambda p$ and $q' = q_0 + \lambda q$; we have $(a, b) = (u - \lambda v)(p, q) + v(p', q')$, so that $g(a, b) \cdot (1, 0) = u - \lambda v$ and $\|g(a, b)\|^2 = (u - v\lambda)^2 + v^2$. It is readily seen that for $|x| \leq 1/2$, we have $\frac{|x|}{\sqrt{x^2+1}} \leq \frac{1}{\sqrt{5}}$. □

Lemma 2. *Let D be a closed parallelogram in $\mathbb{R}^2$, symmetric around the origin, with area $\mu(D) \geq 50$. Either the set $D_{\mathbb{Z}}$ of the integral points in D is 2-isomorphic to a subset of a 2-progression with volume at most equal to $14\mu(D)$ or $D_{\mathbb{Z}}$ lies on a line.*

Proof. We let $\lambda = 2/\sqrt{\mu_D}$. The set λD is a convex body, symmetric with respect to the origin and has an area $\mu(\lambda D) = 4$. By Minkowski's theorem, it contains at least one integral point different from 0. For each such point A, there exists a unique positive number k_A such that the point E satisfying $\overrightarrow{OE} = k_A\overrightarrow{OA}$ lies on the boundary of D. We select a point A such that k_A is maximal and defines F to be the smaller parallelogram that contains D and has its sides respectively parallel to OE and to a (usually the) side of D that contains E. We call $F_1\ F_2$ the side of F that contains E. We have $\mu(F) \leq 2\mu(D)$. Since k_A was chosen to be maximal, the coordinates of A are coprime. We may apply Lemma 1 with $A = (p, q)$ and $(a, b) = \overrightarrow{F_1 F_2}$; we denote by B the point (p', q') and retain the notation of the Lemma for g. Since g induces a bijective linear map from $\mathbb{Z}^2$ to

$\mathbb{Z}^2$, it is a 2-isomorphism. By construction, $g(F)$ is a parallelogram which has two sides parallel to the x-coordinate axis, and these sides have a length of $2k_A$. Since we have $\mu(g(F)) = \mu F$ (because g is unimodular) and $\mu(F) \leq 2\mu(D)$, the height of $g(F)$ is at most $\mu(D)/k_A$. By Lemma 1, the sides of the smaller rectangle with sides parallel to the coordinate axes have length at most $\mu(D)/k_A$ and $2k_A + 2\mu(D)/k_A$. The number of integer points inside this rectangle is thus at most

$$V = (\mu(D)/k_A + 2)(2k_A + 2\mu(D)/k_A + 2).$$

By construction, $k_A \geq 1/\lambda = \sqrt{\mu(D)}/2$; this leads to

$$\begin{aligned} V &\leq 2\mu(D) + 4k_A + 2\Big(\frac{\mu(D)}{k_A}\Big)^2 + 6\frac{\mu(D)}{k_A} + 4 \\ &\leq 10\mu(D) + 4k_A + 12\sqrt{\mu(D)} + 4 . \end{aligned}$$

Since all the points in $D_{\mathbb{Z}}$ are not on a line, and D is symmetric with respect to the origin, we have $2k_A - 2 \leq \mu(D)$, whence the result.

2.2 Simultaneous approximations.

Theorem 3. *Let $\boldsymbol{\alpha} = (\alpha_1, \ldots, \alpha_t)$ be a family of t numbers such that $(1, \alpha_1, \ldots, \alpha_t)$ are $\mathbb{Q}$-linearly independent, $\boldsymbol{\varepsilon} = (\varepsilon_1, \ldots, \varepsilon_t)$ be a family of t numbers in $(0, 1/4)$. When N is sufficiently large, the set*

$$B(\boldsymbol{\alpha}, \boldsymbol{\varepsilon}, N) := \{n; |n| \leq N, \ \|n\alpha_i\| < \varepsilon_i \quad \text{for all} \quad i\}$$

is 2-isomorphic to the set of the integral points of a $(t+1)$-dimensional parallelepiped in $\mathbb{R}^{2t}$.

We notice that a general result in the structure theory of set addition permits to deduce from Theorem 3 that $B(\boldsymbol{\alpha}, \boldsymbol{\varepsilon}, N)$ is 2-isomorphic to a subset of a 2-progression with rank at most $t+1$ and volume at most $c(\boldsymbol{\varepsilon})N$, whereas a straightforward application of Theorem 1 would lead to a rank at most $2^{t+1} - 1$ for the 2-progression.

Let us write $P_{r_i}^{(i)}/Q_{r_i}^{(i)}$ the r_i-th convergent of α_i and let us define $\gamma_{r_i}^{(i)} = \alpha Q_{r_i}^{(i)} - P_{r_i}^{(i)}$. We choose (s_i) such that

$$\frac{1}{2}Q_{s_i-2}^{(i)}Q_{s_i-1}^{(i)} \leq N < \frac{1}{2}Q_{s_i-1}^{(i)}Q_{s_i}^{(i)} \qquad (1 \leq i \leq k).$$

Let us define the maps (f_i) from $\mathbb{Z}^2$ to $\mathbb{Z}$ by $f_i(x, y) = Q_{s_i-1}^{(i)}x + Q_{s_i}^{(i)}y$.

Since $Q_{s_i-1}^{(i)}$ and $Q_{s_i}^{(i)}$ are coprime, f_i is surjective. The kernel of f_i is the set of integral multiples of $\left(Q_{s_i}^{(i)}, -Q_{s_i-1}^{(i)}\right)$. For a real number u, we define $\mathcal{L}^{(i)}(u)$ the line defined by the equation

$$\gamma_{s_i-1}^{(i)}x + \gamma_{s_i}^{(i)}y = u.$$

By the well-known property of consecutive convergents to α_i, we have

$$(*) \qquad \gamma_{s_i-1}^{(i)}Q_{s_i}^{(i)} - \gamma_{s_i}^{(i)}Q_{s_i-1}^{(i)} = -P_{s_i-1}^{(i)}Q_{s_i}^{(i)} + P_{s_i}^{(i)}Q_{s_i-1}^{(i)} = (-1)^{s_i-1} .$$

This implies that the set $S_{\mathbb{Z}}^{(i)}$ of points with integral coordinates lying in the strip $S^{(i)}$ limited by the lines $\mathcal{L}^{(i)}(-\frac{1}{2})$ and $\mathcal{L}^{(i)}(-\frac{1}{2})$ is a set of representatives of $\mathbb{Z}^2/\operatorname{Ker} f_i$. (We may notice that the lines $\mathcal{L}^{(i)}(-\frac{1}{2})$ and $\mathcal{L}^{(i)}(\frac{1}{2})$ contain no point with integral coordinates, since α_i is irrational). We now define

$$S^{(i)}_{\mathbb{Z},\varepsilon_i} = \{(x,y) \in S^i_{\mathbb{Z}} \big/ \|\alpha_i f_i(x,y)\| < \varepsilon_i\},$$

so that $S^{(i)}_{\mathbb{Z},\varepsilon_i}$ consists of the integral points in $\mathbb{R}^2$ lying inside the strip defined by the two lines $\mathcal{L}^{(i)}(-\varepsilon_i)$ and $\mathcal{L}^{(i)}(\varepsilon_i)$. An integral point in $S^{(i)}_{\mathbb{Z}}$ which is mapped by f_i into $[-N,N]$ belongs to the strip $R^{(i)}_N$ lying between the two lines $\mathcal{M}^{(i)}(-N)$ and $\mathcal{M}^{(i)}(N)$, where $\mathcal{M}^{(i)}(v)$ has the equation

$$Q^{(i)}_{s_i-1}x + Q^{(i)}_{s_i}y = v.$$

The map f_i induces a bijection between the two sets $S^{(i)}_{\mathbb{Z},\varepsilon_i} \cap R^{(i)}_N$ and $A^{(i)}(\alpha_i,\varepsilon_i,N)$. We shall show that this bijection is a 2-isomorphism. Since f_i is linear, it is clearly a 2-homomorphism. Let us consider 4 points (x_k,y_k) for $1 \le k \le 4$, such that

$$f_i(x_1,y_1) + f_i(x_2,y_2) = f_i(x_3,y_3) + f_i(x_4,y_4)$$

and let us write $x = x_1 + x_2 - x_3 - x_4$ and $y = y_1 + y_2 - y_3 - y_4$; since f_i is linear, (x,y) belongs to the kernel of f_i. On the other hand, since each (x_k,y_k) belongs to $S^{(i)}_{\mathbb{Z},\varepsilon_i}$ and ε_i is less than $1/4$, the point (x,y) belongs to $S^{(i)}_{\mathbb{Z}}$. Since (x,y) belongs to $S^{(i)}_{\mathbb{Z}}$ and $\operatorname{Ker} f_i$, it is the point $(0,0)$, whence

$$(x_1,y_1) + (x_2,y_2) = (x_3,y_3) + (x_4,y_4).$$

Using relation $(*)$, it is a routine computation to check that the area of the parallelogram in $\mathbb{R}^2$ bounded by the lines $\mathcal{L}^{(i)}(-\varepsilon_i)$, $\mathcal{L}^{(i)}(\varepsilon_i)$, $\mathcal{M}^{(i)}(-N)$ and $\mathcal{M}^{(i)}(N)$ is equal to

$$4\varepsilon_i N \big| Q^{(i)}_{s_i-1}\gamma^{(i)}_{s_i} - Q^{(i)}_{s_i}\gamma^{(i)}_{s_i-1} \big| = 4\varepsilon_i N.$$

So far, we have proved that the sets $A^{(i)}(\alpha_i,\varepsilon_i,N)$ are 2-isomorphic to the set of integral points lying in parallelograms $\mathcal{P}^{(i)} = \mathcal{P}^{(i)}(\alpha_i,\varepsilon_i,N)$ of volume $4\varepsilon_i N$.

Let us consider the parallelepiped

$$\mathcal{P} = \mathcal{P}^{(1)} \otimes \cdots \otimes \mathcal{P}^{(t)} = \{(x_1,y_1,x_2,y_2,\dots,x_t,y_t)/(x_i,y_i) \in \mathcal{P}^{(i)} \quad \text{for all} \quad i\}.$$

The map $f = f_1 \otimes \cdots \otimes f_t$ from $\mathbb{R}^{2t}$ to $\mathbb{R}^t$ that maps $(x_1,\dots,y_t)$ to $(f_1(x_1,y_1),\dots,f_t(x_t,y_t))$ induces a 2-isomorphism between $\mathcal{P}_{\mathbb{Z}}$ and the set of $\mathbf{n} = (n_1,\dots,n_t)$ such that $|n_i| \le N$ and $\|\alpha_i x_i\| < \varepsilon_i$. The set $A(\boldsymbol{\alpha},\boldsymbol{\varepsilon},N)$ is thus 2-isomorphic to the set of those $(x_1,y_1,\dots,x_t,y_t)$ in $\mathcal{P}_{\mathbb{Z}}$ such that

$$f_1(x_1,y_1) = \cdots = f_t(x_t,y_t).$$

We have so far described a 2-isomorphism of the set $A(\boldsymbol{\alpha},\boldsymbol{\varepsilon},N)$ to the set of the integral points in $\mathbb{R}^{2t}$ lying in the intersection of $\mathcal{P}_{\mathbb{Z}}$ and a linear space $\mathcal{H}$ of dimension $t+1$ which is a convex body in $\mathcal{H}$. We now show that this convex body is indeed a parallelepiped. Let us consider the points

$$\widetilde{A}_i = (-1)^{s_i-1}N\Big(-\gamma^{(i)}_{s_i},\gamma^{(i)}_{s_i-1}\Big) \quad (= \mathcal{L}^{(i)}(0) \cap \mathcal{M}^{(i)}(N))$$

and

$$\widetilde{B}_i = (-1)^{s_i-1}\varepsilon_i\Big(Q^{(i)}_{s_i},-Q^{(i)}_{s_i-1}\Big) \quad \Big(= \mathcal{L}^{(i)}(\varepsilon_i) \cap \mathcal{M}^{(i)}(0)\Big).$$

Among the 4^t vertices $(\pm\widetilde{A}_1 \pm \widetilde{B}_1, \dots, \pm\widetilde{A}_t \pm \widetilde{B}_t)$ of $\mathcal{P}$, the 2^{t+1} vertices $(\delta\widetilde{A}_1 \pm \widetilde{B}_1, \dots, \delta\widetilde{A}_t \pm \widetilde{B}_t)$ with $\delta \in \{+1, -1\}$ lie on $\mathcal{H}$. Let us write

$$\begin{aligned} A &= N\Big(-(-1)^{s_1-1}\gamma_{s_1}^{(1)}, (-1)^{s_1-1}\gamma_{s_1-1}^{(1)}, \dots, (-1)^{s_t-1}\gamma_{s_t-1}^{(t)}\Big)\,, \\ B_i &= (-1)^{s_i-1}\varepsilon_i(0,0,\dots,Q_{s_i}^{(i)}, -Q_{s_i-1}^{(i)}, 0\cdots 0) \quad \text{for} \quad 1 \le i \le t\,. \end{aligned}$$

These $(t+1)$ vectors generate $\mathcal{H}$ and it is readily seen that $\mathcal{P} = \{\theta_0 A + \theta_1 B_1 + \dots + \theta_t B_t / -1 \le \theta_i \le 1\}$; thus $\mathcal{P}$ is a $(t+1)$-dimensional parallelepiped. □

2.3 Linear forms.

Theorem 4. *Let $\boldsymbol{\alpha} = (\alpha_1, \dots, \alpha_t)$ be a family of t (irrational) numbers such that $(1, \alpha_1, \dots, \alpha_t)$ are $\mathbb{Q}$-linearly independent and $\boldsymbol{\varepsilon}$ be in $(0, 1/4)$. There exists C such that when N is sufficiently large, the set*

$$C(\boldsymbol{\alpha}, \boldsymbol{\varepsilon}, N) := \{(x_1, \dots, x_t); |x_i| \le N, \|\sum_{i=1}^{t} \alpha_i x_i\| < \boldsymbol{\varepsilon}\}$$

is 2-isomorphic to a subset of a 2-progression of rank at most $2^{t+1}-1$ and cardinality at most $C\boldsymbol{\varepsilon}N^t$.

To view this result in the frame of dynamical systems, we consider $\Omega = (\mathbb{R}/\mathbb{Z})$ equipped with its Haar measure μ. We denote by T_i the transform $z \mapsto z + \alpha_i$, and for $\mathbf{n} = (n_1, \dots, n_t)$ we write

$$T^{\mathbf{n}}0 = T_1^{n_1} \cdots T_t^{n_t} 0 = \alpha_1 n_1 + \dots + \alpha_t n_t.$$

We thus have

$$C(\boldsymbol{\alpha}, \boldsymbol{\varepsilon}, N) = \{\mathbf{n} \in \mathbb{Z}^t / |n_i| \le N \quad \text{and} \quad \|T^{\mathbf{n}}0\| < \boldsymbol{\varepsilon}\}.$$

Since $1, \alpha_1, \dots, \alpha_t$ are $\mathbb{Q}$-linearly independent, we have (cf. [4])

$$|C(\boldsymbol{\alpha}, \boldsymbol{\varepsilon}, N)| \sim 2^{t+1}\boldsymbol{\varepsilon}N^t$$

when $t \to \infty$, and since

$$C(\boldsymbol{\alpha}, \boldsymbol{\varepsilon}, N) + C(\boldsymbol{\alpha}, \boldsymbol{\varepsilon}, N) \subset C(\boldsymbol{\alpha}, 2\boldsymbol{\varepsilon}, 2N),$$

we have when N is sufficiently large

$$|C(\boldsymbol{\alpha}, \boldsymbol{\varepsilon}, N) + C(\boldsymbol{\alpha}, \boldsymbol{\varepsilon}, N)| \le \Big(2^{t+1} + 1/2\Big)|C(\boldsymbol{\alpha}, \boldsymbol{\varepsilon}, x)|.$$

Theorem 4 now follows from Theorem F.

References

[1] Yuri Bilu, "Structure of sets with small sumset". Astérisque 258 (1999), 77–108.

[2] G.A. Freiman, "Foundations of a Structural Theory of Set Addition", Translations of Mathematical Monographs 37, Amer. Math. Soc., Providence, R.I., 1973.

[3] G.A. Freiman, "Structure theory of set addition". Astérisque 258 (1999), 1–33.

[4] L. Kuipers and H. Niederreiter, "Uniform Distribution of Sequences". John Wiley and Sons, New York, 1974.

[5] M.B. Nathanson, "Additive Number Theory. Inverse Theorems and the Geometry of Sumsets". Springer-Verlag, New York, 1994.

Jean-Marc Deshouillers
Statistique Mathématiques, EA 2961
Université Victor Segalen Bordeaux 2
33076 Bordeaux, Cedex, France
and
A2X, UMR 5465, CNRS - Université Bordeaux 1
E-mail address: Jean-Marc.Deshouillers@math.u-bordeaux.fr

Gregory A. Freiman
Sackler Faculty of Exact Sciences
School of Mathematical Sciences
Tel Aviv University
Ramat Aviv, Tel Aviv, Israel 69978
E-mail address: grisha@math.tau.ac.il

Alexander A. Yudin
Department of Mathematics
Vladimir Pedagogical University
600011 vladimir, Russia
E-mail address: aayudin@vgpu.elcom.ru

DIMACS Series in Discrete Mathematics
and Theoretical Computer Science
Volume **64**, 2004

Linear Algebra and Additive Theory

J.A. Dias da Silva

ABSTRACT. We present a description of some Linear Algebra techniques in Additive Number Theory.

1. Introduction

Additive Number Theory has recently been approached using the tools of Linear Algebra. Connections between Linear Algebra and Additive Number Theory have been known for many years. There were two main ways they have appeared. On one hand, linear algebraists have deduced their sought Linear Algebra theorems from well known results in Additive Theory. Articles using this approach have appeared since the beginning of the sixties [**M, M-Sh1, M-Sh2, M-Sh3, S**]. On the other hand Linear Algebra tools have been used to prove results that have an Additive Theory translation.

In this article we are mainly concerned with this second type of connection. It has been approached using the degree of the invariant polynomials of a (diagonal) linear operator as estimator for the cardinality or for the multiplicities of the elements of its spectrum. By using this estimation, it is possible to translate the inequalities between degrees of invariant polynomials of linear operators on finite dimensional vector spaces, as inequalities on cardinalities of finite subsets of a field. Then the goal is to translate results or questions in Additive Theory to a Linear Algebra setting and to derive from them the corresponding additive results.

The first result to be achieved in this way was the classical Cauchy-Davenport Theorem [**C, D1, D2**]. In [**DS-Ha1**] an alternative proof of Cauchy-Davenport was presented using the technique described above. This estimation procedure probably preformed better in proving the Erdös-Heilbroon conjecture [**DS-Ha2**]. Recently some other results in Additive Theory (like Pollard Theorem) were viewed in this way [**Cr-DS1, Cr-DS2**].

2000 *Mathematics Subject Classification.* Primary 11P99; Secondary 15A69.

Key words and phrases. Degree of invariant polynomials, Sumsets, Tensor spaces, Grassmann spaces.

This work was done within the activities of the Centro de Estruturas Lineares e Combinatórias and was partially supported by Fundação para a Ciência e Tecnologia and Programa Ciência, Tecnologia e Inovação do Quadro Comunitário de Apoio.

2. Preliminaries

We denote by $\mathbb{N}_0$ the set of nonnegative integers. We use p to mean the characteristic of the field $\mathbb{F}$, in the case that $\mathbb{F}$ has finite characteristic, and ∞ if $\mathbb{F}$ has characteristic zero (assume the usual conventions on the symbol ∞). If A is a set, $|A|$ denotes the cardinality of A. If f is a linear operator on the finite dimensional vector space V over $\mathbb{F}$, we use $\sigma(f)$ for the spectrum of f (meaning either the family or the set of the roots of the characteristic polynomial of f, in the algebraic closure of $\mathbb{F}$). We use P_f to mean the minimal polynomial of f. We say that f is *diagonal* or *of simple structure* if, for some basis of V, the matrix of f is diagonal.

Let $v \in V$. The subspace spanned by the images of v by the powers of f is called the *f-cyclic subspace* of v and denoted $\mathcal{C}_f(v)$, i.e.

$$\mathcal{C}_f(v) = \langle f^j(v) \mid j \in \mathbb{N}_0 \rangle.$$

The identity operator on V is denoted by I_V.

The following theorems are basic tools for the next sections.

THEOREM 2.1. *If f is a diagonal linear operator on V, the degree of the minimal polynomial of f is equal to the cardinality of its spectrum, i.e.*

$$\deg(P_f) = |\sigma(f)|.$$

THEOREM 2.2. *The degree of the minimal polynomial of f is the maximum of the dimensions of the f-cyclic subspaces of the vectors of V, i.e.,*

$$\deg(P_f) = \max_{v \in V} \dim \mathcal{C}_f(v).$$

3. The Cauchy-Davenport Theorem

Let p be a prime number. We denote by $\mathbb{Z}_p$ the field of integers $\mod p$. If A, B are subsets of an abelian group G (we use additive notation), the sum set $A + B$ is the subset of G,

$$A + B = \{a + b \mid a \in A,\ b \in B\}.$$

The following theorem was proved by Cauchy in 1813 [**C**] and reproved by Davenport [**D1**] in 1935.

THEOREM 3.1. *Let A and B be nonempty subsets of $\mathbb{Z}_p$. Then*

$$|A + B| \geq \min\{p, |A| + |B| - 1\}.$$

In order to get a new proof of Theorem 3.1 with the strategy referred to above the first step to be fulfilled is to find linear operators f, g, H whose spectra are related by the equality

$$\sigma(H) = \sigma(f) + \sigma(g).$$

Basic Linear Algebra provides such operators.

THEOREM 3.2. *Let V and W be nonzero finite dimensional vector spaces over the field $\mathbb{F}$. Let f be a linear operator on V and g be a linear operator on W. The spectrum of the Kronecker sum of f and g,*

$$f \otimes I_W + I_V \otimes g,$$

is equal to the sum of the spectra of f and g, i.e.

$$\sigma(f \otimes I_W + I_V \otimes g) = \sigma(f) + \sigma(g).$$

It is now easy to guess what will be the linear counterpart of the Cauchy-Davenport Theorem.

Until the end of this section V and W are nonzero vector spaces over the field $\mathbb{F}$, f is a linear operator of V and g is a linear operator on W.

THEOREM 3.3 (Linear Cauchy-Davenport). *The integer*

$$\min\{p, \deg P_f + \deg P_g - 1\}$$

is a lower bound for the degree of the minimal polynomial of the Kronecker sum of f and g, i.e.

$$\deg P_{f\otimes I_W + I_V\otimes g} \geq \min\{p, \deg P_f + \deg P_g - 1\}. \tag{3.1}$$

We arrive at this theorem by establishing a similar result for the dimensions of cyclic subspaces (f-cyclic subspaces, g-cyclic subspaces and $(f\otimes I_W + I_V\otimes g)$-cyclic subspaces) and then use Theorem 2.2. The crucial step of the argument for this proof of Cauchy-Davenport Theorem lies in the use of the following result:

PROPOSITION 3.4. *Let $v \in V$ and $w \in W$. Then* $\min\{p, \dim \mathcal{C}_f(v) + \dim \mathcal{C}_g(w) - 1\}$ *is a lower bound for the dimension of the $(f\otimes \mathrm{id}_W + \mathrm{id}_V \otimes g)$-cyclic subspace of $v\otimes w$, i.e.,*

$$\dim \mathcal{C}_{f\otimes \mathrm{id}_W + \mathrm{id}_V\otimes g}(v\otimes w) \geq \min\{p, \dim \mathcal{C}_f(v) + \dim \mathcal{C}_g(w) - 1\}. \tag{3.2}$$

Sketch of the proof. We express the successive "powers" $(f\otimes \mathrm{id}_W + \mathrm{id}_V\otimes g)^k(v\otimes w)$ as linear combinations of the linearly independent family of $V\otimes W$,

$$(f^i(v)\otimes g^j(w))_{\substack{i=0,\ldots,\dim \mathcal{C}_f(v)-1\\ j=0,\ldots,\dim \mathcal{C}_g(w)-1}}.$$

Let $\mathcal{S}_k$ be the support of $(f\otimes \mathrm{id}_W + \mathrm{id}_V\otimes g)^k(v\otimes w)$ in this family. For every $\mathcal{S}_k$, $k = 0, \ldots, \min\{p, \dim \mathcal{C}_f(v) + \dim \mathcal{C}_g(w) - 1\} - 1$ there exists an element in S_k which is not in any of $\mathcal{S}_t$'s for t smaller than k. This obviously implies that

$$\{(f\otimes \mathrm{id}_W + \mathrm{id}_V\otimes g)^k(v\otimes w), \quad k = 0, \ldots, \min\{p, \dim \mathcal{C}_f(v) + \dim \mathcal{C}_g(w) - 1\} - 1\}$$

is a linearly independent set and (3.2) follows.

□

Proof of Linear Cauchy-Davenport. If we choose $v \in V$ such that $\dim \mathcal{C}_f(v) = \deg P_f$ and $w \in W$ such that $\dim \mathcal{C}_g(w) = \deg P_g$ (recall Theorem 2.2), we get, from equality (3.2),

$$\deg P_{f\otimes \mathrm{id}_W + \mathrm{id}_V\otimes g} \geq \min\{p, \deg P_f + \deg P_g - 1\}.$$

□

Now we can easily derive the Cauchy Davenport Theorem. Let p be a prime number. Let A and B be nonempty subsets of $\mathbb{Z}_p$ of cardinality r and s respectively. Let f be a diagonal linear operator on an r-dimensional vector space V over $\mathbb{Z}_p$ whose spectrum is A. Let g be a linear operator on an s-dimensional space over $\mathbb{Z}_p$ whose spectrum is B. Using Theorem 2.1 and replacing the degrees of the minimal polynomials of f, g and $f\otimes I_W + I_V\otimes g$ by the cardinality of their spectra in (3.1) (recall that if f and g are diagonal then $f\otimes \mathrm{id}_W + \mathrm{id}_V\otimes g$ is diagonal), we get Cauchy-Davenport Theorem.

4. Erdös-Heilbronn conjecture

The Erdös-Heilbroon conjecture is another additive theorem that has been successively fitted into the procedure described above. Let us start by introducing some more terminology and notation. We say *m-set* to mean a set of cardinality m. Let A be a nonempty finite subset of $\mathbb{F}$. We denote the set of the sums of the m-subsets of A (*restricted sums*) by $\wedge^m A$. In 1964 Erdös and Heilbronn [**E-G**] made the following conjecture.

CONJECTURE 4.1. *Let p be a prime number and A be a nonempty subset of $\mathbb{Z}_p$. The set of the sums of the 2-subsets of A has cardinality at least* $\min\{p, 2|A| - 3\}$, *i.e.*

$$|\wedge^2 A| \geq \min\{p, 2|A| - 3\}.$$

In fact, we may consider a more general problem:

PROBLEM 4.2. *Let A be a nonempty subset of $\mathbb{Z}_p$. Find a lower bound for the cardinality of the set of the m-subsets of A.*

To make the linear approach to this problem, we have to find linear operators f and H such that the spectrum of H is the set of the sums of the m-subsets of the spectrum of f. As before, Linear Algebra provides those linear operators .

Let f be a linear operator on V. Consider the linear operator $D(f)$ on the mth exterior product of V, $\wedge^m V$, defined (by the values assumed on decomposable exterior tensors) by the equalities [**B**, AIII, pg 129],

$$\begin{aligned} D(f)(v_1 \wedge \cdots \wedge v_m) &= f(v_1) \wedge v_2 \wedge \cdots \wedge v_m + v_1 \wedge f(v_2) \wedge \cdots \wedge v_m + \cdots \\ &\qquad \cdots + v_1 \wedge v_2 \wedge \cdots \wedge f(v_m), \\ &\qquad\qquad v_1, \ldots, v_m \in V \end{aligned}$$

The following theorem is an easy consequence of the definition of $D(f)$.

THEOREM 4.3. *Let f be a diagonal linear operator. Then $D(f)$ is diagonal and the spectrum of $D(f)$ is the set of sums of the m subsets of $\sigma(f)$, i.e.*

$$\sigma(D(f)) = \wedge^m \sigma(f).$$

To pursue the described procedure, the next task is to give a lower bound for the dimension of the $D(f)$-cyclic subspaces. With this objective, we express the powers of $D(f)$ as linear combinations of a basis of $\wedge^m V$ (designed to fit in this problem). In order to prepare the tools for this, we need to make a combinatorial digression.

A *partition of* m is a decreasing (nonincreasing) sequence of nonnegative integers whose terms sum is equal to m. We say that a partition of m has *length* s (we write $\ell(\lambda) = s$) if the number of positive terms of the sequence is s. We denote by $\mathcal{P}_{m,s}$ the set of partitions of m of length at most s, and by $\mathcal{P}_s = \cup_{i \in \mathbb{N}} \mathcal{P}_{i,s}$ the set of the partitions of length at most s. To each partition $\lambda = (\lambda_1, \ldots, \lambda_t)$ of m we consider the associated Young tableau $[\lambda]$ of λ which consists of m boxes placed in t rows each starting in the same column, where the ith row has λ_i boxes, $i = 1, \ldots, t$. For instance, the Young tableau associated with the partition $(5, 3, 1)$ is

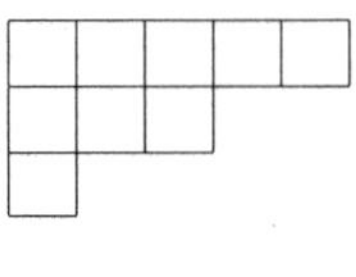

Let λ be a partition of m. The (i,j)-*hook* of $[\lambda]$ is the subset of boxes of $[\lambda]$ consisting of the (i,j)-box of $[\lambda]$ (the box in the ith row and jth column of $[\lambda]$) together with the boxes in the same row to the right and the boxes in the same column under it. We denote the (i,j)-hook of $[\lambda]$ by H_{ij}^{λ} and the cardinality of H_{ij}^{λ} by h_{ij}^{λ}. We identify partitions of m that only differ by zero terms. Then, for a partition λ when we write $\lambda = (\lambda_1, \ldots, \lambda_m)$ we are considering it a sequence with m terms, adding the necessary zeros.

It is possible to express the powers of $D(f)$ as linear combinations of the basis

$$f^{\lambda_m}(v) \wedge f^{\lambda_{m-1}+1}(v) \wedge \cdots \wedge f^{\lambda_1+m-1}(v) \mid \lambda \in \mathcal{P}_m, \lambda_1 \leq \dim \mathcal{C}_f(v) - m \tag{4.1}$$

of the mth exterior power of $\mathcal{C}_f(v)$.

THEOREM 4.4.
$$D(f)^t(v \wedge f(v) \wedge \cdots \wedge f^{m-1}(v)) =$$
$$= \sum_{\lambda \in \mathcal{P}_{t,m}} \frac{t!}{\prod_{i,j} h_{ij}^{\lambda}} f^{\lambda_m}(v) \wedge f^{\lambda_{m-1}+1}(v) \wedge \cdots \wedge f^{\lambda_1+m-1}(v)$$

We now have the tools needed to prove a proposition that, in connection with Problem 1, plays a role similar to the Proposition 3.4. In fact, if we use Theorem 4.4 to express the successive powers of $D(f)$ as linear combination of the basis (4.1), we can again use a support argument to conclude that $\{D(f)^t \mid t = 0, \ldots, \min\{p, m(\dim \mathcal{C}_f(v) - m) + 1)\} - 1\}$ is linearly independent, i.e. to conclude the following proposition:

PROPOSITION 4.5. *Let $v \in V$. The integer* $\min\{p, m(\dim \mathcal{C}_f(v) - m) + 1\}$ *is a lower bound for the dimension of $D(f)$-cyclic subspace of $v \wedge f(v) \wedge \cdots \wedge f^{m-1}(v)$, i.e.*

$$\dim \mathcal{C}_{D(f)}(v \wedge f(v) \wedge \cdots \wedge f^{m-1}(v)) \geq \min\{p, m(\dim \mathcal{C}_f(v) - m) + 1\}.$$

Using this proposition and arguments similar to the ones used to prove the Linear Cauchy-Davenport we can establish the following theorem:

THEOREM 4.6. *Let f be a linear operator on V. The degree of the minimal polynomial of $D(f)$ is bounded from below by* $\min\{p, m(\dim \mathcal{C}_f(v) - m) + 1\}$, *i.e.,*

$$\deg P_{D(f)} \geq \min\{p, m(\deg P_f - m) + 1\}.$$

Considering a diagonal linear operator with spectrum A and the line of arguments presented after the proof of linear Cauchy-Davenport, we can easily derive the following theorem that resolves Problem 1 and also implies, by making $m = 2$, the inequality conjectured by Erdös and Heilbronn (Conjecture 1).

THEOREM 4.7. *Let A be a nonempty subset of $\mathbb{Z}_p$. Then*

$$|\wedge^m A| \geq \min\{p, m(|A| - m) + 1\}.$$

5. The Pollard Theorem

Given a real number b, we denote by $\lfloor b \rfloor$ the greatest integer less than or equal to b.

Let A, B be finite nonempty subsets of $\mathbb{F}$. For each element $c \in A+B$, $\nu_c(A,B)$, the *multiplicity of c in $A+B$*, is the cardinality of the set of pairs (a,b) such that

$a + b = c$. If i is a positive integer, we denote by $\mu_i(A, B)$, or briefly, by μ_i the integer

$$\mu_i(A, B) := |\{c \in A + B \mid \nu_c(A, B) \geq i\}|.$$

Recall that if $A = \{a_1, \ldots, a_n\}$, then the set $\wedge^2 A$ is defined as follows:

$$\wedge^2 A = \{a_j + a_k \mid 1 \leq j < k \leq n\}.$$

Following the preceding notation, if $c \in \wedge^2 A$, denote the cardinality of the set of the pairs (j, k) satisfying $1 \leq j < k \leq n$ and $a_j + a_k = c$ by

$$\nu_c^{(R)}(A).$$

The integer $\nu_c^{(R)}(A)$ is called the *multiplicity of* c *in* $\wedge^2 A$. If i is a positive integer, $\mu_i^{(R)}(A)$ denotes the integer

$$\mu_i^{(R)}(A) := |\{c \in \wedge^2 A \mid \nu_c^{(R)}(A) \geq i\}|.$$

In 1974 J. M. Pollard proved a result that estimates the average of the number of elements of $A + B$ whose multiplicity is less than or equal to a certain integer t. He proved that if A and B are nonempty subsets of $\mathbb{Z}_p$, then for $t = 1, 2, ..., \min\{|A|, |B|\}$we have

$$\sum_{i=1}^{t} \mu_i \geq t \min\{p, |A| + |B| - t\} \tag{5.1}$$

The aim of this section is to use this technique to prove the following two theorems:

THEOREM 5.1. *Let A and B be finite nonempty subsets of $\mathbb{F}$. Then, for $t = 1, 2, ..., \min\{|A|, |B|\}$,*

$$\sum_{i=1}^{t} \mu_i \geq t \min\{p, |A| + |B| - t\} \tag{5.2}$$

THEOREM 5.2. *Let $A \subseteq \mathbb{F}$ and $1 \leq t \leq \lfloor \frac{|A|}{2} \rfloor$. Assume that $|A| \geq 2$. Then we have*

$$\sum_{i=1}^{t} \mu_i^{(R)} \geq t \min\{p, 2(|A| - t) - 1\}. \tag{5.3}$$

The first one of these theorems is an extension of Pollard Theorem to arbitrary fields. Theorem 5.2 is a statement on restricted sums, analogous to Pollard Theorem.

Let f be a linear operator on the vector space V of dimension n over $\mathbb{F}$. We denote the invariant polynomials of f by

$$\alpha_{f,1}, \cdots, \alpha_{f,n} = P_f, \qquad (\alpha_{f,1} | \cdots | \alpha_{f,n}).$$

Next we will see that there is a generalization of the concept of f-cyclic subspace for the family of invariant polynomials that extends inequality (3.2).

Let $x_1, \ldots, x_t$ be linearly independent vectors of V, and f a linear operator on V. We call *generalized f-cyclic subspace associated to $x_1, \ldots, x_t$* the subspace

$$C_f(x_1, \ldots, x_t) = \langle f^i(x_j) \ : \ i \in \mathbb{N} \cup \{0\} \quad j = 1, \ldots, t \rangle.$$

If the set

$$\mathcal{I} = \bigcup_{i=1}^{t} \{x_i, f(x_i), f^2(x_i), \ldots, f^{s_i-1}(x_i)\}$$

is linearly independent, we say that $\mathcal{I}$ is an *f-nice independent set* or $(x_1, \ldots, x_t, f)$-*nice independent set*, and the integers $s_1, \ldots, s_t$ will be called the *indices of $\mathcal{I}$*. Observe that if $v \neq 0$ is a vector of V and $t \leq \dim \mathcal{C}_f(v)$, then $\{v, f(v), \ldots, f^{t-1}(v)\}$ is a (v, f)-nice independent set of index t.

Let $a = (a_1, \ldots, a_n)$ and $b = (b_1, \ldots, b_n)$ be sequences of nonnegative integers. Denote by $(\overline{a}_1, \ldots, \overline{a}_n)$ and $(\overline{b}_1, \ldots, \overline{b}_n)$ the reordering, in a nonincreasing way, of a and b respectively. We say that a *weakly-dominates* b, and we write

$$a \sqsupseteq b$$

if

$$\sum_{i=1}^{k} \overline{a}_i \geq \sum_{i=1}^{k} \overline{b}_i, \quad k = 1, \ldots, n.$$

The next theorem is proved in [**Cr-DS2**] and is an extension of equality (3.2). In fact, bearing in mind the remark following the definition of nice independent set, we can see that inequality (3.2) is equivalent to $\deg P_f \geq t$ for every (v, f)-nice independent set of index t.

THEOREM 5.3. *Let f be a linear operator on V. Let $s_1, \ldots, s_t$ be positive integers. If there exist linearly independent, vectors $v_1, \ldots, v_t$ such that*

$$\bigcup_{i=1}^{t} \{v_i, f(v_i), f^2(v_i), \ldots, f^{s_i-1}(v_i)\}$$

is linearly independent then the following condition holds:

$$(s_1, \ldots, s_t) \sqsubseteq (\deg(\alpha_{f,n}), \ldots, \deg(\alpha_{f,n-t+1})).$$

We are now able to prove the linear counterpart of Pollard Theorem (Linear Pollard Theorem).

THEOREM 5.4 (Linear Pollard Theorem). *Let V and W be nonzero finite dimensional vector spaces over the field $\mathbb{F}$. Let f be a linear operator on V and g be a linear operator on W. Then we have, for $1 \leq t \leq \min\{\deg P_f, \deg P_g\}$,*

$$\sum_{i=1}^{t} \deg(\alpha_{f \otimes I_W + I_V \otimes g, mn-i+1}) \geq t \min\{p, \deg(P_f) + \deg(P_g) - t\}. \tag{5.4}$$

Sketch of proof. Let $\ell = \min\{p, \deg P_f + \deg P_g - t\}$. Let v be a vector of V satisfying $\dim \mathcal{C}_f(v) = \deg(P_f)$ and w a vector of W satisfying $\dim \mathcal{C}_g(w) = \deg(P_g)$. We prove that

$$\mathcal{I} = \bigcup_{i=1}^{t} \{f^i(v) \otimes w, (f \otimes I_W + I_V \otimes g)(f^i(v) \otimes w), \ldots, (f \otimes I_W + I_V \otimes g)^{t-1}(f^i(v) \otimes w)\}$$

is a $(v \otimes w, f(v) \otimes w, \ldots, f^{t-1}(v) \otimes w)$-nice independent set with indices $(\ell, \ell, , \ldots, \ell)$. Therefore, using Theorem 5.3, we have

$$\begin{aligned}\sum_{i=1}^{t} \deg(\alpha_{f\otimes I_W + I_V \otimes g, mn-i+1}) &\geq t\ell \\ &\geq t \min\{p, \deg P_f + \deg P_g - t\}.\end{aligned}$$

□

Let $D(f)$ be the restriction of $f \otimes I_V + I_V \otimes f$ to the second exterior power of V, $\wedge^2 V$. Using the same line of arguments, we can prove the following theorem:

THEOREM 5.5 (Linear version of Theorem 5.2). *Let V be a nonzero m-dimensional vector space over the field $\mathbb{F}$. Let f be a linear operator on V. Assume that* $\deg(P_f) \geq 2$. *Then, for $1 \leq t \leq \lfloor \deg(P_f)/2 \rfloor$, we have*

$$\sum_{i=1}^{t} \deg(\alpha_{D(f), \binom{m}{2}-i+1}) \geq t \min\{p, 2\deg(P_f) - t) - 1\}.$$

Now, we are able to prove Theorems 5.1 and 5.2.

The main argument to prove Theorems 5.1 and 5.2 is to make an extension of a technique used in sections 3 and 4. This technique relies strongly upon the equality between the degree of the minimal polynomial and the cardinality of the spectrum of diagonal linear operators on finite dimensional vector spaces. We show the following generalization of this equality. Let $V \neq 0$ be an n-dimensional vector space over the field $\mathbb{F}$, and let f be a diagonal linear operator on V. Let i be a positive integer. Denote by $m_i(f)$ the cardinality of the elements of $\sigma(f)$ whose algebraic multiplicity is greater than or equal to i. Then for $j = 1, \ldots, n$,

$$m_1(f) + \cdots + m_j(f) = \sum_{i=1}^{j} \deg(\alpha_{f,n-i+1}). \tag{5.5}$$

Let A and B be nonempty finite subsets of $\mathbb{F}$ ($|A| = n$, $|B| = m$). Let V and W be vector spaces over $\mathbb{F}$ of dimensions n and m respectively. Let f be a diagonal linear operator on V with spectrum $\sigma(f) = A$ and g be a diagonal linear operator on W with spectrum $\sigma(g) = B$. Then, using Theorem 3.2, for $j = 1, \ldots, \min\{|A|, |B|\}$ we have

$$m_1(f \otimes I_W + I_V \otimes g) + \cdots + m_j(f \otimes I_W + I_V \otimes g) = \mu_1(A, B) + \cdots + \mu_j(A, B). \tag{5.6}$$

Since f and g are diagonal, $f \otimes I_W + I_V \otimes g$ is diagonal, and then we use equalities (5.5) and (5.6) to replace the left hand side in (5.4) by $\sum_{i=1}^{t} \mu_i$, and we use the equality between the degree of the minimal polynomial of a linear operator and the cardinality of its spectrum to replace the right hand side by $t \min\{p, |A| + |B| - t\}$, resulting in Theorem 5.1.

A similar reasoning can be used to prove Theorem 5.2.

Acknowledgment-The author thanks Joanne Haderer for her careful review of the english of this paper.

References

[B] N. Bourbaki, *Elements de Mathématique, Algèbre I*, Hermann, Paris, (1970).

[C] A. Cauchy, Recherches sur les nombres, *J. École Polytech.* 9:99-116(1813).

[Cr-DS1] Cristina Caldeira and J. A. Dias da Silva, The Invariant Polynomials Degrees of the Kronecker sum of Two Linear Operators and Additive Theory, *Linear Algebra and Appl* 315:125-138(2000).

[Cr-DS2] Cristina Caldeira and J. A. Dias da Silva, A Pollard Type Result for Restricted Sums, *Journal of Number Theory* 72:153-173(1998).

[D1] H. Davenport, On the addition of residue classes, *J. London Math. Soc.* 10:30-32(1935).

[D2] H. Davenport, A Historical Note, *J. London Math. Soc.* 22:100-101(1947).

[DS-Ha1] J.A. Dias da Silva and Y. O. Hamidoune, A note on the minimal polynomial of the Kronecker sum of two linear operators, *Linear Algebra and Appl.* 141:283-287(1990).

[DS-Ha2] J.A. Dias da Silva and Y. O. Hamidoune, Cyclic Spaces for Grassmann Derivatives and Additive Theory, *Bull. London Math. Soc.* 26:140-146(1994).

[E-G] P. Erdös and R. L. Graham, *Old and New Problems and Results in Combinatorial Additive Theory*, L'Enseignement Mathématique, Genève, 1980.

[L] Serge Lang, *Algebra*, Addison-Wesley Publishing Company Inc., New York, 1993.

[M] M. Marcus, The Minimal Polynomial of a Commutator, *Portugaliae Math.* 25:73-76(1964).

[M-Sh1] M. Marcus and M. Shafqat Ali, On the degree of the minimal polynomial of a commutator operator, *Pacific J. Math.* 37:361-565(1971).

[M-Sh2] M. Marcus and M. Shafqat Ali, Minimal Polynomials of Additive Commutators and Jordan Products, *J. Algebra* 22:12-33(1972).

[M-Sh3] M. Marcus and M. Shafqat Ali, On the Degree of the Minimal Polynomial of the Lyapunov Operator, *Monatshefte für Mathematik* 78:229-236(1974).

[Me] Melvyn B. Nathanson, *Additive Number Theory: 1. Inverse Problems and the Geometry of Sumsets*, Springer Verlag, Berlin, 1996.

[P] J. M. Pollard, A generalization of a theorem of Cauchy and Davenport, *J. London Math. Soc.* 8:460-462(1974).

[S] Renato Spiegler, An Application of Group Theory to Matrices and Ordinary Differential Equations, *Linear Algebra Appl.* 44:143-151(1982).

Centro de Estruturas Lineares e Combinatórias, Departamento de Matemática, Universidade de Lisboa, Av Gama Pinto 2, 1649-003 Lisboa PORTUGAL

DIMACS Series in Discrete Mathematics
and Theoretical Computer Science
Volume **64**, 2004

On the Distribution of the Power Generator Modulo a Prime Power

John B. Friedlander, Jan S. dT. Hansen, and Igor E. Shparlinski

ABSTRACT. We study the multidimensional distribution of the power generator of pseudorandom numbers modulo a high power of a fixed prime number. These results complement some recently obtained results about the power generator modulo a product of two distinct primes in which case the generator is of great value for many cryptographic applications. The case of a prime power modulus, although it does not have any immediate cryptography related applications, may still be of interest for other applications which require quality pseudorandom numbers. Moreover, in this case new effects arise which allow us to apply some recent bounds for exponential sums with sparse polynomials to study the multidimensional distribution. In the case of moduli which are the product of two primes such results are known only for power generators with small exponents.

1. Introduction

Let $e \geq 2$, $m \geq 1$ and ϑ be integers such that $\gcd(\vartheta, m) = 1$. Then one can define the sequence (u_n) by the recurrence relation

$$u_n \equiv u_{n-1}^e \pmod m, \quad 0 \leq u_n \leq m-1, \qquad n = 1, 2, \dots, \tag{1.1}$$

with the *initial value* $u_0 = \vartheta$.

This sequence is known as the *power generator* of pseudorandom numbers and in the case when the modulus m is a product of two distinct primes it has numerous applications to cryptography, see [**1, 4, 18, 28**].

2000 *Mathematics Subject Classification.* Primary 11B83, 11L07.

Key words and phrases. exponential sums, pseudorandom number generators, sparse polynomials.

J. F. was supported in part by NSERC and by an NEC grant to the Institute for Advanced Study.

J. H. was supported in part by an APA Scholarship.

I. S. was supported in part by an ARC grant.

It is obvious that the sequence (1.1) eventually becomes periodic with some period τ. In this paper we shall assume that $\gcd(e, \varphi(m)) = 1$; from this it follows that the sequence (u_n) is *purely periodic*.

Although various properties have been studied in a number of works, see [**1, 2, 3, 4, 6, 9, 11, 14, 18, 21, 27, 28**], very few unconditional results are known. Amongst them are results of [**3, 9, 10**] about the distribution of the sequence produced by this generator in the case where the modulus m is a product of two distinct primes. On the other hand, although in the each of the aforementioned papers [**3, 9, 10**] a different method is used, none of them can be applied to the case of multidimensional distributions, for example even to study the distribution of pairs (u_n, u_{n+1}). The only exception is the case of a very small exponent e, say $e = 2$, for which some results have been obtained in [**10**].

Here we show that the method of [**10**] combined with bounds for exponential sums with sparse polynomials from [**26**] allows us to study the multidimensional distribution of the power generator of pseudorandom numbers modulo a high power of a small prime number p.

Several other results about non-linear pseudorandom number generators have been obtained in the series of papers [**12, 13, 22, 23, 24, 25**]. However these apply to generators of the form $u_n \equiv f(u_{n-1}) \pmod m$ where f is a polynomial or a rational function of small degree, while in this paper we do not impose any restrictions on the size of e.

Throughout the paper the implied constants in symbols 'O', '$\gg$' and '$\ll$' may depend on the prime p, the integer s and on the small positive parameter ε (we recall that $A \ll B$ and $B \gg A$ are equivalent to $A = O(B)$).

Also, $\log z$ denotes the logarithm to the base p.

2. Exponential sums

Let p be a fixed prime number. For an integer vector $\mathbf{a} = (a_0, \ldots, a_{s-1}) \in \mathbb{Z}^s$ we define the exponential sum

$$S_k(\mathbf{a}) = \sum_{n=1}^{\tau} \mathbf{e}_k \left(\sum_{i=0}^{s-1} a_i u_{n+i} \right).$$

where

$$\mathbf{e}_r(z) = \exp(2\pi i z / p^r).$$

We obtain a non-trivial upper bound for the sums $S_k(\mathbf{a})$ and derive (see Theorem 3.2) the uniformity of distribution modulo $m = p^r$ of the elements u_n, $n = 1, \ldots, \tau$.

For a t-element set $\mathcal{R} = \{r_1, \ldots, r_t\} \subseteq \mathbb{Z}$ denote by $\Delta(\mathcal{R})$ the following determinant

$$\Delta(\mathcal{R}) = \det\left(\binom{r_i}{j}\right)_{i,j=1,\ldots,t},$$

where for integers $m \ge 0$ and k we set

$$\binom{k}{m} = \frac{k(k-1)\ldots(k-m+1)}{m!}.$$

It is not difficult to check that

$$\Delta(\mathcal{R}) = \prod_{\nu=1}^{t} \frac{r_\nu}{\nu!} \prod_{1 \le i < j < t} (r_j - r_i). \tag{2.1}$$

Let $\operatorname{ord}_p z$ denote the p-adic order of $z \in \mathbb{Z}$.

LEMMA 2.1. *Let p be a prime and let $\alpha \ge 1$ be an integer. Then, for any set $\mathcal{R} = \{r_1, \ldots, r_t\} \subseteq \mathbb{Z}$ and any integers $A_1, \ldots, A_t$ with $\gcd(A_1, \ldots, A_t, p) = 1$ the bound*

$$\left| \sum_{\substack{x=1 \\ \gcd(x,p)=1}}^{p^\alpha} \mathbf{e}_\alpha \left(\sum_{1 \le j \le t} A_j x^{r_j} \right) \right| \le C(p,t) p^{\alpha(1-1/t+\varepsilon)+\gamma}$$

holds, where $\gamma = \operatorname{ord}_p \Delta(\mathcal{R})$ and the constant $C(p,t)$ depends only on p and t.

PROOF. The necessary arguments are already in [**26**]; see the proof of Lemma 5 in that paper. More precisely we proceed as in the proof of formula (8) there but here we want the result uniform in α whereas there it needed to be uniform in p. In fact this makes it a little simpler in our case, the bound (4) in that paper not being needed here. □

We say that an integer g is *regular* modulo a prime p with $\gcd(p,g) = 1$ if $g^{p-1} \not\equiv 1 \pmod{p^2}$ for odd p and $g \equiv 5 \pmod 8$ for $p = 2$.

The following statement is a partial case of a more general fact; see [**15**], Remark after Lemma 1 and [**16**], beginning of Section 1. It also follows immediately from Lemma P of Section 3.2.1.2 of [**17**].

LEMMA 2.2. *Let the integer g be regular modulo the prime p and let T_r be the multiplicative order of g modulo p^r. Then $T_r = T_1 p^{r-1}$, $r = 1, 2, \ldots$, if p is odd, and $T_r = 2^{r-2}$, $r = 2, 3, \ldots$, if $p = 2$.*

In particular Lemma 2.2 implies that if g is regular modulo the prime p then

$$\operatorname{ord}_p (g^k - 1) \le \operatorname{ord}_p k + 2. \tag{2.2}$$

LEMMA 2.3. *Let $s \ge 1$ and $k \ge s$ be integers and let*

$$\mathcal{R} = \{1, \ldots, e^{s-1}, e^k, \ldots, e^{k+s-1}, \}.$$

If e is regular modulo p then

$$\operatorname{ord}_p \Delta(\mathcal{R}) \le 8s^2 + s\log k.$$

PROOF. We have

$$\Delta(\mathcal{R}) = e^f \prod_{\nu=1}^{2s} \frac{1}{\nu!} \prod_{0\le i<j\le s-1} \left(e^{j-i}-1\right)^2 \prod_{i,j=0}^{s-1} \left(e^{k+j-i}-1\right)$$

for some exponent f depending on s and k.

From (2.1) and (2.2) we derive, ignoring the product of factorials in the denominator, and since $\operatorname{ord}_p e = 0$,

$$\begin{aligned}\operatorname{ord}_p \Delta(\mathcal{R}) &\le 2\sum_{0\le i<j\le s-1} (\operatorname{ord}_p (j-i)+2) + \sum_{i,j=0}^{s-1} (\operatorname{ord}_p (k+j-i)+2)\\ &\le 2\sum_{0\le i<j\le s-1} \operatorname{ord}_p (j-i) + \sum_{i,j=0}^{s-1} \operatorname{ord}_p (k+j-i) + 4s^2 - 2s.\end{aligned}$$

Using $\lfloor a+b\rfloor - \lfloor a\rfloor \le \lceil b\rceil$, for any integers $h, w \ge 0$ and recalling that all logarithms are to base p, we have

$$\begin{aligned}\sum_{j=1}^{w} \operatorname{ord}_p (h+j) &= \sum_{\nu=1}^{\lfloor\log(h+w)\rfloor} \left\lfloor \frac{h+w}{p^\nu}\right\rfloor - \sum_{\mu=1}^{\lfloor\log h\rfloor} \left\lfloor \frac{h}{p^\mu}\right\rfloor\\ &= \sum_{\nu=1}^{\lfloor\log(h+w)\rfloor} \left(\left\lfloor \frac{h+w}{p^\nu}\right\rfloor - \left\lfloor \frac{h}{p^\nu}\right\rfloor\right)\\ &\le \sum_{\nu=1}^{\lfloor\log(h+w)\rfloor} \left\lceil \frac{w}{p^\nu}\right\rceil \le \sum_{\nu=1}^{\lfloor\log(h+w)\rfloor} \left(\frac{w}{p^\nu}+1\right)\\ &\le w\sum_{\nu=1}^{\infty} \frac{1}{2^\nu} + \lfloor\log(h+w)\rfloor \le w + \log(h+w).\end{aligned}$$

Therefore,

$$\sum_{0\le i<j\le s-1} \operatorname{ord}_p (j-i) = \sum_{j=1}^{s-1}\sum_{i=1}^{j} \operatorname{ord}_p i \le \sum_{j=1}^{s-1} (j+\log j) \le \frac{s^2}{2} + s\log s$$

and

$$\begin{aligned}\sum_{i,j=0}^{s-1} \operatorname{ord}_p (k+j-i) &= \sum_{i=0}^{s-1}\sum_{j=0}^{s-1} \operatorname{ord}_p (k+j-i)\\ &= \sum_{i=0}^{s-1}\sum_{j=1}^{s} \operatorname{ord}_p ((k-i-1)+j)\\ &\le \sum_{i=0}^{s-1} \{s + \log(k-i-1+s)\}\\ &\le s^2 + s\log 2k \le s^2 + s + s\log k.\end{aligned}$$

Collecting the above estimates together, we obtain

$$\operatorname{ord}_p \Delta(\mathcal{R}) \le 6s^2 + 2s\log s + s\log k - s$$

and, taking into account that $\log s \le s$, we obtain the required result. □

Now we are prepared to formulate our main estimate.

THEOREM 2.4. *Let $e \ge 2$ and $m = p^r$ where p is a prime such that e is regular modulo p, Assume that the sequence (u_n), given by (1.1) is periodic with period τ. Fix $\varepsilon > 0$. Then, for every integer s, and every vector $\mathbf{a} = (a_0, \dots, a_{s-1}) \in \mathbb{Z}^s$ with $\gcd(a_0, \dots, a_{s-1}, p^r) = \mu$, we have*

$$|S_{\mathbf{a}}| \ll \tau^{1/2} m^{1/2} (m/\mu)^{-1/4s(s+1)+\varepsilon}$$

where the implied constant depends at most on p, s and ε.

PROOF. Let $\mathcal{K} = \{1, 2, \dots, K\}$, so $|\mathcal{K}| = K \in \mathbb{Z}^+$. For any nonnegative integer k we have

$$\begin{aligned} S_{\mathbf{a}} &= \sum_{n=1}^{\tau} \mathbf{e}_r\left(\sum_{i=0}^{s-1} a_i u_{n+k+i}\right) \\ &= \sum_{n=1}^{\tau} \mathbf{e}_r\left(\sum_{i=0}^{s-1} a_i \vartheta^{e^{n+k+i}}\right), \end{aligned}$$

and so

$$KS_{\mathbf{a}} = \sum_{k\in\mathcal{K}} \sum_{n=1}^{\tau} \mathbf{e}_r\left(\sum_{i=0}^{s-1} a_i \vartheta^{e^{n+k+i}}\right).$$

Let $\mu = p^\rho$.

Since ϑ^{e^n}, $n = 1, \dots, \tau$, are distinct modulo p^r, the Cauchy-Schwarz inequality gives

$$\begin{aligned} K^2|S_{\mathbf{a}}|^2 &\le \tau \sum_{n=1}^{\tau} \left| \sum_{k\in\mathcal{K}} \mathbf{e}_r\left(\sum_{i=0}^{s-1} a_i \vartheta^{e^{n+k+i}}\right)\right|^2 \\ &\le \tau \sum_{\substack{x=0\\ \gcd(x,p)=1}}^{p^r} \sum_{k,l\in\mathcal{K}} \mathbf{e}_r\left(\sum_{i=0}^{s-1} a_i \left(x^{e^{k+i}} - x^{e^{l+i}}\right)\right) \\ &\le 2\tau \sum_{\substack{k,l\in\mathcal{K}\\ l<k}} \sum_{\substack{x=0\\ \gcd(x,p)=1}}^{p^r} \mathbf{e}_r\left(\sum_{i=0}^{s-1} a_i \left(\left(x^{e^l}\right)^{e^{k-l+i}} - \left(x^{e^l}\right)^{e^i}\right)\right) + \tau K p^r \\ &= 2\tau \sum_{\substack{k,l\in\mathcal{K}\\ l<k}} \sum_{\substack{x=0\\ \gcd(x,p)=1}}^{p^r} \mathbf{e}_r\left(\sum_{i=0}^{s-1} a_i \left(x^{e^{k-l+i}} - x^{e^i}\right)\right) + \tau K p^r \\ &= 2\tau p^\rho \sum_{\substack{k,l\in\mathcal{K}\\ l<k}} \sum_{\substack{x=0\\ \gcd(x,p)=1}}^{p^{r-\rho}} \mathbf{e}_{r-\rho}\left(\sum_{i=0}^{s-1} b_i \left(x^{e^{k-l+i}} - x^{e^i}\right)\right) + \tau K p^r \end{aligned}$$

where $a_i = b_i p^\rho$ and $\gcd(b_0, \ldots, b_{s-1}, p) = 1$.

Applying Lemma 2.1 and Lemma 2.3, we derive

$$\begin{aligned} K^2|S_{\mathbf{a}}|^2 &\ll \tau p^\rho p^{(r-\rho)(1-1/2s+\varepsilon)} \sum_{\substack{l,k\in\mathcal{K}\\ l<k}} p^{8s^2+s\log(k-l)} + \tau K p^r \\ &\ll \tau p^\rho p^{(r-\rho)(1-1/2s+\varepsilon)} K^{s+2} + \tau K p^r \\ &\ll \tau m^{1-1/2s+\varepsilon} \mu^{1/2s-\varepsilon} K^{s+2} + \tau m K. \end{aligned}$$

Thus

$$|S_{\mathbf{a}}|^2 \ll \tau m \left((m/\mu)^{-1/2s+\varepsilon} K^s + K^{-1} \right).$$

Balancing the two terms in the above estimate (up to the $(m/\mu)^\varepsilon$) by selecting $K = \lfloor (m/\mu)^{1/2s(s+1)} \rfloor$ we obtain the result claimed. □

If for example $\mu = 1$ then for any $\delta > 0$ the bound of Theorem 2.4 is nontrivial for $\tau > m^{1-1/2s(s+1)+\delta}$ provided that r is sufficiently large in terms of p, s and δ.

3. Discrepancy

For a sequence of N points

$$\Gamma = (\gamma_{1,n}, \ldots, \gamma_{s,n})_{n=1}^N \tag{3.1}$$

in the half-open box $[0,1)^s$, denote by Δ_Γ its *discrepancy*, that is,

$$\Delta_\Gamma = \sup_{B\subseteq[0,1)^s} \left| \frac{T_\Gamma(B)}{N} - |B| \right|,$$

where $T_\Gamma(B)$ is the number of points of the sequence Γ which hit the box

$$B = [\alpha_1, \beta_1) \times \ldots \times [\alpha_s, \beta_s) \subseteq [0,1)^s$$

and the supremum is taken over all such boxes.

For an integer vector $\mathbf{a} = (a_1, \ldots, a_s) \in \mathbb{Z}^s$ we put

$$|\mathbf{a}| = \max_{i=1,\ldots,s} |a_i|, \qquad h(\mathbf{a}) = \prod_{i=1}^s \max\{|a_i|, 1\}.$$

This discrepancy of a sequence of points in the s-dimensional unit cube can be estimated by the *Erdös–Turán–Koksma inequality* (see Theorem 1.21 of [**5**]) which we present in the following form.

LEMMA 3.1. *There exists a constant $C_s > 0$ depending only on the dimension s such that, for any integer $M \geq 1$, for the discrepancy of a sequence of points (3.1) the bound*

$$\Delta_\Gamma < C_s \left(\frac{1}{M} + \frac{1}{N} \sum_{0<|\mathbf{a}|\leq M} \frac{1}{h(\mathbf{a})} \left| \sum_{n=1}^N \exp\left(2\pi i \sum_{j=1}^s a_j \gamma_{j,n} \right) \right| \right)$$

holds, where the sum is taken over all integer vectors

$$\mathbf{a} = (a_1, \dots, a_s) \in \mathbb{Z}^s$$

with $0 < |\mathbf{a}| \le M$.

Let D_s denote the discrepancy of the points

$$\left(\left\{\frac{u_n}{p^r}\right\}, \dots, \left\{\frac{u_{n+s-1}}{p^r}\right\}\right), \qquad n = 1, \dots, \tau.$$

THEOREM 3.2. *Assume that the sequence* (u_n), *given by (1.1) with* $m = p^r$ *where* p *is a prime such that* e *is regular modulo* p, *is periodic with period* τ. *Fix* $\varepsilon > 0$. *Then for every positive integer* s

$$D_s \ll \tau^{-1/2} m^{1/2 - 1/4s(s+1) + \varepsilon}$$

where the implied constant depends at most on p, s *and* ε.

PROOF. From Theorem 2.4 and Lemma 3.1, applied with $M = m$ we see

$$\begin{aligned}
D_s &\ll \frac{1}{m} + \frac{1}{\tau} \sum_{\rho=0}^{r} \sum_{\substack{0<|\mathbf{a}|\le p^r \\ \gcd(a_0,\dots,a_{s-1},p^r)=p^\rho}} \frac{1}{h(\mathbf{a})} \tau^{1/2} m^{1/2} (m/p^\rho)^{-1/4s(s+1)+\varepsilon/2} \\
&\ll \frac{1}{m} + \tau^{-1/2} m^{1/2-1/4s(s+1)+\varepsilon/2} \sum_{\rho=0}^{r} p^{\rho(1/4s(s+1)-\varepsilon/2)} \\
&\qquad\qquad \times \sum_{\substack{0<|\mathbf{a}|\le p^r \\ \gcd(a_0,\dots,a_{s-1},p^r)=p^\rho}} \frac{1}{h(\mathbf{a})} \\
&\ll \frac{1}{m} + \tau^{-1/2} m^{1/2-1/4s(s+1)+\varepsilon/2} \sum_{\rho=0}^{r} p^{-\rho(1+\varepsilon/2-1/4s(s+1))} \sum_{0<|\mathbf{a}|\le p^{r-\rho}} \frac{1}{h(\mathbf{a})} \\
&\ll \frac{1}{m} + \tau^{-1/2} m^{1/2-1/4s(s+1)+\varepsilon/2} \sum_{\rho=0}^{r} p^{-\rho(1+\varepsilon/2-1/4s(s+1))} \left(\log p^{r-\rho}\right)^s
\end{aligned}$$

and after simple calculations we derive the desired statement. □

We may remark that, for any $\delta > 0$, the bound of the theorem is $O\left(m^{-\delta/2}\right)$ provided that $\tau > m^{1-1/2s(s+1)+\delta}$ and r is sufficiently large in terms of p, s and δ.

4. Remarks

There is no particular reason for choosing the base p to be a prime number. Although this seems the most natural choice, the methods here work more generally. For example, a similar result holds for moduli m which are products of high powers of several fixed primes. Moreover these primes need not be kept fixed (nor need s) but could grow slowly with m although this would require a little more work. Also one could try to study the distribution not of consecutive $s-$tuples of the sequence but rather $s-$tuples $u_{n+g(i)}$, $1 \le i \le s$, for a fixed function g taking integer values distinct modulo τ. This will certainly work for certain simple functions g.

However it seems that neither the method of this work nor the method of [**7, 10**] can be extended to arbitrary composite moduli m. On the other hand, in [**8**], an appropriate modification of the method of [**9, 20**] has been developed. This variant of the method of [**9, 20**] has led to nontrivial upper bounds of the exponential sums involved in studying the uniformity of distribution of the power generator modulo a composite and to a number of other results.

Studying other characteristics of the power generator (1.1) with prime power moduli $m = p^r$ is of interest as well. In particular, it would be interesting to obtain analogues of the results of [**11, 27**] on the *linear complexity* of this generator modulo RSA moduli $m = pl$, where p and l are two distinct primes.

It would also be interesting to extend the results of this paper to the case of the *exponential generator*

$$v_n \equiv g^{v_{n-1}} \pmod m, \quad 0 \le v_n \le m-1, \qquad n = 1, 2, \dots ,$$

which also has numerous cryptographic applications [**4, 18**].

References

[1] L. Blum, M. Blum and M. Shub, 'A simple unpredictable pseudo-random number generator', *SIAM J. Comp.*, **15** (1986), 364–383.

[2] J. J. Brennan and B. Geist, 'Analysis of iterated modular exponentiation: The orbit of $x^\alpha \bmod N$', *Designs, Codes and Cryptography*, **13** (1998), 229–245.

[3] T. W. Cusick, 'Properties of the $x^2 \bmod N$ pseudorandom number generator', *IEEE Trans. Inform. Theory*, **41** (1995), 1155–1159.

[4] T. W. Cusick, C. Ding and A. Renvall, *Stream Ciphers and Number Theory*, Elsevier, Amsterdam, 1998.

[5] M. Drmota and R.F. Tichy, *Sequences, discrepancies and applications*, Springer-Verlag, Berlin, 1997.

[6] R. Fischlin and C. P. Schnorr, 'Stronger security proofs for RSA and Rabin bits', *J. Cryptology*, **13** (2000), 221–244.

[7] J. B. Friedlander, J. Hansen and I. E. Shparlinski, 'On character sums with exponential functions', *Mathematika*, **47** (2000), 75–85.

[8] J. B. Friedlander, S. V. Konyagin and I. E. Shparlinski, 'Some doubly exponential sums over $\mathbb{Z}_m$', *Acta Arith.*, **105** (2002), 349–370.

[9] J. B. Friedlander, D. Lieman and I. E. Shparlinski, 'On the distribution of the RSA generator', *Proc. Intern. Conf. on Sequences and their Applications (SETA '98), Singapore*, Springer-Verlag, London, 1999, 205–212.

[10] J. B. Friedlander and I. E. Shparlinski, 'On the distribution of the power generator', *Math. Comp.* **70** (2001), 1575–1589.

[11] F. Griffin and I. E. Shparlinski, 'On the linear complexity profile of the power generator', *Trans. IEEE on Information Theory*, **46** (2000), 2159–2162.

[12] F. Griffin, H. Niederreiter and I. E. Shparlinski, 'On the distribution of nonlinear recursive congruential pseudorandom numbers of higher orders', *Proc. the 13th Symp. on Appl. Algebra, Algebraic Algorithms, and Error-Correcting Codes*, Hawaii, 1999, Lect. Notes in Comp. Sci., Springer-Verlag, Berlin, **1719** (1999), 87–93.

[13] J. Gutierrez, H. Niederreiter and I. E. Shparlinski, 'On the multidimensional distribution of inversive congruential pseudorandom numbers in parts of the period', *Monatsh. Math.*, **129** (2000), 31–36.

[14] J. Håstad and M. Näslund, 'The security of individual RSA bits', *Proc 39th IEEE Symp. on Foundations of Comp. Sci.*, 1998, 510–519.

[15] N. M. Korobov, 'Exponential sums with exponential functions and the distribution of digits in periodic fractions', *Matem. Zametki*, **8** (1970), 641–652. *Matem. Notes*, **8** (1970), 831–837.

[16] N. M. Korobov, 'On the distribution of digits in periodic fractions', *Matem. Sbornik*, **89** (1972), 654–670. *Mat. USSR-Sb.*, **18** (1972), 659–676.

[17] D. Knuth, *The art of computer programming, vol. 2*, Addison-Wesley, Reading, MA, 1998.

[18] J. C. Lagarias, 'Pseudorandom number generators in cryptography and number theory', *Proc. Symp. in Appl. Math.*, Amer. Math. Soc., Providence, RI, **42** (1990), 115–143.

[19] R. Lidl and H. Niederreiter, *Finite fields*, Cambridge University Press, Cambridge, 1997.

[20] D. Lieman and I. E. Shparlinski, 'On a new exponential sum', *Canad. Math. Bull.*, **44** (2001), 87–92.

[21] A. J. Menezes, P. C. van Oorschot and S. A. Vanstone, *Handbook of Applied Cryptography*, CRC Press, Boca Raton, FL, 1996.

[22] H. Niederreiter and I. E. Shparlinski, 'On the distribution and lattice structure of nonlinear congruential pseudorandom numbers', *Finite Fields and Their Appl.*, **5** (1999), 246–253.

[23] H. Niederreiter and I. E. Shparlinski, 'On the distribution of inversive congruential pseudorandom numbers modulo a prime power', *Acta Arithm.*, **92** (2000), 89–98.

[24] H. Niederreiter and I. E. Shparlinski, 'On the distribution of pseudorandom numbers and vectors generated by inversive methods', *Appl. Algebra in Engin., Commun. and Computing*, **10** (2000), 189–202.

[25] H. Niederreiter and I. E. Shparlinski, 'On the distribution of inversive congruential pseudorandom numbers in parts of the period', *Math. Comp.*, **70** (2001), 1569–1574.

[26] I. E. Shparlinski, 'On exponential sums with sparse polynomials', *J. Number Theory*, **60** (1996), 233–244.

[27] I. E. Shparlinski, 'On the linear complexity of the power generator', *Designs, Codes and Cryptography*, **23** (2001), 5–10.

[28] D. R. Stinson, *Cryptography: Theory and Practice*, CRC Press, Boca Raton, FL, 1995.

[29] I. M. Vinogradov, *Elements of Number Theory*, Dover Publ., NY, 1954.

Dept. of Mathematics, University of Toronto,Toronto, Ontario M5S 3G3, Canada

E-mail address: `frdlndr@math.toronto.edu`

Dept. of Computing, Macquarie University, Sydney, NSW 2109, Australia

E-mail address: `jhansen@ics.mq.edu.au`

Dept. of Computing, Macquarie University, Sydney, NSW 2109, Australia

E-mail address: `igor@ics.mq.edu.au`

DIMACS Series in Discrete Mathematics
and Theoretical Computer Science
Volume **64**, 2004

Number Theoretical Error Estimates in a Quantization Scheme for Bandlimited Signals

C. Sinan Güntürk

Abstract. Sigma-delta quantization is a way of representing bandlimited signals (functions with compactly supported Fourier transforms) by $\{0,1\}$ sequences for each sampling density such that convolving these sequences with appropriately chosen filters produces approximations of the original signals. Approximations are refined by increasing the sampling density; this is what makes such a scheme fundamentally different from more conventional quantization schemes, where the sampling density is not varied. We present various examples of how tools from analytic number theory are employed in sharpening the error estimates in sigma-delta systems.

1. Introduction

Consider the problem of representing real numbers in $[0,1]$ by binary sequences in the following translation invariant manner: Each $x \in [0,1]$ is mapped to a sequence $q \in \{0,1\}^{\mathbb{Z}}$ such that for some appropriate sequence $h \in l^1(\mathbb{Z})$, called the reconstruction *filter*, one has

$$q * h = x, \tag{1.1}$$

where $*$ denotes the additive convolution of two sequences, and the symbol x also denotes the constant sequence $(\dots, x, x, \dots)$. A natural normalization for h is that $\sum h(n) = 1$, so that the number 1 is necessarily represented by the sequence $q = (\dots, 1, 1, \dots)$. As we shall show in Section 2, this problem is too strict in terms of the reconstruction formula (1.1) to be solvable for all x: a solution exists if and only if x is rational, and the solution q is necessarily periodic. An alternative approach is to ask for a sequence of filters $(h_\lambda)_{\lambda>0}$ such that

$$q * h_\lambda \to x, \tag{1.2}$$

uniformly, or at least pointwise, as $\lambda \to \infty$. The normalization condition can also be relaxed to the weaker form $\sum h_\lambda(n) \to 1$ as $\lambda \to \infty$. Clearly, there would not be any gain in introducing this alternative for a choice of sequence (h_λ) that converges in

2000 *Mathematics Subject Classification.* Primary 94A20, 11K06; Secondary 11L07, 41A25.
Key words and phrases. one-bit A/D conversion, sigma-delta quantization, uniform distribution, discrepancy, exponential sums.
The author's research was supported by a Francis Robbins Upton honorific graduate fellowship from Princeton University.

l^1, since the problem would then be immediately reduced to the case (1.1). Indeed, a typical choice is $h_\lambda(n) = \frac{1}{\lambda}\chi_{[0,1)}(\frac{n}{\lambda})$, which converges to 0 uniformly as $\lambda \to \infty$, but not in l^1. It turns out that in this new formulation, the problem has plenty of solutions for *all* x. Yet another possibility is to let the binary representation also vary with λ and ask for

$$q_\lambda * h_\lambda \to x. \tag{1.3}$$

The last two settings are quite flexible and the main question consists in finding efficient representations in the sense that (1.2) or (1.3) converges rapidly in λ, where the filters h_λ are scaled versions of an averaging window as in the example we have just given. It may also be desirable to determine the exact rate of convergence for particular schemes that have other features of interest.

In this more general context, it is also possible to formulate the problem for more general functions. Let $x(\cdot)$ be a function on $\mathbb{R}$, taking values in $[0, 1]$. Given a sequence u, we define the measure $\mu_\lambda(u)$ by

$$\mu_\lambda(u) = \frac{1}{\lambda}\sum_{n\in\mathbb{Z}} u(n)\delta_{n/\lambda}, \tag{1.4}$$

where δ_a denotes the Dirac mass at the point a. Then, for each function x in some appropriate class $\mathcal{C}$, the problem is to find a family (q_λ) of binary representations such that a pre-chosen filter φ (or a sequence (φ_λ) of filters) in $L^1(\mathbb{R})$ can decode x in the sense that

$$(\mu_\lambda(q_\lambda) * \varphi)(\cdot) = \frac{1}{\lambda}\sum_{n\in\mathbb{Z}} q_\lambda(n)\varphi(\cdot - \tfrac{n}{\lambda}) \quad \longrightarrow \quad x(\cdot) \tag{1.5}$$

in a given functional sense, as $\lambda \to \infty$. Note that (1.3) is already contained in (1.5) if we define the discrete filter h_λ by $h_\lambda(n) = \frac{1}{\lambda}\varphi(\frac{n}{\lambda})$, and restrict our attention to constant functions. However, the analogy between (1.3) and (1.5) is not exact due to the difference between the two settings.

Functions for which there exist solutions to the above problem include *bandlimited* functions, defined by

$$\mathcal{B}_\Omega = \{x : \mathbb{R} \to \mathbb{R} \mid \hat{x} \text{ is a finite Borel measure supported on } [-\Omega, \Omega]\}. \tag{1.6}$$

Here, $\hat{x}$ denotes the Fourier transform of x. Such a function is the restriction to $\mathbb{R}$ of an entire function of exponential type. Perhaps, the most important property of bandlimited functions is that they can be recovered from their samples taken at or above the critical density π/Ω, called the *Nyquist density.* This is called the *sampling theorem.* In our discussion, we assume $\Omega = \pi$ to ease the notation; otherwise, the analysis can be transposed by a rescaling of the argument. In this case, if the filter φ is such that $\hat{\varphi}$ is a continuous cut-off function satisfying

$$\hat{\varphi}(\xi) = \begin{cases} 1 & \text{if } |\xi| \leq \pi, \text{ and} \\ 0 & \text{if } |\xi| > \lambda\pi, \end{cases} \tag{1.7}$$

then one has the reconstruction formula

$$x(t) = \frac{1}{\lambda}\sum_{n\in\mathbb{Z}} x(\tfrac{n}{\lambda})\varphi(t - \tfrac{n}{\lambda}), \tag{1.8}$$

where we assume $\lambda > 1$. The equality holds pointwise, at least for all values of t for which the right hand side converges, and in general everywhere if the summation method uses Cesàro means. Typically, $\hat{\varphi}$ is chosen to be smooth so that the corresponding fast decay of φ enables an almost "local" reconstruction, which removes any concern about the method of summation. (The formula holds also in the L^2 sense when $x \in L^2(\mathbb{R})$, including the case $\lambda = 1$. However, in this critical sampling case, a smooth $\hat{\varphi}$ cannot be chosen; $\hat{\varphi} = \chi_{[-\pi,\pi]}$ is the only candidate.)

We shall search solutions to (1.5) for the class $\mathcal{C} = \{x : \mathbb{R} \to [0,1] \mid x \in \mathcal{B}_\pi\}$. An important class of solutions are generated by the so-called $\Sigma\Delta$ *quantization* (or $\Sigma\Delta$ *modulation*), which transforms the sequence of samples $(x(\frac{n}{\lambda}))$ into the sequence of output bits $(q_\lambda(n))$. The *first order* $\Sigma\Delta$ quantizer operates according to a very simple principle: Given the sequence of samples $(x(\frac{n}{\lambda}))$ taking values in $[0,1]$, a binary sequence q_λ is constructed such that

$$\sum_{n_1}^{n_2} x(\tfrac{n}{\lambda}) \sim \sum_{n_1}^{n_2} q_\lambda(n), \tag{1.9}$$

for all n_1 and n_2. This is done by the following procedure: Define the sequences X_λ, Q_λ and q_λ (which we denote by X, Q, and q when x is constant) by

$$X_\lambda(n) = \sum_{m=1}^{n} x(\tfrac{m}{\lambda}), \tag{1.10}$$

$$Q_\lambda(n) = \lfloor X_\lambda(n) \rfloor, \quad \text{and} \tag{1.11}$$

$$q_\lambda(n) = Q_\lambda(n) - Q_\lambda(n-1). \tag{1.12}$$

Since x takes values in $[0,1]$, we have $q_\lambda(n) \in \{0,1\}$; and at the same time (1.9) is satisfied, up to an error less than 1. X_λ can be defined naturally for negative indices as well, by integrating backwards.

The property (1.9) implies an immediate estimate for constants; if the resulting sequence q is filtered using a rectangular averaging window of size λ, i.e. $h_\lambda(n) = \frac{1}{\lambda}\chi_{[0,1)}(\frac{n}{\lambda})$, then

$$\|q * h_\lambda - x\|_{l^\infty} \leq \frac{1}{\lambda}. \tag{1.13}$$

It turns out, as shown in [**5**], that the same estimate holds (in the sense of (1.5)) for the whole class of bandlimited functions, using a smooth window that satisfies (1.7). One of the main results we shall present in this paper is that the exponent of λ in this estimate can be improved using number theoretical tools, both for the special case of constants and the general case of bandlimited functions. After a discussion of general considerations in Section 2, we shall present these number theoretical estimates in Sections 3 and 4.

The summation, truncation and differencing steps in (1.10) to (1.12) explain the name "$\Sigma\Delta$ quantization". The reason for the name "first order" is that only first order sums and differences have been used; a k-th order scheme involves k-th order sums and differences. The first construction of stable $\Sigma\Delta$ quantization schemes for all orders is due to Daubechies and DeVore [**5**]. We shall return to this briefly in Section 5. Let us also mention the references [**1**] and [**2**] which cover the history as well as recent advances in the theory and applications of $\Sigma\Delta$ quantization in electrical engineering.

2. General Considerations for Constants

Let us return to the problem (1.1). We claimed that a solution exists if and only if x is rational. This claim will follow simply as a corollary to a theorem of Szegö. We first recall the definition of spectral set for bounded sequences.

DEFINITION 2.1. Let a be a sequence in l^∞. Then the spectral set $\sigma(a)$ is the set of all $\xi \in \mathbb{T}$ such that $b * a = 0$ ($b \in l^1$) implies $\sum_n b(n)e^{-in\xi} = 0$.

The spectral set of a sequence a in l^1 is precisely the closed support of its Fourier transform $\hat{a}(\xi) = \sum_n a(n)e^{-in\xi}$. Definition 2.1 extends this notion for arbitrary sequences in l^∞, whose Fourier transforms in general are not functions. It is always true that $\sigma(a)$ is a closed subset of $\mathbb{T}$.

Szegö's theorem, as stated in the following form in [**7, 8**] by Helson, deals with spectral sets of sequences whose terms come from finite sets:

THEOREM 2.2 (Szegö-Helson). *Let a be a sequence whose terms are all drawn from a finite set S of complex numbers. Unless the sequence is periodic, its spectral set fills $\mathbb{T}$.*

Let us apply this powerful theorem to the sequence $q - x$. Certainly, the terms of this sequence come from the finite set $\{-x, 1-x\}$. On the other hand, $h * x = x$, since $\sum h(n) = 1$. Thus, (1.1) is a restatement of $(q - x) * h = 0$. According to the conclusion of the theorem, a nonperiodic q (hence a nonperiodic $q - x$) would mean $\sigma(q - x) = \mathbb{T}$. This implies $\hat{h}(\xi) = \sum h(n)e^{-in\xi} = 0$ for all $\xi \in \mathbb{T}$, and hence $h \equiv 0$, a contradiction. Hence, only periodic solutions of (1.1) may exist.

Now, consider a periodic solution q, whose period is N. Then $q * h$ is also periodic and its period divides N. It is a simple calculation to show that

$$\sum_{n=1}^{N} (q * h)(n) = \sum_{n=1}^{N} q(n). \tag{2.1}$$

Since the left hand side is Nx and the right hand side is an integer $0 \leq M \leq N$, it follows that $x = M/N$, i.e., x is rational. This proves the "only if" part of the assertion. The "if" part follows from a trivial construction. Consider a rational number $x = M/N$ and let h be the rectangular averaging window of length N. If q is such that it has period N and exactly M of $\{q(k) : 1 \leq k \leq N\}$ are equal to 1, then it is clear that $q * h = x$. In summary:

THEOREM 2.3. *There is a solution to (1.1) if and only if x is rational. The solution q is necessarily periodic and its period is a multiple of the denominator of x in its reduced form.*

The rectangular filter of size N would work for all $x = M'/N'$, where N' divides N. (So, if the size of the window is chosen to be l.c.m.$(1, \ldots, N)$, then all numbers in the Farey sequence $\mathcal{F}_N$ can be decoded with it.[1]) Now, let us consider the implications of (2.1) on the filter h. If a rational number $x = r/s$ (in its reduced form) is represented by a sequence q of period N, then s must divide N and precisely $M = Nr/s$ of $\{q(k) : k = 1, \ldots, N\}$ are equal to 1. Let the corresponding indices be $l_1, \ldots, l_M$ and define $P(\xi) = \sum_{j=1}^{M} e^{-il_j\xi}$. Then, a straightforward calculation shows that $q * h = x$ implies $P(\xi)\hat{h}(\xi) = 0$ for all $\xi = 2\pi p/N$, $p = 1, \ldots, N-1$. If $M \neq N$ (i.e., $x \neq 1$), then $\hat{h}(\xi)$ must vanish at least at one of these points. Indeed,

[1] $\mathcal{F}_N$ is the set of all rationals in $[0, 1]$ with denominators less than or equal to N.

for almost any choice of P, almost all of these roots would belong to $\hat{h}$. This leads us to conjecture the following:

CONJECTURE 2.4. *It is impossible to construct a "universal" filter that can decode all rationals simultaneously. Equivalently, there is no collection of sequences in* $\{0,1\}^{\mathbb{Z}}$ *such that for every* $x \in \mathbb{Q} \cap [0,1]$, *there is a sequence* q *in this collection for which* $q * h = x$, *where* h *is fixed.*

Let us consider the alternative formulation (1.2). We have already stated in the introduction that it is now possible to find solutions for all $x \in [0,1]$. Even stronger, *universal* filter sequences (h_λ) can be employed. Before looking at $\Sigma\Delta$ quantization more closely as a scheme that generates such solutions, let us state the following result which is merely an extension of Theorem 2.3.

THEOREM 2.5. *A solution* q *to (1.2) for an irrational* x *is necessarily non-periodic.*

PROOF. Suppose (1.2) is satisfied for some x and a periodic $q \in \{0,1\}^{\mathbb{Z}}$. Let N be the period of q. Then, as in (2.1), we have

$$\sum_{n=1}^{N} (q * h_\lambda)(n) \to \sum_{n=1}^{N} q(n). \tag{2.2}$$

Combined with (1.2), this implies $x = \frac{1}{N} \sum_{n=1}^{N} q(n)$, i.e. $x \in \mathbb{Q}$. □

3. $\Sigma\Delta$ Quantization: Constants

Let us look at the very simple first order $\Sigma\Delta$ system (given in (1.10) to (1.12)) for constant input $x \in [0,1]$. Define the auxiliary variable $u(n)$ to be $X(n) - Q(n)$. It is equal to the fractional part of $X(n)$, which we denote by $\langle X(n) \rangle$. In practice, neither $X(n)$ nor $Q(n)$ are computed in an electronic circuit, since these variables are unbounded. However, the sequence u is bounded and satisfies the recursion relation

$$u(n) - u(n-1) = x - q(n), \qquad u(0) = 0. \tag{3.1}$$

In fact, this recursion is taken as the starting point in practice. One asks for a bounded solution u of (3.1) such that $q \in \{0,1\}^{\mathbb{Z}}$. The particular construction of q we have considered is just one of the solutions of (3.1).

We are now in the setting (1.2). For a given filter h_λ, let us derive an estimate for the error $e_\lambda = x - q * h_\lambda$, where h_λ obeys the scaling relation $h_\lambda(n) = \frac{1}{\lambda}\varphi(\frac{n}{\lambda})$ for some function φ. For $a \in l^1$, define $S(a) = \sum a(n)$. The error may be bounded by a sum of two contributions:

$$|e_\lambda(n)| \leq |x(1 - S(h_\lambda))| + \left| \sum_k (x - q(k)) h_\lambda(n-k) \right|. \tag{3.2}$$

Let us call the two terms e_λ^1 and e_λ^2. It is possible to choose φ such that the first error term is zero for all λ. For instance, $\varphi = \chi_{[0,1]}$ has this property for all integer λ; on the other hand, $\varphi \in BV \cap \mathcal{B}_\pi$ with $\hat{\varphi}(0) = 1$ has it for all real $\lambda > 1$ (which may easily be seen using Poisson's summation formula). Assume a choice of φ with this property, which leaves us with $e_\lambda^2 = (x - q) * h_\lambda$.

THEOREM 3.1. *For all* λ, $\|e_\lambda\|_{l^\infty} \leq \frac{1}{\lambda}\mathrm{Var}(\varphi)$.

PROOF. Let Δ denote the difference operator acting on sequences, defined by $\Delta u(n) = u(n) - u(n-1)$. Using the recursion relation (3.1), and that u is bounded by 1, it follows that

$$\begin{aligned} \|e_\lambda\|_{l^\infty} &= \|\Delta u * h_\lambda\|_{l^\infty}, \\ &= \|u * \Delta h_\lambda\|_{l^\infty}, \qquad (3.3) \\ &\leq \|u\|_{l^\infty}\|\Delta h_\lambda\|_{l^1}, \\ &\leq \frac{1}{\lambda}\mathrm{Var}(\varphi). \qquad (3.4) \end{aligned}$$

□

We call this the "basic estimate", in the sense that only boundedness of u was used in the derivation. We shall improve the exponent of λ by examining the expression (3.3) more closely.

First we note that $u * \Delta h_\lambda = (u-c) * \Delta h_\lambda$ for any constant c. Next, we consider a particular filter, the triangle function $\varphi(t) = (1-|t|)\chi_{[-1,1]}(t)$. Then,

$$(u * \Delta h_\lambda)(n) = \frac{1}{\lambda^2}\sum_{k=0}^{\lambda-1}\left(u(n+k) - \frac{1}{2}\right) - \frac{1}{\lambda^2}\sum_{k=1}^{\lambda}\left(u(n-k) - \frac{1}{2}\right) \qquad (3.5)$$

In the above expression, the value of c was chosen to be $1/2$ in order to exploit the fact that the state variable $u(n) = \langle X(n)\rangle = \langle nx\rangle$ forms a uniformly distributed sequence in $[0,1]$ for all irrational x. Using well known results in the theory of uniform distribution, we now prove the following improved estimate:

THEOREM 3.2. *Let $\epsilon > 0$ be given. Then for almost every $x \in [0,1]$, one has the estimate $\|e_\lambda\|_{l^\infty} \leq C_x \lambda^{-2}\log^{2+\epsilon}\lambda$, using the triangular filter.*

PROOF. *Koksma's inequality*[2] reduces the problem to considering the discrepancy values for the two sequences $u(n-\lambda), \dots, u(n-1)$ and $u(n), \dots, u(n+\lambda-1)$. The discrepancy can be bounded using the *Erdős-Turán inequality*[3]. In our case, these two inequalities result in

$$\left|\frac{1}{\lambda}\sum_{k=a+1}^{a+\lambda} u(k) - \frac{1}{2}\right| \leq \inf_K C\left(\frac{1}{K} + \sum_{k=1}^{K}\frac{1}{k}\left|\frac{1}{\lambda}\sum_{m=1}^{\lambda} e^{2\pi i k m x}\right|\right), \qquad (3.8)$$

uniformly in a and for all λ. The precise behaviour of this quantity depends on the behaviour of the continued fraction expansion of x (see, e.g. [**11, 12, 13**]). We shall not pursue such a detailed analysis, but rather use a metric result due to Khinchine, which yields the bound $O(\lambda^{-1}\log^{2+\epsilon}\lambda)$ for almost every (Lebesgue) x (see [**11**, pp. 131]). Let us note that the involved constant in general depends on x. This, together with (3.5) gives us the desired estimate. □

---

[2]For any function $f \in BV([0,1])$ and a finite sequence of points $x_1, \dots, x_N$ in $[0,1]$,

$$\left|\frac{1}{N}\sum_{n=1}^{N} f(x_n) - \int_0^1 f(t)dt\right| \leq \mathrm{Var}(f)D_N, \qquad (3.6)$$

where D_N denotes the *discrepancy* of the sequence $x_1, \dots, x_N$ and $\mathrm{Var}(f)$ is the variation of f.

[3]The discrepancy D_N of any real numbers $x_1, \dots, x_N$ is bounded by

$$D_N \leq C\left(\frac{1}{K} + \sum_{k=1}^{K}\frac{1}{k}\left|\frac{1}{N}\sum_{m=1}^{N} e^{2\pi i k x_m}\right|\right) \qquad (3.7)$$

for any positive integer K, where C is an absolute constant.

Let us use the notation $e_{\lambda,x}$ to denote the dependence of the error e_λ on the input value x. We assume that the triangular filter is used, so that (3.5) and (3.8) are in effect. We have already mentioned above that the precise behaviour of $e_{\lambda,x}$ can only be described by means of the continued fraction expansion of x. However, the mean behaviour is simpler. Let us consider the mean squared error (MSE)

$$\text{MSE}(e_\lambda) = \int_0^1 \|e_{\lambda,x}\|_{l^\infty}^2 dx \tag{3.9}$$

A straightforward bound for $\text{MSE}(e_\lambda)$ can be found directly from (3.8). Let $P_{\lambda,k}(x)$ denote the trigonometric polynomial $\lambda^{-1}\sum_{m=1}^{\lambda} e^{2\pi ikmx}$. Note that $\|P_{\lambda,k}\|_{L^2([0,1])} = \lambda^{-1/2}$. Then,

$$\begin{aligned}
\text{MSE}(e_\lambda) &\leq \frac{C}{\lambda^2}\int_0^1 \inf_K \Big(\frac{1}{K} + \sum_{k=1}^{K} k^{-1}|P_{\lambda,k}(x)|\Big)^2 dx \\
&\leq \frac{C'}{\lambda^2}\inf_K \int_0^1 \left(\frac{1}{K^2} + \Big(\sum_{k=1}^{K} k^{-1}|P_{\lambda,k}(x)|\Big)^2\right) dx \\
&\leq \frac{C'}{\lambda^2}\inf_K \left(\frac{1}{K^2} + \sum_{k=1}^{K}\sum_{l=1}^{K}\frac{1}{kl}\int_0^1 |P_{\lambda,k}(x)||P_{\lambda,l}(x)|dx\right) \\
&\leq \frac{C'}{\lambda^2}\inf_K \left(\frac{1}{K^2} + \frac{1}{\lambda}\log^2 K\right),
\end{aligned}$$

where we have used the Cauchy-Schwarz inequality in the last step. Finally, by choosing $K \sim \lambda^{1/2}$, we arrive at the bound

$$\text{MSE}(e_\lambda) \leq C\lambda^{-3}\log^2\lambda. \tag{3.10}$$

The exponent of λ in this estimate is optimal. Indeed, using number theoretical tools, it is shown in [**14**] that

$$C_1\lambda^{-3} \leq \int_0^1 |e_{\lambda,x}(0)|^2 dx \leq C_2\lambda^{-3}. \tag{3.11}$$

It is natural to conjecture that this is also true for $\| \int_0^1 |e_{\lambda,x}(\cdot)|^2 dx\|_{l^\infty}$, a quantity smaller than $\text{MSE}(e_\lambda)$. It is shown in [**3**] that the quantity

$$\int_0^1 \frac{1}{2N+1}\sum_{n=-N}^{N} |e_{\lambda,x}(n)|^2 dx \tag{3.12}$$

(which is yet even smaller) also behaves as $O(\lambda^{-3})$ as $N \to \infty$.

4. ΣΔ Quantization: Arbitrary Bandlimited Inputs

In this section, we generalize the results of the previous section to arbitrary bandlimited functions. The difference equation now reads as

$$u_\lambda(n) - u_\lambda(n-1) = x_\lambda(n) - q_\lambda(n), \qquad u(0) = 0; \tag{4.1}$$

where $x_\lambda(n)$ and $u_\lambda(n)$ are defined to be $x(\frac{n}{\lambda})$ and $X_\lambda(n) - Q_\lambda(n) = \langle X_\lambda(n)\rangle$, respectively. Clearly, u_λ takes its values in $[0,1]$. Our setting is (1.5), and in general we allow φ to depend on λ, which will be denoted by φ_λ.

The Basic Estimate. We start with the corresponding "basic estimate" given in [**5**], where it is also possible to use a fixed filter φ for all λ.

THEOREM 4.1 ([**5**]). *Let $x \in \mathcal{B}_\pi$ with $0 \le x(t) \le 1$ for all t, and $\varphi \in BV(\mathbb{R})$ satisfying (1.7) for some fixed $\lambda_0 > 1$. Then, $\|x - \mu_\lambda(q_\lambda) * \varphi\|_{L^\infty} \le \frac{1}{\lambda}\mathrm{Var}(\varphi)$ for all $\lambda \ge \lambda_0$.*

PROOF. The sampling theorem states that $x = \mu_\lambda(x_\lambda) * \varphi$ for all $\lambda \ge \lambda_0$. Let Δ_η be the difference operator whose action on a measure is given by $\Delta_\eta \mu(\cdot) = \mu(\cdot) - \mu(\cdot - \eta)$ and let $\mathbf{1}$ denote the constant sequence of 1's. Then,

$$\begin{aligned} x - \mu_\lambda(q_\lambda) * \varphi &= \mu_\lambda(x_\lambda - q_\lambda) * \varphi, \\ &= \mu_\lambda(\Delta u_\lambda) * \varphi, \\ &= \Delta_{1/\lambda}\mu_\lambda(u_\lambda) * \varphi, \\ &= \mu_\lambda(u_\lambda) * \Delta_{1/\lambda}\varphi, \end{aligned} \tag{4.2}$$

so that

$$\begin{aligned} \|x - \mu_\lambda(q_\lambda) * \varphi\|_{L^\infty} &\le \|\mu_\lambda(\mathbf{1}) * |\Delta_{1/\lambda}\varphi|\,\|_{L^\infty}, \\ &\le \frac{1}{\lambda}\mathrm{Var}(\varphi). \end{aligned} \tag{4.3}$$

In the last step, we made use of the identity $\mu_\lambda(\mathbf{1}) * f = \frac{1}{\lambda}\sum f(\cdot - \frac{n}{\lambda})$. □

An Improved Estimate. We shall apply the ideas of the previous section to prove the following theorem, which is an improvement of the above basic estimate:

THEOREM 4.2. *For all $\eta > 0$, there exists a family $\{\varphi_\lambda\}_{\lambda \ge 1}$ of filters such that, for all x in Theorem 4.1, and all t for which $x'(t)$ does not vanish, we have*

$$|x(t) - (\mu_\lambda(q_\lambda) * \varphi_\lambda)(t)| \le C\lambda^{-4/3+\eta} \tag{4.4}$$

for some constant $C = C(\eta, x'(t))$.

Note that bandlimited functions are analytic, so that the derivative x' of a non-constant bandlimited function x has at most countably many zeros, with no accumulation point. It is also possible to carry out a higher order analysis at the zeros of x', but we shall not touch upon this here.

PROOF. We divide the proof into a number of steps.
1. Fix t (for which $x'(t) \ne 0$). For each λ, let $N_\lambda = \lfloor \lambda t \rfloor$, and define the sequence U_λ by

$$U_\lambda(n) - U_\lambda(n-1) = u_\lambda(n) - \frac{1}{2}, \qquad U_\lambda(N_\lambda) = 0. \tag{4.5}$$

Let also $t_\lambda = N_\lambda/\lambda$ and $\delta_\lambda = t - t_\lambda$. Note that $|\delta_\lambda| \le 1/\lambda$. Now, (4.2) can be written as

$$\begin{aligned} x(t) - (\mu_\lambda(q_\lambda) * \varphi_\lambda))(t) &= \frac{1}{\lambda}\sum_n \Delta U_\lambda(n)\Delta_{1/\lambda}\varphi(t - \tfrac{n}{\lambda}), \\ &= \frac{1}{\lambda}\sum_n U_\lambda(n)\Delta_{1/\lambda}^2\varphi(t - \tfrac{n}{\lambda}), \\ &= \frac{1}{\lambda}\sum_n U_\lambda(N_\lambda + n)\Delta_{1/\lambda}^2\varphi(-\tfrac{n}{\lambda} + \delta_\lambda), \end{aligned} \tag{4.6}$$

for any φ that decays sufficiently fast. Denote the error expression by $e_\lambda(t)$.

2. Our purpose is to find non-trivial bounds for $U_\lambda(N_\lambda + n)$ by accounting for the cancellations in

$$U_\lambda(N_\lambda + n) = \sum_{m=1}^{n} \Big(u_\lambda(N_\lambda + m) - \frac{1}{2}\Big), \tag{4.7}$$

where we have assumed $n > 0$, the other case being essentially the same. Note that the trivial bound is $|n|/2$. We shall prove the following estimate:

$$|U_\lambda(N_\lambda + n)| \le C_1\Big(\lambda^{2/3} + \frac{\lambda^{1/2}}{|x'(t)|^{1/2}}\Big), \tag{4.8}$$

for all $n \le C_2|x'(t)|\lambda$ and $\lambda > C_3|x'(t)|^{-1}$. [4]

The inequalities of Koksma and Erdős-Turán result in the bound

$$|U_\lambda(N_\lambda + n)| \le \inf_K C\left(\frac{n}{K} + \sum_{k=1}^{K}\frac{1}{k}\Big|\sum_{m=1}^{n} e^{2\pi i k u_\lambda(N_\lambda+m)}\Big|\right), \tag{4.9}$$

which reduces our task to analyzing the behaviour of the exponential sums

$$S_{\lambda,k}(n) := \sum_{m=1}^{n} e^{2\pi i k X_\lambda(N_\lambda+m)}, \tag{4.10}$$

since $u_\lambda(n) = \langle X_\lambda(n)\rangle$.

3. We will use the stationary phase methods of van der Corput to estimate the exponential sums given in (4.10). The following well-known theorems serve well for this purpose:

THEOREM 4.3 (Truncated Poisson, [**9**]). *Let f be a real-valued function and suppose that f' is continuous and increasing on $[a, b]$. Put $\alpha = f'(a)$, $\beta = f'(b)$. Then*

$$\sum_{a\le m\le b} e^{2\pi i f(m)} = \sum_{\alpha-1\le\nu\le\beta+1} \int_a^b e^{2\pi i (f(\tau)-\nu\tau)}\,d\tau + O(\log(2+\beta-\alpha)). \tag{4.11}$$

(If f' is decreasing, then by taking the complex conjugate of the above expression applied to $-f$, one finds the same expression with α and β switched.)

THEOREM 4.4 (van der Corput, [**10**]). *Suppose ϕ is real-valued and smooth in (a, b), and that $|\phi^{(r)}(t)| \ge \mu$ for all $t \in (a, b)$ and for a positive integer r. If $r = 1$, suppose additionally that ϕ' is monotonic. Then there exists an absolute constant c_r such that*

$$\Big|\int_a^b e^{i\phi(t)}\,dt\Big| \le c_r \mu^{-1/r}. \tag{4.12}$$

---

[4] As is customary, we shall use the notations $C, C', C_1, C_2, \ldots$ for generic constants that may change value from one proof to another; constants of different values occurring in the same argument will be distinguished by different indices.

In our case, X_λ is initially only defined on the integers; however, (1.10) immediately yields an (analytic) interpolation of X_λ. We call this new function X_λ as well, and show in the Appendix that for $\lambda \geq C|x'(t)|^{-1}$, and all real τ in the range $0 \leq \tau \leq C_3|x'(t)|\lambda$, one has

$$C_1 \frac{|x'(t)|}{\lambda} \leq |X''_\lambda(N_\lambda + \tau)| \leq C_2 \frac{|x'(t)|}{\lambda}, \tag{4.13}$$

where C, C_1, C_2, and C_3 are absolute numerical constants.

In Theorem 4.3, we set $f = kX_\lambda$, $a = N_\lambda + 1$, and $b = N_\lambda + n$, and assume $n \leq C_3|x'(t)|\lambda$. It follows from (4.13) that the number of integral terms in the right hand side of (4.11) is bounded by

$$\begin{aligned} |\beta - \alpha + 3| &\leq 3 + k(n-1) \sup_{1 \leq \tau \leq n} |X''_\lambda(N_\lambda + \tau)| \\ &\leq 3 + k(n-1) C_2 \frac{|x'(t)|}{\lambda}. \end{aligned} \tag{4.14}$$

On the other hand, using Theorem 4.4 (for $r = 2$) with (4.13), each exponential integral term in (4.11) is bounded by $C(k|x'(t)|/\lambda)^{-1/2}$. Combining this with the bound on the number of terms that we have just found, we get

$$|S_{\lambda,k}(n)| \leq C_1 n \Big(\frac{k}{\lambda}\Big)^{1/2} + C_2 \Big(\frac{k}{\lambda}\Big)^{-1/2} |x'(t)|^{-1/2} + O(\log(2+k)), \tag{4.15}$$

where in the first term we have made use of the fact that $\|x'\|_{L^\infty} \leq \pi$ (which follows from Bernstein's inequality[5]), and in the logarithmic term that $|\beta - \alpha| \leq k$ for the given range of n. Note that, for small k, this bound significantly improves the trivial bound n. Now, if in (4.9), one chooses $K \sim \lambda^{1/3}$, then (4.15) yields our desired estimate (4.8).

4. We finish the proof of Theorem 4.4 by bounding (4.6) for a particular family of filters which we construct next. For this, we fix a filter φ such that $\hat{\varphi}$ is C^∞, $\text{supp}(\hat{\varphi}) \subset [-c_0\pi, c_0\pi]$ for some small fixed $c_0 > 1$, and $\hat{\varphi}(\xi) = 1$ on $[-\pi, \pi]$. Then φ is a Schwartz function, i.e., φ has rapidly decreasing derivatives: there are constants $C_N^{(l)}$ for all $N \geq 0$ and $l \geq 0$ such that

$$|\varphi^{(l)}(t)| \leq \frac{C_N^{(l)}}{(1+|t|)^N}. \tag{4.16}$$

For a small $\eta > 0$, we set $\Omega_\lambda = \lambda^{\eta/2}$ and define φ_λ by

$$\varphi_\lambda(t) = \Omega_\lambda \varphi(\Omega_\lambda t) \tag{4.17}$$

for $\lambda \geq 1$. Then $\hat{\varphi}_\lambda(\xi) = \hat{\varphi}(\xi/\Omega_\lambda)$ and hence $\{\varphi_\lambda\}$ is an admissible family of reconstruction filters. We turn back to the expression (4.6). For small n (i.e., for $|n| \leq c|x'(t)|\lambda$ for a sufficiently small constant c), we will use the estimate (4.8) in the form $O(|x'(t)|^{-1/2}\lambda^{2/3})$, and for large n, the trivial estimate $|n|/2$. Thus,

$$\begin{aligned} |e_\lambda(t)| &\leq \frac{1}{\lambda}\Big(O(|x'(t)|^{-1/2}\lambda^{2/3}) \sum_{|n| \leq c|x'(t)|\lambda} |\Delta^2_{1/\lambda}\varphi_\lambda(-\tfrac{n}{\lambda} + \delta_\lambda)| \\ &\quad + \sum_{|n| > |x'(t)|c\lambda} \frac{|n|}{2} |\Delta^2_{1/\lambda}\varphi_\lambda(-\tfrac{n}{\lambda} + \delta_\lambda)| \Big) \end{aligned} \tag{4.18}$$

[5] If $f \in \mathcal{B}_\Omega$, then $\|f^{(s)}\|_{L^\infty} \leq \Omega^s \|f\|_{L^\infty}$.

The first sum term can easily be bounded by

$$2\lambda^{-1}\|\varphi_\lambda''\|_{L^1} = 2\lambda^{-1+\eta}\|\varphi''\|_{L^1}, \tag{4.19}$$

and the second sum term by

$$\sum_{|n|>c|x'(t)|\lambda} \frac{|n|}{2}\cdot\frac{2}{\lambda^2}\cdot\frac{\Omega_\lambda^3 C_N^{(2)}}{(1+\Omega_\lambda|n|\lambda^{-1})^N} \leq C(N,|x'(t)|)\Omega_\lambda^{-N+3}, \tag{4.20}$$

for all N. We choose N such that $(N-3)\eta/2 > 1/3$. Putting together, this results in the estimate

$$\epsilon_\lambda(t) \leq C(\eta,|x'(t)|)\lambda^{-4/3+\eta}, \tag{4.21}$$

concluding the proof. □

5. Other Results

In the introduction, we mentioned higher order $\Sigma\Delta$ schemes. In a k-th order $\Sigma\Delta$ scheme, one is interested in solutions $q \in \{0,1\}^{\mathbb{Z}}$ of the difference equation

$$\Delta^k u = x - q, \tag{5.1}$$

with $u \in l^\infty$ (such a scheme is called *stable*). Again, we use the notations u_λ, x_λ and q_λ when x is an arbitrary bandlimited function sampled at the rate λ.

The second order case is best understood among all higher order schemes, and there are many construction strategies for q that are stable. We shall not get into any specific details of these schemes, but state only the error estimates. The basic estimate is now $O(\lambda^{-2})$ both for constants and general bandlimited functions. However, in this case, the exponent can be improved only for constants [**15**]. A notable difficulty arises in analyzing a particular family of piecewise affine dynamical systems in the plane, whose attracting invariant sets turn out to tile the plane with the action of the integer lattice. Using similar number theoretical techniques (such as mean values of Gauss sums and higher dimensional versions of discrepancy and the inequalities of Koksma and Erdős-Turán), the MSE(e_λ) is shown to be bounded by $O(\lambda^{-4.5})$. A similar improvement of the exponent for the uniform error is also found using the corresponding metric results.

The basic estimate for a stable k-th order scheme is $O(\lambda^{-k})$, which makes higher order schemes more interesting in terms of their approximation properties. However in practice, the first order scheme is widely used for its many attractive features, such as its robustness to hardware imperfections. There are various (mostly ad hoc) constructions in the electrical engineering literature of schemes of higher (single digit) orders ([**1, 2**]), without proof of stability. A recent mathematical achievement in this direction has been the construction of a family of arbitrary order stable $\Sigma\Delta$ schemes in [**5**].

For a given order k, let C_k be the constant hidden in the $O(\lambda^{-k})$ estimate. Then, it is natural to let the order depend on λ and look for the best possible decay of the error in λ. For the scheme in [**5**], the authors show that $C_k \sim c^{k^2}$, which results in an $O(c^{-\log^2\lambda})$ type decay using the optimal choice of order for each λ. On the other hand, a lower bound can easily shown to be $2^{-\lambda}$ using Kolmogorov entropy. It is an unsolved problem to determine whether exponential decay of error can be achieved.[6] Another step towards this lower bound is taken in [**17**] for the constant input case, in which a purely number theoretical construction yields an $O(c^{-\sqrt{\lambda}})$ type of error decay.

6. Appendix

In this section, we present the construction of the analytic interpolation of the sequence X_λ, which was stated in the proof of Theorem 4.4. Using the Taylor expansion of x about the point t_λ, we proceed as follows:

$$\begin{aligned} X_\lambda(N_\lambda + m) &= X_\lambda(N_\lambda) + \sum_{l=1}^{m} x(t_\lambda + \tfrac{l}{\lambda}) \\ &= X_\lambda(N_\lambda) + \sum_{s=0}^{\infty} \frac{x^{(s)}(t_\lambda)}{\lambda^s s!} \sum_{l=1}^{m} l^s \\ &= X_\lambda(N_\lambda) + \sum_{s=0}^{\infty} \frac{x^{(s)}(t_\lambda)}{\lambda^s s!} \sum_{j=1}^{s+1} P_{s,j} m^j \qquad (6.1) \\ &= X_\lambda(N_\lambda) + \sum_{j=1}^{\infty} m^j \sum_{s=j-1}^{\infty} \frac{x^{(s)}(t_\lambda)}{\lambda^s s!} P_{s,j}, \qquad (6.2) \end{aligned}$$

where $P_{s,j}$ are related to Bernoulli numbers. We use this last expression to define

$$X_\lambda(N_\lambda + \tau) = X_\lambda(N_\lambda) + \sum_{j=1}^{\infty} a_j \tau^j \qquad (6.3)$$

for all $\tau \geq 0$, where a_j is the sum term appearing in (6.2). The simple bound $P_{s,j} \leq s!/j!$ and Bernstein's inequality easily yields

$$|a_j| < \frac{2}{j!}\left(\frac{\pi}{\lambda}\right)^{j-1} \qquad (6.4)$$

for $\lambda > 2\pi$. Let us show that

$$X'_\lambda(N_\lambda + \tau) = x(t_\lambda + \tfrac{\tau}{\lambda}) + R_\lambda(\tau), \qquad (6.5)$$

where $R_\lambda(\tau)$ is small compared to $x(t_\lambda + \frac{\tau}{\lambda})$ for $\tau = O(\lambda)$. We start with noting that $P_{s,s+1} = 1/(s+1)$ for all s. Then, starting from (6.1),

$$\begin{aligned} X'_\lambda(N_\lambda + \tau) &= \sum_{s=0}^{\infty} \frac{x^{(s)}(t_\lambda)}{\lambda^s s!} \sum_{j=1}^{s+1} P_{s,j} j \tau^{j-1} \\ &= \sum_{s=0}^{\infty} \frac{x^{(s)}(t_\lambda)}{\lambda^s s!} \left(\tau^s + \sum_{j=1}^{s} P_{s,j} j \tau^{j-1}\right) \\ &= x(t_\lambda + \tfrac{\tau}{\lambda}) + R_\lambda(\tau), \qquad (6.6) \end{aligned}$$

[6]**Note added in October 2003:** This problem is now solved. In [**16**], we construct $\Sigma\Delta$ families which collectively achieve the error bound $\|e_\lambda\|_{L^\infty} = O(2^{-0.07\lambda})$ for arbitrary π-bandlimited functions. The best achievable constant in the exponent is still unknown.

where

$$\begin{aligned} R_\lambda(\tau) &= \sum_{s=0}^{\infty} \frac{x^{(s)}(t_\lambda)}{\lambda^s s!} \sum_{j=1}^{s} P_{s,j} j \tau^{j-1} \\ &= \sum_{j=1}^{\infty} j\tau^{j-1} \sum_{s=j}^{\infty} \frac{x^{(s)}(t_\lambda)}{\lambda^s s!} P_{s,j} \\ &=: \sum_{j=1}^{\infty} j\tau^{j-1} b_j. \end{aligned} \tag{6.7}$$

A similar estimate for b_j is

$$|b_j| < \frac{2}{j!}\left(\frac{\pi}{\lambda}\right)^j, \tag{6.8}$$

which, through (6.6) and (6.7), provides us the estimate

$$|X_\lambda''(N_\lambda + \tau) - \tfrac{1}{\lambda}x'(t_\lambda + \tfrac{\tau}{\lambda})| \le 2\left(\tfrac{\pi}{\lambda}\right)^2 e^{\tau\pi/\lambda}. \tag{6.9}$$

Now,

$$|x'(t_\lambda + \tfrac{\tau}{\lambda}) - x'(t)| \le \frac{(\tau+1)}{\lambda}\pi^2, \tag{6.10}$$

so that

$$\frac{1}{2}|x'(t)| \le |x'(t_\lambda + \tfrac{\tau}{\lambda})| \le \frac{3}{2}|x'(t)|, \tag{6.11}$$

for all $0 \le \tau \le C\lambda|x'(t)|$, where C is a sufficiently small absolute constant. Hence, from (6.9) and (6.11), it follows that

$$\begin{aligned} |X_\lambda''(N_\lambda + \tau)| &\ge \frac{1}{\lambda}|x'(t_\lambda + \tfrac{\tau}{\lambda})| - 2e^{\pi\tau/\lambda}\left(\tfrac{\pi}{\lambda}\right)^2 \\ &\ge C_1 \frac{|x'(t)|}{\lambda} \end{aligned} \tag{6.12}$$

if $\lambda \ge C'|x'(t)|^{-1}$. It follows similarly that

$$|X_\lambda''(N_\lambda + \tau)| \le C_2 \frac{|x'(t)|}{\lambda} \tag{6.13}$$

for the same range of λ and τ.

7. Acknowledgments

The author would like to thank I. Daubechies for her support and helpful advice during this work, and P. Sarnak for suggesting **[9]** for number-theoretic results of possible relevance to this problem. The author would also like to thank W. Schlag for a remark that simplified an earlier version of the proof of Theorem 4.4 and which led to a generalization of the class of filters that can be used.

References

[1] S.R. Norsworthy, R. Schreier, G.C. Temes, eds., *Delta-Sigma Data Converters*, IEEE Press, 1997.

[2] J.C. Candy, G.C. Temes, eds., *Oversampling Delta-Sigma Data Converters*, IEEE Press, 1992.

[3] R.M. Gray, "Spectral Analysis of Quantization Noise in a Single-Loop Sigma-Delta Modulator with dc Input", *IEEE Transactions on Communications, Vol. 37, June 1989.*

[4] R.M. Gray, W. Chou, P.W. Wong, "Quantization Noise in Single-Loop Sigma-Delta Modulation with Sinusoidal Inputs", *IEEE Transactions on Communications, Vol. 37, Sept. 1989.*

[5] I. Daubechies, R. DeVore, "Reconstructing a Bandlimited Function From Very Coarsely Quantized Data: A Family of Stable Sigma-Delta Modulators of Arbitrary Order", *Ann. of Math.*, vol. 158, no. 2, pp. 679–710, Sept. 2003.

[6] C.S. Güntürk, "Approximating a Bandlimited Function Using Very Coarsely Quantized Data: Improved Error Estimates in Sigma-Delta Modulation", *J. Amer. Math. Soc.*, posted on August 1, 2003, PII S 0894-0347(03)00436-3 (to appear in print).

[7] H. Helson, *Harmonic Analysis*, Addison-Wesley, 1983.

[8] H. Helson, "On a theorem of Szegö", *Proc. Amer. Math. Soc.* 6 (1955), 235-242.

[9] H.L. Montgomery, *Ten Lectures on the Interface Between Analytic Number Theory and Harmonic Analysis*, AMS, 1994.

[10] E.M. Stein, *Harmonic Analysis: Real-Variable Methods, Orthogonality, and Oscillatory Integrals*, Princeton University Press, 1993.

[11] L. Kuipers and H. Niederreiter, *Uniform Distribution of Sequences*, Wiley, 1974

[12] S. Lang, *Introduction to Diophantine Approximation*, Springer-Verlag, 1995.

[13] M. Drmota, R.F. Tichy, *Sequences, Discrepancies and Applications*, Springer, 1997.

[14] C.S. Güntürk, J.C. Lagarias, and V.A. Vaishampayan, "On the robustness of single loop sigma-delta modulation," *IEEE Transactions on Information Theory.*, 47 (5):1735–1744, July 2001.

[15] C.S. Güntürk and N.T. Thao, "Refined Analysis of Error in Second Order Sigma-Delta Modulation with DC Inputs," submitted to IEEE Transactions on Information Theory, in revision.

[16] C.S. Güntürk, "One-Bit Sigma-Delta Quantization with Exponential Accuracy," *Comm. Pure Appl. Math.*, vol. 56, pp. 1608–1630, no. 11, 2003.

[17] S. Konyagin, private communication, 1999.

PROGRAM IN APPLIED AND COMPUTATIONAL MATHEMATICS, PRINCETON UNIVERSITY, FINE HALL, WASHINGTON ROAD, PRINCETON, NJ, 08544.

Current address: Courant Institute of Mathematical Sciences, New York University, 251 Mercer Street, New York, NY, 10012.

E-mail address: `gunturk@cims.nyu.edu`

DIMACS Series in Discrete Mathematics
and Theoretical Computer Science
Volume **64**, 2004

Random walks with badly approximable numbers

Doug Hensley and Francis Edward Su

ABSTRACT. Using the discrepancy metric, we analyze the rate of convergence of a random walk on the circle generated by d rotations, and establish sharp rates that show that badly approximable d-tuples in $\mathbf{R}^d$ give rise to walks with the fastest convergence.

1. Introduction

Fix $\bar{\alpha} = (\alpha_1, \alpha_2, ..., \alpha_d)$, a d-tuple of real numbers, not all rational. Consider a random walk on the circle S^1 which proceeds as follows: at each step one of the α_i is chosen (with probability $1/d$) and the walk moves forwards or backwards (with probability $1/2$) on the circle by an angle $2\pi\alpha_i$. The distribution of this walk converges to the uniform (Haar) measure on S^1; in this paper, we investigate how quickly this convergence can take place.

To quantify this, identify S^1 with the unit circle in $\mathbf{R}^2$, and let Q be the probability measure supported on $\{e^{\pm 2\pi i \alpha_j}\}$, $j = 1, ..., d$, with equal weights $1/2d$. If at time $k = 0$ the random walk starts at 1, then the convolution power Q^{*k} represents the probability distribution of the random walk at time k. In this framework, we study how quickly Q^{*k} approaches U, the uniform (Haar) measure on the unit circle. Define the *discrepancy* $D(P)$ of a probability measure on S^1 by

$$D(P) = \sup_I |P(I) - U(I)| \tag{1}$$

where I represents any connected interval on S^1. The discrepancy $D(P)$ measures how uniform P is, and metrizes weak-* convergence to the uniform distribution on S^1. The usual definition of discrepancy found in the study of uniform distribution of sequences mod 1 (see [**6**]) is a special case of our definition, by letting P be an *equally weighted* measure on a sequence.

We seek bounds on how quickly $D(Q^{*k})$ diminishes as a function of the step size k and the numbers α_i. Clearly the convergence will not occur if all α_i are rational.

The case $d = 1$ reduces to a random walk generated by one irrational rotation; this problem was posed by Diaconis in [**3**, p. 34]. Su [**13**] obtained discrepancy bounds for this walk, showing that it converges most quickly when α is a quadratic irrational. In that case,

$$\frac{C_1}{\sqrt{k}} \le D(Q^{*k}) \le \frac{C_2}{\sqrt{k}}$$

2000 *Mathematics Subject Classification.* Primary 11K60, 60B15; Secondary 11J68, 11J71, 11K38, 60B10, 60G50.

Key words and phrases. random walk on a circle, discrepancy, badly approximable vector, Erdös-Turan inequality, rate of convergence.

Francis Su is grateful to the School of Operations Research at Cornell University for their hospitality during a sabbatical in which this was completed.

for constants C_1, C_2 which can be determined given α. We investigate whether the inclusion of additional generators can speed up the convergence of this walk. In particular we show that if $\bar{\alpha}$ is a *badly approximable* vector in $\mathbf{R}^d$, then the random walk generated by the $\bar{\alpha}$ converges as fast as possible for a set of d generators. In Theorem 1 we obtain matching upper and lower bounds:

$$\frac{C_1}{k^{d/2}} \leq D(Q^{*k}) \leq \frac{C_2}{k^{d/2}}.$$

We also show how the constants depend on $\bar{\alpha}$.

This result is reminiscent of results in the theory of uniform distribution of sequences mod 1; however, there is a notable lack of any log terms in the upper and lower bounds. To compare for $d = 1$, it is known (e.g., see [**6**]) that for a fixed irrational $\alpha \in \mathbf{R}$ the discrepancy of the sequence $\{\alpha, 2\alpha, 3\alpha, \ldots, k\alpha\}$ in $\mathbf{R}$ mod 1 falls as $\log k/k$ up to constant factors. For our corresponding random walk, it is not a surprise that the exponent gets halved, but it is quite surprising that the log term disappears.

While the total variation metric is more frequently used to study random walk convergence (e.g., see [**3**]), it is not appropriate for this walk, because Q^{*k} is a finitely supported measure and its total variation distance from U remains at 1 for all k. Moreover, we favor the use of discrepancy over other common metrics on probabilities (e.g., Wasserstein, Prokhorov) because bounding techniques for such metrics are not well developed, and the use of Fourier bounds for the discrepancy metric admits the possibility of sharp analysis. An analogous discrepancy metric may be used to study random walks on other spaces, e.g., [**9, 12, 14, 15**]. For a survey of bounds relating various metrics, see [**5**].

We also remark that while the literature contains many results on rates of convergence for random walks on finite groups (e.g., see [**3**]), and a few results for walks on infinite compact groups ([**8**],[**10**]), very little has been done for *discrete* walks on infinite compact groups, mainly due to lack of bounding techniques for appropriate metrics. Our analysis of the d-generator random walk on the circle reveals that bounds for the discrepancy metric are refined enough to yield sharp rates of convergence; this and [**13**] are the only sharp results we are aware of in this direction.

2. Random Walk Bounds

Throughout this paper, let $\|\mathbf{x}\|$ denote the Euclidean distance of $\mathbf{x} \in \mathbf{R}^d$ to the nearest integer lattice point (the dimension inherent in the notation $\|\cdot\|$ is to be understood by context). Given an arbitrary d-tuple $\bar{\alpha} = (\alpha_1, \alpha_2, ..., \alpha_d)$, the Dirichlet Approximation Theorem [**4**, p.68] implies that there exists a constant $B = B(\bar{\alpha})$ such that for any q, there exists a positive integer $N \leq q^d$ such that

$$\|N\bar{\alpha}\| < \frac{B}{q} \leq \frac{B}{N^{1/d}}. \tag{2}$$

In fact for any d-tuple $\bar{\alpha}$, the above bound holds for $B = \sqrt{d}$.

The d-tuple $\bar{\alpha}$ is said to be *badly approximable* if there exists a constant $\beta = \beta(\bar{\alpha}) > 0$ such that for all positive integers N,

$$\|N\bar{\alpha}\| \geq \frac{\beta}{N^{1/d}}. \tag{3}$$

We refer to B and β as *approximation constants* for $\bar{\alpha}$. The approximation constant β is only defined for badly approximable $\bar{\alpha}$.

Our main result is the following:

THEOREM 1. *Suppose $\bar{\alpha} = (\alpha_1, \alpha_2, ..., \alpha_d)$ is badly approximable. Let Q denote the generating measure of the random walk on S^1 generated by the d generators α_i. Then the discrepancy of the k-th step probability distribution of the walk satisfies*

$$\frac{C_1}{k^{d/2}} \leq D(Q^{*k}) \leq \frac{C_2}{k^{d/2}} \tag{4}$$

with constants that depend on d and $\bar{\alpha}$:

$$\begin{aligned} C_1 &= 0.0947\ (\sqrt{d}/5B)^d \\ C_2 &= 19.857\ (d\sqrt{d}/\beta)^d \end{aligned}$$

where B, β are approximation constants for $\bar{\alpha}$. The lower bound holds for walks generated by an arbitrary d-tuple.

Therefore, among all d-tuples, walks generated by badly approximable d-tuples converge the fastest. Note that the constants C_1, C_2 depend on the approximation constants of $\bar{\alpha}$. However since one may choose $B = \sqrt{d}$, it follows that $C_1 \geq 0.0947(1/5)^d$.

PROOF. We establish the lower bound using an inequality of Su [**14**] for the discrepancy of an arbitrary probability measure P on S^1:

$$D(P) \ \geq \ \left(\frac{2}{\pi^2} \sum_{m=1}^{\infty} \frac{|\widehat{P}(m)|^2}{m^2} \right)^{1/2}. \tag{5}$$

Here $\widehat{P}(m)$ denotes the m-th Fourier coefficient of P (viewed on $\mathbf{R}$ mod $\mathbf{Z}$). Since every term is positive, we can use a dominant term in the sum as a lower bound.

The Fourier coefficients for the d-generator walk are

$$\widehat{Q}(m) = \sum_{l=1}^{d} \frac{1}{2d}(e^{2\pi i m\alpha_l} + e^{-2\pi i m \alpha_l}) = \frac{1}{d}\sum_{l=1}^{d} \cos(2\pi m \alpha_l). \tag{6}$$

Since $\cos(2\pi x) = \cos(2\pi \|x\|) \geq 1 - 2\pi^2 \|x\|^2$, for any m we have

$$\widehat{Q}(m) \geq 1 - \frac{2\pi^2}{d} \|m\bar{\alpha}\|^2 \tag{7}$$

The inequality $(1-x)^k \geq 1 - kx$ holds for $k \geq 1$ and $x \leq 1$. Then

$$|\widehat{Q}(m)|^k \geq 1 - \frac{2\pi^2 k}{d} \|m\bar{\alpha}\|^2$$

as long as $2\pi^2 k \|m\bar{\alpha}\|^2/d < 1$. We ensure this by setting $Z_1 < 1$ (to be specified shortly) and let $q = (2\pi^2 B^2 k / Z_1 d)^{1/2}$. Then (2) implies there exists an integer $m \leq q^d$ such that $\|m\bar{\alpha}\| < B/q$, which yields $2\pi^2 k \|m\bar{\alpha}\|^2/d < Z_1 < 1$, as desired. (Note that q was chosen to ensure $|\widehat{Q}(m)|^k \geq 1 - Z_1$.)

For this m we use just the m-th term in (5), and since $m < q^d$ and $\widehat{Q^{*k}}(m) = \widehat{Q}^k(m)$, we obtain

$$D(Q^{*k}) \geq \frac{\sqrt{2}|\widehat{Q}(m)|^k}{\pi m} \geq C_1\ k^{-d/2} \tag{8}$$

with

$$C_1 = \frac{\sqrt{2}(1-Z_1)}{\pi}\left(\frac{Z_1 d}{2\pi^2 B^2}\right)^{d/2}.$$

Choosing $Z_1 = 2\pi^2/25 < 1$, we recover the lower bound (4) of the theorem.

The upper bound is trickier to compute. We start with the Erdös-Turán inequality [**7**]: given a probability measure P on S^1, for any integer M,

$$D(P) \leq \frac{4}{M+1} + \frac{4}{\pi}\sum_{m=1}^{M}\frac{|\widehat{P}(m)|}{m} \tag{9}$$

where $\widehat{P}$ represents the Fourier transform of P. Note that one may choose M in the Erdös-Turán inequality so as to optimize the bound obtained.

Since $|\cos(2\pi x)| \leq 1 - 4\|2x\|^2$ for all $x \in \mathbf{R}$, it follows from (6) that

$$\begin{aligned} |\widehat{Q}(m)| &\leq \frac{1}{d}\sum_{l=1}^{d}|\cos(2\pi m\alpha_l)| \\ &\leq 1 - \frac{4}{d}\sum_{l=1}^{d}\|2m\alpha_l\|^2 \\ &\leq \exp(-\frac{4}{d}\|2m\bar{\alpha}\|^2). \end{aligned}$$

In light of the Erdös-Turán inequality and $\widehat{Q^{*k}}(m) = \widehat{Q}^k(m)$, we need to estimate a sum of the form

$$\sum_{m=1}^{M}\frac{|\widehat{Q}^k(m)|}{m} \leq \sum_{m=1}^{M}\frac{1}{m}\exp(-\frac{4k}{d}\|2m\bar{\alpha}\|^2) =: S.$$

Since M may be chosen freely, choose an integer M such that

$$M \leq \frac{1}{2}(\beta^2 k/d^3)^{d/2} < M+1. \tag{10}$$

Recall that β is an approximability constant for $\bar{\alpha}$ from (3) and k is the number of steps in the walk. The reason for this choice of M will be evident later. Choose an integer J such that

$$2^{J-1} \leq M \leq 2^J - 1. \tag{11}$$

The sum in S may be grouped into J cohorts of integers $m \in [2^{j-1}, 2^j - 1]$ for $j = 1, ..., J$. Within each such cohort, the use of inequality (3) yields $\|2m\bar{\alpha}\| \geq \beta/(2m)^{1/d} \geq \beta/2^{(j+1)/d}$. This says that the points of the sequence $\{2m\bar{\alpha}\}$ (mod 1) in the unit cube in $\mathbf{R}^d$ are bounded away from the corners of the unit cube. In fact, they are bounded away from each other, since if $m_1, m_2 \in [2^{j-1}, 2^j - 1]$, then

$$\|2(m_1 - m_2)\bar{\alpha}\| \geq \beta/2^{(j+1)/d} \tag{12}$$

as well. Therefore if $s = \beta/\sqrt{d}\,2^{(j+1)/d}$, then any box of side length s can contain at most one of the multiples from the sequence $\{2m\bar{\alpha}\}$ (mod 1), $m \in [2^{j-1}, 2^j - 1]$, with no such multiples in boxes containing the origin.

So divide up the unit d-cube into disjoint boxes of side length s and sides parallel to the axes. In the worst case, all M points are distributed in the boxes in the 2^d corners of the unit cube nearest the origin. The nearest points in such boxes (under the L_1 metric) are at integral multiples of s and extend out in layers to distance at most $M^{1/d}$. A crude upper bound for the number of boxes whose nearest point is at L_1-distance ns from the origin is $(n+1)^{d-1}$, and in the Euclidean metric the nearest point in such boxes is at least distance $ns/\sqrt{d}$ from the origin. Hence we can bound S by grouping first by cohorts, then by corners and layers:

$$\begin{aligned} S &\leq \sum_{j=1}^{J} \sum_{m=2^{j-1}}^{2^j-1} \frac{1}{m} \exp(-\frac{4k}{d}\|2m\bar{\alpha}\|^2) \\ &\leq \sum_{j=1}^{J} 2^d \sum_{n=1}^{M^{1/d}} \frac{(n+1)^{d-1}}{2^{j-1}} \exp(-\frac{4k}{d}\frac{\beta^2 n^2}{d \cdot 2^{2(j+1)/d}}) \\ &\leq \sum_{j=1}^{J} \frac{2^d}{2^{j-1}} \sum_{n=1}^{\infty} (n+1)^{d-1} \exp(-4n^2 d 2^{2(J-j-1)/d}). \end{aligned}$$

The second inequality used the bound $\|2m\bar{\alpha}\| \geq ns/\sqrt{d}$ and the definition of s. The third inequality follows by noting $k \geq d^3 2^{2J/d}/\beta^2$ from the definitions of J and M in (10) and (11).

Using $j \leq J$, the log derivative of the expression of the innermost sum with respect to n can be bounded:

$$\frac{d-1}{n+1} - 8nd2^{2(J-j-1)/d} \leq \frac{d-1}{n+1} - 8nd2^{-2/d} \leq \frac{d-1}{2} - d \leq -1$$

for all choices of $n \geq 1$ and $d \geq 1$. Hence the expression in the inner sum decreases geometrically (by at least ratio e^{-1}) as j increases, and so the inner sum is bounded by the first term (at $n = 1$) times the constant $Z_2 = 1/(1-e^{-1}) \approx 1.5820$. Thus

$$S \leq \sum_{j=1}^{J} Z_2\, 2^{2d-j} \exp(-4d2^{2(J-j-1)/d}).$$

This sum may be bounded by noting that the largest term occurs when $j = J$. For $j \leq J$, the log derivative of the terms with respect to j is $\ln 2(-1+8\cdot 2^{2(J-j-1)/d}) \geq \ln 2$ for $d \geq 1$. Thus the sum decreases geometrically (with at least ratio 1/2) as $j \leq J$ decreases, so the sum is therefore bounded by twice the final term at $j = J$:

$$\begin{aligned} S &\leq 2Z_2\, 2^{2d-J} \exp(-4d2^{-2/d}) \\ &\leq \frac{Z_2 2^{2d+1} \exp(-4d2^{-2/d})}{M+1} \leq \frac{4.6559}{M+1} \end{aligned} \tag{13}$$

where the second inequality follows from (11), and the final inequality from noting that the numerator is greatest for $d = 1$. Using the Erdös-Turán inequality we obtain the upper bound

$$D(Q^{*k}) \leq \frac{4}{M+1} + \frac{4}{\pi} S \leq \frac{4 + (4/\pi)4.6559}{M+1} \leq \frac{9.9281}{M+1}$$

for all $d \geq 1$. An application of (10) produces

$$D(Q^{*k}) \leq 19.857\, d^{3d/2} \beta^{-d} k^{-d/2}$$

which establishes the constant C_2 in the statement of the theorem. □

We remark that this argument simplifies the proof given in [**13**] for the case $d = 1$. For specific d, the constants C_1, C_2 can be improved by adjusting the derivations of the constants Z_1, Z_2 and the bound for the last inequality in (13).

3. Choosing Generators

Badly approximable d-tuples are plentiful. Cassels [**1**] shows that there uncountably many badly approximable d-tuples in $\mathbf{R}^d$. Moreover, results of Schmidt [**11**, pp. 53-59] imply that the Hausdorff dimension of the set of badly approximable vectors is positive. For some concrete examples, it can be shown (see, e.g., [**4**, p. 68]) that if $1, \alpha_1, ..., \alpha_d$ are linearly independent over $\mathbf{Z}$, and if the degree of the extension $[\mathbf{Q}(\alpha_1, ..., \alpha_d) : \mathbf{Q}] = d+1$, then $\bar{\alpha}$ is badly approximable.

We have shown that badly approximable d-tuples in $\mathbf{R}^d$ give rise to random walks with the fastest convergence, established sharp rates of convergence, and exhibited how the constants depend on the approximation constants β, B of the given d-tuple. However, just among badly approximable d-tuples, can we say which random walks converge "the fastest"? By Theorem 1, this amounts to identifying d-tuples with the largest possible approximation constant β.

For $d = 1$, the number $\phi = \frac{\sqrt{5}-1}{2}$ satisfies (3) and is therefore badly approximable; in fact, it is known to be the "most" badly approximable number in the sense that its approximation constant β is larger than for any other number.

For $d = 2$, we conjecture that the "most" badly approximable 2-tuple is the vector $\mathbf{v} = (\gamma^{-2}, \gamma^{-1})$, where $\gamma \approx 1.3247$ is the unique real root of $x^3 - x - 1$. For this $\mathbf{v}$, we believe (based on heuristic arguments and numerical evidence) that for any $\beta < 0.54850...$, there is a sufficiently large N such that $n > N$ implies that

$$\|n\mathbf{v}\| > \beta\, n^{-1/2}. \tag{14}$$

Furthermore, no other vector can have a much larger approximation constant β, because the work of Davenport and Mahler [**2**] implies that if $\beta > (2/\sqrt{23})^{1/2} \approx .64577...$, then there is no vector $\mathbf{v}$ in $\mathbf{R}^2$ for which (14) holds for all n.

Thus the pair $(\gamma^{-2}, \gamma^{-1})$ yields a random walk on the circle with 2 generators whose convergence rate appears close to fastest possible over all badly approximable pairs. Is it fastest? For larger d the question is also open.

References

[1] J. W. S. Cassels, *Simultaneous Diophantine Approximation II*, Proc. London Math. Soc. (3) **5** (1955), 435-448.

[2] H. Davenport and K. Mahler, *Simultaneous Diophantine Approximation*, Duke Math. J. **13** (1946), 105-111.

[3] P. Diaconis, *Group Representations in Probability and Statistics*, Institute of Mathematical Statistics Lecture Notes, Vol. 11, Hayward, CA, 1988.

[4] M. Drmota and R. F. Tichy, *Sequences, Discrepancies and Applications*, Lecture Notes in Math. 1651, Springer-Verlag, 1997.

[5] A. Gibbs and F. E. Su, *On choosing and bounding probability metrics.* Internat. Statist. Review **70** (2002), 419-435.

[6] L. Kuipers and H. Niederreiter, *Uniform Distribution of Sequences*, Wiley, New York, 1974.

[7] H. Niederreiter and W. Philipp, *Berry-Esseen bounds and a theorem of Erdös and Turán on uniform distribution mod 1*, Duke Math. J. **40** (1973), 633-649.

[8] U. Porod, *The cut-off phenomenon for random reflections*, Ann. Probab. **24** (1996), 74-96.

[9] T. Prescott and F. E. Su, *Random walks on the torus with several generators*, preprint, `arXiv:math.PR/0309011`.

[10] J. S. Rosenthal, *Random rotations: characters and random walks on SO(n)*, Ann. Probab. **22** (1994), 398-423.

[11] W. M. Schmidt, *Diophantine Approximation*, Lecture Notes in Math. 785, Springer-Verlag, 1980.

[12] F. E. Su *Methods of Quantifying Rates of Convergence for Random Walks on Groups*, Ph.D. Thesis, Harvard University, 1995.

[13] F. E. Su, *Convergence of random walks on the circle generated by an irrational rotation*, Trans. Amer. Math. Soc. **350** (1998), 3717-3741.

[14] F. E. Su, *A LeVeque-type lower bound for discrepancy*, in *Monte Carlo and Quasi-Monte Carlo Methods 1998*, H. Niederreiter and J. Spanier (eds.), Springer-Verlag, 2000, pp. 448-458.

[15] F. E. Su, *Discrepancy convergence for the drunkard's walk on the sphere.* Electron. J. Probab. **6** (2001) no. 2, 1-20.

DEPARTMENT OF MATHEMATICS, TEXAS A&M UNIVERSITY, COLLEGE STATION, TX 77843
E-mail address: Doug.Hensley@math.tamu.edu

DEPARTMENT OF MATHEMATICS, HARVEY MUDD COLLEGE, CLAREMONT, CA 91711
E-mail address: su@math.hmc.edu

DIMACS Series in Discrete Mathematics
and Theoretical Computer Science
Volume **64**, 2004

Distribution of Values of Real Quadratic Zeta Functions

Joshua Holden

ABSTRACT. The author has previously extended the theory of regular and irregular primes to the setting of arbitrary totally real number fields. It has been conjectured that the Bernoulli numbers, or alternatively the values of the Riemann zeta function at odd negative integers, are evenly distributed modulo p for every p. This is the basis of a well-known heuristic, given by Siegel in [**16**], for estimating the frequency of irregular primes. So far, analyses have shown that if $\mathbf{Q}(\sqrt{D})$ is a real quadratic field, then the values of the zeta function $\zeta_D(1-2m) = \zeta_{\mathbf{Q}(\sqrt{D})}(1-2m)$ at negative odd integers are also distributed as expected modulo p for any p. However, it has proven to be very computationally intensive to calculate these numbers for large values of m. In this paper, we present the alternative of computing $\zeta_D(1-2m)$ for a fixed value of D and a large number of different m.

1. Introduction, Conjectures, and Previous Results

Siegel, in [**16**], conjectured that the numerators of the Bernoulli numbers B_{2m} were evenly distributed modulo p for any odd prime p. This hypothesis was used by Johnson ([**14**]) and independently by Wooldridge ([**20**]) to predict the density of primes with a given index of irregularity, that is such that p divides a given number of the Bernoulli numbers $B_2, \ldots, B_{p-3}$. It also comes in handy for predicting many other values that are related to irregular primes, such as the order of magnitude of the first prime of a given index of irregularity. (See, for example, [**18**].)

Since $B_{2m} = -\zeta(1-2m)(2m)$, it is equivalent to say that the values of $\zeta(1-2m)$ are evenly distributed modulo p, where $\zeta(s)$ is the Riemann zeta function. Of course, the Riemann zeta function can be generalized to any number field k to get a zeta function $\zeta_k(s)$ associated with the number field k. If k is a totally real number field, there are a number of situations where the values $\zeta_k(s)$ can be seen as analogous to the Bernoulli numbers, including generalizations of Kummer's Criterion for predicting when p divides the class number of the p-th cyclotomic field and cases where the equation $x^p + y^p = z^p$ of Fermat's Last Theorem can be shown not to have solutions in k. (See for example, [**5, 6, 7**], [**10**], [**9**], [**10**], [**11**], and [**15**].)

2000 *Mathematics Subject Classification.* Primary 11Y40, 11Y60, 11R42; Secondary 11B68, 11R29, 11R18.

Key words and phrases. Bernoulli numbers, irregular primes, zeta functions, quadratic extensions, cyclotomic extensions, class groups.

Little or no progress has been made on proving Siegel's hypothesis, but a great deal of data has been collected, especially in regard to the prediction of Johnson and Wooldridge. Specifically, this prediction says that as $p \to \infty$, the probability that p has index of irregularity r goes to

$$\left(\frac{1}{2}\right)^r \frac{e^{-1/2}}{r!} \ .$$

(In addition to the original sources, the details may be found in Section 5.3 of [**19**].) Note that this prediction does not rely on the full strength of Siegel's hypothesis, but merely on the weaker hypothesis that the Bernoulli numbers are 0 modulo p with probability $1/p$. The investigations focused on in this paper relate only to predictions about indices of irregularity based on this weaker hypothesis.

Wagstaff, in [**18**], computed $u_r(x)$, the fraction of primes not exceeding x with index r of irregularity for each r between 0 and 2 and for all $r \geq 3$ grouped together, and compared this distribution to the predicted distribution for each multiple x of 1000 up to 125000. The result of the chi-squared test "fluctuated usually between 0.1 and 1.0 and had the value 0.29 at $x = 125000$. It was 0.03 at $x = 8000$" [**18**]. These results correspond to significance levels of .992, .801, .962, and .999, respectively. (The significance levels used in this paper correspond roughly to the probability that the agreement between the observed results and the predicted results is *not* due to chance. Statisticians consider the threshold for considering a result to be not due to chance to be a significance level of .9 to .95. Since we are not actually conducting a valid statistical study in this paper, all of the statistical results should be taken with a very large grain of salt.)

Buhler, Crandall, Ernvall, and Metsänkylä hold the record for computations with irregular primes, having found all the irregular primes below four million as described in [**2**]. They do not seem to have done a chi-squared analysis, but they tabulate the values of $u_r(x)$ for $x = 4000000$ and r between 0 and 7. A chi-squared test using the same methodology as before has the result 1.02, for a significance level of .796. Earlier, in [**3**], Buhler, Crandall, and Sompolski tabulated the same data for $x = 1000000$. The result of the same chi-squared test is 0.78, for a significance level of .854.

Unfortunately, the only way to collect data to test Siegel's hypothesis is to investigate B_{2m} for larger and larger m, which is very computationally intensive. (See [**1**] or [**8**] for details.)

However, in the more general number field case, there are many more dimensions to the problem. We start by restricting our attention to the case of k an abelian totally real number field. Then we know that

$$\zeta_k(s) = \prod_{\chi \in \hat{G}} L(s, \chi)$$

where $\hat{G}$ is the character group of $G = \mathrm{Gal}(k/\mathbf{Q})$ and $L(s,\chi)$ is the L-function associated with the character χ. Note that $L(s,1) = \zeta(s)$, so the Riemann zeta function is a factor of the zeta function for k. (See [**4**], e.g., for more details.) Certainly it seems likely that for a fixed (totally real) number field k and character χ the values of the numerator of $L(1-2m,\chi)$ are evenly distributed modulo p as m varies. (It is known that these values are rational numbers.) We also hypothesize that these values for different χ are independent, which implies that the numerators of $\zeta_k(1-2m)$ are distributed modulo p like the product of $|G|$ independent integer variables, each of which is evenly distributed modulo p. We will refer to this as the "product distribution", for lack of a better term. However, it also is reasonable to conjecture that for a fixed m the values of $\zeta_k(1-2m)$ are distributed according to the product distribution modulo p as k varies. More precisely, if we fix m and the degree of k we expect the values to be distributed according to the product distribution modulo p as the discriminant of k varies. Alternatively, if we fix m and the discriminant of k we expect the values to be distributed according to the product distribution modulo p as the degree varies.

In this paper we will be considering the former situation, with the degree fixed at 2, making k a real quadratic field. We let $k = \mathbf{Q}(\sqrt{D})$ (where D is the discriminant of k) and $\zeta_D(s) = \zeta_k(s)$. In this case

$$\zeta_D(s) = L(s,1)L(s,\chi) = \zeta(s)L(s,\chi)$$

where $\chi(s) = \left(\frac{D}{s}\right)$, the Kronecker symbol, where appropriate.

We make the following definitions:

DEFINITION 1. Let $k = \mathbf{Q}(\sqrt{D})$ be a real quadratic number field with discriminant D. We say that an odd prime p is *k-regular* (or *D-regular*) if p is relatively prime to $\zeta_k(1-2m)$ for all integers m such that $2 \leq 2m \leq \delta - 2$ and also p is relatively prime to $p\zeta_k(1-\delta)$, where $\delta = p-1$ unless $D = p$, in which case $\delta = (p-1)/2$. The number of such zeta-values that are divisible by p will be the *index of k-irregularity* (or *index of D-irregularity*) of p.

Further, we will say that p is *χ-regular* if p is relatively prime to $L(1-2m,\chi)$ for all integers m such that $2 \leq 2m \leq \delta - 2$ and also p is relatively prime to $pL(1-2m,\chi)$. The number of such L-values that are divisible by p will be the *index of χ-irregularity* of p.

(See [**11**], [**12**] or [**13**] for an explanation of why the definition has exactly this form.)

Saying that the values of $\zeta_k(1-2m)$ are distributed according to the product distribution and that the values of $\zeta(1-2m)$ are evenly distributed is the same as saying that the values of $L(1-2m,\chi)$ are evenly distributed modulo p. Then we can make the same prediction about the indices of *chi*-irregularity that Johnson and Wooldridge made about the indices of irregularity in the rational case. We briefly investigated this issue in [**11**], where there are tables of the analogue of $u_r(x)$ (using the index of χ-irregularity) for $x = 1000$, r from 0 to 4, and $D = 5, 8, 12$, and 13. The chi-squared test results are not included, but using the methodology discussed earlier they are 3.32, 1.74, 1.15, and 2.54. The corresponding significance levels are .345, .628, .765, and .469, respectively. We could total the values of (the analogue of) $u_r(x)$ for the four values of D and compare them to the predicted values; we might expect that this would give us a better significance level because of the larger "sample size". However, in this case the chi-squared result is 3.53 and the significance level is .316, which is worse than any of the results for the values of D taken separately! We could instead average the values for the four values of D and compare them to the predicted values; in this case the chi-squared result is 0.884 and the significance level is .829, which is quite good. However, it is not clear to us what the actual meaning of these averages are. Let the reader beware.

The papers [**11**] and [**13**] investigate some algorithms, based on formulas given by Siegel in [**17**], for calculating values of $\zeta_D(1-2m)$. We show there that while the best-known algorithms for calculating $\zeta_D(1-2m)$ letting m vary take amortized time polynomial in m, it is possible to calculate $\zeta_D(1-2m)$ with m fixed and D varying in amortized time subpolynomial in D. We will use such an algorithm in the following.

2. New Results

In the course of testing the algorithms in [**13**], we collected more data in addition to that above. Table 1 shows the number of primes less than 5000 which have χ-index of irregularity r for various values of r and $D = 5$. We compared the observed and predicted distributions, using the methodology above, for primes below x where x was 1000, 2000, 3000, 4000, and 5000, and found chi-squared values of 3.32, 5.03, 2.51, 1.73, and 2.10 and significance levels of .344, .170, .473, .630, and .552, respectively.

TABLE 1. Results for $D = 5$ and $p < 5000$

r	number	predicted number	predicted fraction
0	422	405.16	.606531
1	186	202.58	.303265
2	51	50.65	.075816
3	7	8.44	.012636
4	2	1.06	.001580

Other data was obtained using the philosophy, described above, of computing the values of $L(1-2m,\chi)$ for large numbers of D and relatively small values of m. As in the discussion of $D = 5, 8, 12$, and 13 above, we present both the total and the average across the different discriminants. Table 2 presents the data for all $D < 5000$ and $p < 100$. The chi-squared value for the totals is 81.1 and the significance level is .000. The chi-squared value for the averages is 0.053 and the significance level is .997. As before, the actual meaning of these numbers is not clear.

TABLE 2. Results for $D < 5000$ and $p < 100$

r	total number	predicted total number	average number	predicted average number	predicted fraction
0	21864	22068.01	14.42	14.56	.606531
1	11596	11034.01	7.65	7.28	.303265
2	2529	2758.50	1.67	1.82	.075816
3	347	459.75	0.23	0.30	.012636
4	41	57.47	0.03	0.04	.001580
5	7	5.75	0.005	0.004	.000158

Finally, we computed the values of $L(-1,\chi)$ and $L(-3,\chi)$ for the 303957 discriminants D less than one million, and calculated the indices of χ-irregularity for the primes 3 and 5. In this case, since the p involved are very small, we computed the expected number of primes with given index of irregularity directly from Siegel's hypothesis, rather than using the limit as $p \to \infty$. This takes into account, for instance, the fact that for $p = 3$ the index of irregularity cannot be more than 1 and for $p = 5$ it cannot be more than 2. The results are shown in Table 3. The chi-squared value for the totals is 24636 and the significance level is .000. (In this case the categories used for the chi-squared test are $r = 0$, $r = 1$, and $r = 2$.) The chi-squared value for the averages is 0.081 and the significance level is .960. Nevertheless, in view of the large amount of data there seems to be a significant discrepancy between the predicted results and the observed results, even when the averages are considered. Coupled with our doubts about the meaning of the averaged numbers, this leads us to believe that something other than a straightforward extension of Siegel's hypothesis is influencing the distribution of the values of $L(1-2m,\chi)$ as the discriminant varies. We hope to make the nature of this influence clearer in the future.

TABLE 3. Results for $D < 1000000$ and $p = 3$ or $p = 5$

r	total number	predicted total number	average number	predicted average number	predicted fraction
0	338966	397170.48	1.115177	1.306667	0.653333
1	252832	198585.24	0.831802	0.653333	0.326667
2	16116	12158.28	0.053021	0.040000	0.020000

3. Other Conjectures and Future Work

A number of other conjectures about irregular primes could be extended to the setting which we have presented. One such, mentioned by Wagstaff in [**18**] without attribution, is that the irregular primes are evenly distributed across the possible residue classes modulo n for every positive integer n. To be exact, one expects that the ratio of irregular primes in a residue class to all odd primes in the class to be the same for each possible residue class. Wagstaff investigated this for primes below 125000 and $3 \leq n \leq 37$ (and also $n = 59$ and 67) and gave tables for $n = 3, 4$, and 5. He does not give the chi-squared results, but they are 0.107, 0.060, and 2.420, respectively, with significance levels of .744, .806, and .490. We conjecture that the χ-irregular primes are also distributed evenly, and intend to investigate this using the data we have collected.

Wagstaff also mentions a conjecture due to Wooldridge about the distribution of the numbers $2k/p$ for which $(p, 2k)$ is an irregular pair; i.e. p divides B_{2k}. Wooldridge conjectures that these numbers have a uniform distribution in the interval $(0, 1)$. We also intend to investigate whether these numbers are uniformly distributed for χ-irregular pairs.

Two other conjectures about irregular primes which are widely believed cannot really be checked using statistical methods. It is thought that there are primes with arbitrarily large indices of irregularity; the largest observed so far is 7, for $p = 3238481$ (see [**2**]). The largest index of χ-irregularity yet observed is 5, which occurred 7 times among the primes less than 100 using discriminants less than 5000. It is also likely that there are Bernoulli numbers divisible by arbitrarily large powers of p. Although as yet p^2 has not been seen to divide B_{2m}, as Washington says, "there does not seem to be any reason to believe this in general" [**19**]. On the other hand, powers as large as 3^7 have been observed to divide $L(1-2m,\chi)$; this happens at $D = 3869$ and again at $D = 3937$.

Studies such as [**2**] also frequently investigate applications such as Fermat's Last Theorem, Vandiver's conjecture, and Iwasawa's cyclotomic invariants. All of these could be modified to the situation explored here. Additional data would have to be collected in order to study these applications; we are not yet certain whether new algorithms would have to be implemented.

References

[1] Eric Bach. The complexity of number-theoretic constants. *Inform. Process. Lett.*, 62:145–152, 1997.

[2] J. Buhler, R. Crandall, R. Ernvall, and T. Metsänkylä. Irregular primes and cyclotomic invariants up to four million. *Math. Comp.*, 61:151–153, 1993.

[3] J. P. Buhler, R. E. Crandall, and R. W. Sompolski. Irregular primes to one million. *Math. Comp.*, 59:717–722, 1992.

[4] J. W. S. Cassels and A. Fröhlich, editors. *Algebraic Number Theory.* Academic Press, 1986. Reprint of the 1967 original.

[5] Reijo Ernvall. Generalized Bernoulli numbers, generalized irregular primes, and class number. *Ann. Univ. Turku. Ser. A I*, 178, 1979. 72 pp.

[6] Reijo Ernvall. Generalized irregular primes. *Mathematika*, 30:67–73, 1983.

[7] Reijo Ernvall. A generalization of Herbrand's theorem. *Ann. Univ. Turku. Ser. A I*, 193, 1989. 15 pp.

[8] Sandra Fillebrown. Faster computation of Bernoulli numbers. *J. Algorithms*, 13:431–445, 1992.

[9] Ralph Greenberg. A generalization of Kummer's criterion. *Invent. Math.*, 21:247–254, 1973.

[10] Fred H. Hao and Charles J. Parry. Generalized Bernoulli numbers and m-regular primes. *Math. Comp.*, 43:273–288, 1984.

[11] Joshua Holden. Irregularity of prime numbers over real quadratic fields. In J. P. Buhler, editor, *Algorithmic Number Theory: Third International Symposium; Proceedings*, volume 1423 of *Springer Lecture Notes in Computer Science*, pages 454–462. Springer-Verlag, 1998.

[12] Joshua Holden. On the Fontaine-Mazur Conjecture for number fields and an analogue for function fields. *J. Number Theory*, 81:16–47, 2000.

[13] Joshua Holden. Comparison of algorithms to calculate quadratic irregularity of prime numbers. *Mathemathics of Computation*, 71:863–871, 2002.

[14] Wells Johnson. Irregular primes and cyclotomic invariants. *Math. Comp.*, 29:113–120, 1975.

[15] Aichi Kudo. On a generalization of a theorem of Kummer. *Mem. Fac. Sci. Kyushu Univ. Ser. A*, 29(2):255–261, 1975.

[16] Carl Ludwig Siegel. Zu zwei bemerkungen Kummers. *Nachr. Akad. Wiss. Göttingen Math.-Phys. Kl. II*, 6:51–57, 1964.

[17] Carl Ludwig Siegel. Bernoullische Polynome und quadratische Zahlkörper. *Nachr. Akad. Wiss. Göttingen Math.-Phys. Kl. II*, 2:7–38, 1968.

[18] Samuel S. Wagstaff, Jr. The irregular primes to 125000. *Math. Comp.*, 32:583–591, 1978.

[19] Lawrence C. Washington. *Introduction to Cyclotomic Fields*, volume 83 of *Graduate Texts in Mathematics*. Springer-Verlag, second edition, 1997.

[20] K. Wooldridge. *Some Results in Arithmetical Functions Similar to Euler's Phi-Function.* PhD thesis, University of Illinois at Urbana-Champaign, 1975.

Department of Mathematics, Rose-Hulman Institute of Technology Terre Haute, IN 47803-3999, USA

E-mail address: holden@rose-hulman.edu

DIMACS Series in Discrete Mathematics
and Theoretical Computer Science
Volume **64**, 2004

Standardizing Nonstandard Methods For Upper Banach Density Problems

Renling Jin

ABSTRACT. Many results in [**J1, J2**] about the addition of sets of positive upper Banach density are proven here using standard methods. These standard methods are translated from the nonstandard methods used in [**J1, J2**].

1. Introduction

There are many interesting theorems about Shnirel'man density or lower asymptotic density (see [**HR, N1, N2**]) in additive number theory. There are a few interesting results about upper Banach density (see [**B, F**]) in combinatorial number theory. However, not very many results about upper Banach density involving addition of sets can be found in literature. Recently, I obtained some results of this kind in [**J1, J2**] using nonstandard analysis. Some of these results were presented at the DIMACS workshop *Unusual Applications of Number Theory*. Since some people who are interested in the subject may not be comfortable reading the proofs using nonstandard analysis, I would like to translate them into the standard proofs in order to reach a wider range of readers.

I denote by $\mathbb{N}$ the set of all natural numbers including 0 and by $\mathbb{Z}$ the set of all integers. For any two integers $m \leqslant n$, I write $[m,n]$ exclusively for the interval of integers between m and n including m and n. The upper case letters $A, B, C, \ldots$ except X and Y, which I reserve for something else, will always denote the subsets of $\mathbb{N}$ and the lower case letters $a, b, c, \ldots$ will always denote the elements of $\mathbb{N}$ unless specified. For a finite set A, let $|A|$ be the number of elements in A. The notion $A(n)$ is an abbreviation of $|A \cap [1,n]|$ and the notion $A(m,n)$ with $m \leqslant n$ is an abbreviation of $|A \cap [m,n]|$. For any sets $A_1, A_2, \ldots, A_h$, let $\sum_{i=1}^{h} A_i$ denote the

2000 *Mathematics Subject Classification.* Primary: 11B05, 11B13, 11B75; Secondary: 11P70, 11P99, 28D05

Key words: upper Banach density, Shnirel'man density, lower asymptotic density, additive number theory.

The author was supported in part by Ralph E. Powe Junior Faculty Enhancement Award from Oak Ridge Associated Universities and by Faculty Research and Development Summer Grant from College of Charleston. The author would also like to thank DIMACS and the organizing committee for inviting him to the workshop *Unusual Applications of Number Theory.*

set

$$\{\sum_{i=1}^{h} a_i : a_i \in A_i \text{ for } i = 1, 2, \ldots, h\}.$$

If all A_i's are same, say $A_i = A$ for $i = 1, 2, \ldots, h$, then we write simply hA for $\sum_{i=1}^{h} A_i$. The Shnirel'man density $\sigma(A)$, the lower asymptotic density $\underline{d}(A)$, the upper asymptotic density $\bar{d}(A)$, and the upper Banach density $BD(A)$ of a set A are defined as the following.

$$\sigma(A) = \inf_{n \geq 1} \frac{A(n)}{n}$$

$$\underline{d}(A) = \liminf_{n \to \infty} \frac{A(n)}{n}$$

$$\bar{d}(A) = \limsup_{n \to \infty} \frac{A(n)}{n}$$

$$BD(A) = \lim_{n \to \infty} \sup_{m-l=n} \frac{A(l, m)}{m - l + 1}.$$

It is easy to see that for any set A,

$$0 \leqslant \sigma(A) \leqslant \underline{d}(A) \leqslant \bar{d}(A) \leqslant BD(A) \leqslant 1.$$

2. Sumset Phenomenon

The results in this section were proven in [**J1**] using the methods from nonstandard analysis. Here we give standard alternatives of those proofs.

A set A is called syndetic if there is a positive integer k such that $A \cap [n, n+k] \neq \emptyset$ for every $n \in \mathbb{N}$. Equivalently, a set A is syndetic iff $(A \cup \{0\}) + [0, k] = \mathbb{N}$ for some positive integer k. There is a well-known result [**F**, Proposition 3.19 (a)] in combinatorial number theory, which says that if a set A has positive upper Banach density, then $A - A = \{a - a' : a, a' \in A \text{ and } a \geqslant a'\}$ is syndetic. It is natural to ask what can be said for $A + A$. One can easily construct a set A with $BD(A) = 1$ and $A + A$ is not syndetic.

Essentially, upper Banach density is a concept of measure and syndeticity is a concept of "order-topology" on $\mathbb{N}$. So one can ask the question in a more abstract way. What kind of order-topological properties one can have for $A + A$ when A has positive measure? A result called Steinhaus's theorem in real analysis says that if two sets of real numbers, X and Y, have positive Lebesgue measure, then $X + Y$ covers a non-empty open interval. Inspired by this result, a question in nonstandard analysis was asked [**KL**, Problem 9.13] (the reader doesn't need to know the definitions of the concepts in nonstandard analysis mentioned below for reading the rest of the article).

> Whether are there any cut U in a hyperfinite Loeb probability space H for which there exist sets $A, B \subseteq H$ of positive Loeb measure such that A plus B is U–meager?

Note that a set in a topological space is called meager if it is a countable union of nowhere dense sets.

Above question was answered negatively in [**J1**].

Since a set A has upper Banach density $\geqslant \alpha$ if and only if the nonstandard copy of A has Loeb measure $\geqslant \alpha$ in a hyperfinite Loeb probability space, the set $A + A$ should bear some properties of the order structure of $\mathbb{N}$. In fact, if one translate the U–meagerness in a Loeb space into the standard world, it corresponds non-piecewise syndeticity.

A set A is called thick if A contains k consecutive positive integers for every positive integer k. A set A is called piecewise syndetic if $A + [0, k]$ is thick for some $k \in \mathbb{N}$. Thick sets, syndetic sets and piecewise syndetic sets are the objects studied in combinatorial number theory (see [**B, F**]). By deriving the consequences of the main result in [**J1**], I obtained several theorems of an interesting general phenomenon in the standard world, which says that if X and Y are large in terms of "measure", then $X+Y$ is not small in terms of "order–topology". Note that there are usually "nowhere dense" set of positive "measure". For example, Steinhaus's theorem is one of these consequences. The next theorem is another consequence.

THEOREM 2.1. *For any sets A and B, if $BD(A) > 0$ and $BD(B) > 0$, then $A + B$ is piecewise syndetic.*

Before proving Theorem 2.1, I would like to introduce two lemmas first.

LEMMA 2.2. *Let A be a set, $a \leqslant b$ be two positive integers, and $\alpha = \frac{A(a,b)}{b-a+1}$. For any real number ϵ between 0 and 1, there is an $a' \in [a, b]$ such that $b - a' + 1 \geqslant \epsilon(b - a + 1)$ and*

$$\inf_{a' \leq i \leq b} \frac{A(a', i)}{i - a' + 1} > \alpha - \epsilon. \tag{2.1}$$

PROOF. Let $a' = a$ if

$$\inf_{a \leq i \leq b} \frac{A(a, i)}{i - a + 1} > \alpha - \epsilon.$$

Otherwise, we construct an increasing finite sequence $a_0 < a_1 < \cdots < a_k$ inductively with $a_0 = a - 1$. Suppose a_i is defined for $i \geqslant 0$. If the set

$$S = \{h \in [a_i + 1, b] : \frac{A(a_i + 1, h)}{h - a_i} \leqslant \alpha - \epsilon\}$$

is not empty, then let $a_{i+1} = \min S$. Otherwise the construction is terminated. Let $a' = a_k + 1$, where a_k is the last term of the sequence. Next we want to show that $b - a' + 1 \geqslant \epsilon(b - a + 1)$. Since

$$\begin{aligned}
\alpha &= \frac{A(a, b)}{b - a + 1} \\
&= \frac{\sum_{i=0}^{k-1} A(a_i + 1, a_{i+1}) + A(a', b)}{b - a + 1} \\
&= \frac{1}{b - a + 1} \sum_{i=0}^{k-1} \frac{A(a_i + 1, a_{i+1})}{a_{i+1} - a_i}(a_{i+1} - a_i) + \frac{A(a', b)}{b - a + 1} \\
&\leqslant \alpha - \epsilon + \frac{b - a' + 1}{b - a + 1},
\end{aligned}$$

then $\frac{b-a'+1}{b-a+1} \geqslant \epsilon$. Hence $b - a' + 1 \geqslant \epsilon(b - a + 1)$. Now 2.1 is true because a_k is the last term of the sequence and $a' = a_k + 1$.

□

LEMMA 2.3. *Let $A, B \subseteq \mathbb{N}$ and $[a,b], [c,d] \subseteq \mathbb{N}$ such that*

$$\frac{A(a,b)}{b-a+1} = \alpha > \frac{1}{2} \text{ and } \frac{B(c,d)}{d-c+1} = \beta > \frac{1}{2}.$$

Then there is an interval $[s,t] \subseteq (A+B) \cap [a+c, b+d]$ such that $t-s+1 \geqslant \epsilon \cdot \min\{b-a+1, d-c+1\}$, where $\epsilon = \min\{\alpha - \frac{1}{2}, \beta - \frac{1}{2}\}$.

PROOF. By Lemma 2.2, there are $a' \in [a,b]$ and $c' \in [c,d]$ such that $b-a'+1 \geqslant \epsilon(b-a+1)$, $d-c'+1 \geqslant \epsilon(d-c+1)$,

$$\inf_{a' \leq i \leq b} \frac{A(a',i)}{i-a'+1} > \frac{1}{2} \tag{2.2}$$

and

$$\inf_{c' \leq i \leq d} \frac{A(c',i)}{i-c'+1} > \frac{1}{2}. \tag{2.3}$$

Let $s = a'+c'$ and $t = s + \min\{b-a', d-c'\}$. Then

$$t-s+1 \geqslant \epsilon \cdot \min\{b-a+1, d-c+1\}.$$

Obviously, $[s,t] \subseteq [a+c, b+d]$. For each $m \in [s,t]$, we need to show $m \in A+B$. Note that $a'+(m-s) \leqslant b$ and $c'+(m-s) \leqslant d$. Then $A(a', a'+(m-s)) > \frac{1}{2}(m-s+1)$ by 2.2 and $B(c', c'+(m-s)) > \frac{1}{2}(m-s+1)$ by 2.3. Let

$$A' = A \cap [a', a'+(m-s)]$$

and

$$B' = \{m-i : i \in B \cap [c', c'+(m-s)]\}.$$

Then $A', B' \subseteq [a', a'+(m-s)]$. Also $|A'| > \frac{1}{2}(m-s+1)$ and

$$|B'| = B(c', c'+(m-s)) > \frac{1}{2}(m-s+1).$$

So $A' \cap B' \neq \emptyset$. Let $k \in A' \cap B'$. Then $k = x = m-y$ for some $x \in A$ and $y \in B$. Hence $m = x+y \in A+B$.

□

Now we prove Theorem 2.1.

PROOF. Assume the theorem is not true. We need to derive a contradiction. First define two numbers α and β such that

$$\alpha = \sup\{BD(X) : X \subseteq \mathbb{N} \text{ and there is a } Y \subseteq \mathbb{N} \text{ such that } BD(Y) > 0 \text{ and } X+Y \text{ is not piecewise syndetic.}\}$$

and

$$\beta = \sup\{BD(Y) : Y \subseteq \mathbb{N} \text{ and there is a } X \subseteq \mathbb{N} \text{ such that } BD(X) > (23/24)\alpha \text{ and } X+Y \text{ is not piecewise syndetic.}\}.$$

By the assumption that the theorem is not true, we have $\beta > 0$.

Claim 2.1.1: $\beta \leqslant \alpha$ and $\beta < \frac{7}{12}$.

Proof It is obvious that $\beta \leqslant \alpha$ due to the way they are defined. Suppose $\beta \geqslant \frac{7}{12}$. Then $\alpha \geqslant \frac{7}{12}$. Choose A and B such that $BD(A) > (23/24)\alpha$ and $BD(B) > (23/24)\beta$. Then $BD(A) > (23/24) \cdot (7/12) > \frac{1}{2}$ and $BD(B) > \frac{1}{2}$. Let

$\langle [a_n, b_n] : n \in \mathbb{N}\rangle$ and $\langle [c_n, d_n] : n \in \mathbb{N}\rangle$ be two sequences of intervals such that $\lim_{n\to\infty}(b_n - a_n) = \infty$, $\lim_{n\to\infty}(d_n - c_n) = \infty$,

$$\frac{A(a_n, b_n)}{b_n - a_n + 1} \geqslant \frac{1}{2} + \epsilon,$$

and

$$\frac{B(c_n, d_n)}{d_n - c_n + 1} \geqslant \frac{1}{2} + \epsilon$$

for some fixed positive number ϵ. Let M be any positive number and choose an n such that $\epsilon(b_n - a_n + 1) > M$ and $\epsilon(d_n - c_n + 1) > M$. By Lemma 2.3, there is an interval $[s, t] \subseteq A + B$ such that $t - s + 1 > M$. This showed that $A + B$ is thick, which contradicts that $A + B$ is not piecewise syndetic. □(Claim 2.1.1)

Let's fix two sets A and B such that $BD(A) > \frac{23}{24}\alpha$, $BD(B) > \frac{23}{24}\beta$, and $A + B$ is not piecewise syndetic. For each $k \in \mathbb{N}$, clearly

$$(A + B) + [0, k] = A + (B + [0.k])$$

is not piecewise syndetic. By the supremality of β, we have

$$BD(B + [0, k]) \leqslant \beta$$

for every $k \in \mathbb{N}$. Let $\langle [a_n, b_n] : n \in \mathbb{N}\rangle$ and $\langle [c_n, d_n] : n \in \mathbb{N}\rangle$ be two sequences of intervals such that for each $n \in \mathbb{N}$,

$$\frac{A(a_n, b_n)}{b_n - a_n + 1} > \frac{23}{24}\alpha,$$
$$\frac{B(c_n, d_n)}{d_n - c_n + 1} > \frac{23}{24}\beta,$$
$$\frac{(B + [0, 2n])(c_n, d_n)}{d_n - c_n + 1} < \frac{25}{24}\beta,$$

and $d_n - c_n + 1 > n^2$. For each $n > 40$, let

$$I_n = \{[c_n + in, c_n + (i + 1)n - 1] : i = 0, 1, \ldots, l_n\}$$

where $l_n = [\frac{d_n - c_n + 1}{n}]$. If

$$|\{i \in [0, l_n - 1] : B(c_n + in, c_n + (i + 1)n - 1) = \emptyset\}| \leqslant \frac{1}{3} l_n,$$

which means that there are at least two third of the intervals in I_n containing elements from B, then

$$\begin{aligned} |(B + [0, 2n])(c_n, d_n)| &\geqslant (2/3)l_n n - n \\ &\geqslant (2/3)(d_n - c_n + 1 - n) - n \\ &\geqslant (2/3)(d_n - c_n + 1) - 2n. \end{aligned}$$

So

$$\frac{(B + [0, 2n])(c_n, d_n)}{d_n - c_n + 1} \geqslant \frac{2}{3} - \frac{2}{n} > \frac{2}{3} - \frac{1}{20} = \frac{37}{60}.$$

On the other hand, we have

$$\frac{(B + [0, 2n])(c_n, d_n)}{d_n - c_n + 1} < \frac{25}{24}\beta \leqslant \frac{25}{24} \cdot \frac{7}{12} < \frac{37}{60}.$$

So we have a contradiction. Hence, for every $n > 40$,

$$|\{i \in [0, l_n - 1] : B(c_n + in, c_n + (i + 1)n - 1) = \emptyset\}| > \frac{1}{3} l_n$$

Claim 2.1.2: For each $n > \max\{40, \frac{12}{\beta}\}$, There exists an $i_n \in [0, l_n]$ such that

$$\frac{B(c_n + i_n n, c_n + (i_n + 1)n - 1)}{n} \geqslant \frac{13}{12}\beta.$$

Proof Suppose the claim is not true. Then

$$\begin{aligned}\frac{B(c_n, d_n)}{d_n - c_n + 1} &= \frac{\sum_{i=0}^{l_n - 1} B(c_n + in, c_n + (i+1)n - 1) + B(c_n + l_n n, d_n)}{d_n - c_n + 1} \\ &= \frac{n}{d_n - c_n + 1} \sum_{i=0}^{l_n - 1} \frac{B(c_n + in, c_n + (i+1)n - 1)}{n} + \frac{B(c_n + l_n n, d_n)}{d_n - c_n + 1} \\ &\leqslant \frac{n}{d_n - c_n + 1} \cdot l_n \cdot \frac{2}{3} \cdot \frac{13}{12}\beta + \frac{n}{d_n - c_n + 1} \\ &\leqslant (26/36)\beta + \frac{1}{n} < (29/36)\beta.\end{aligned}$$

This contradicts that

$$\frac{B(c_n, d_n)}{d_n - c_n + 1} > \frac{23}{24}\beta > \frac{29}{36}\beta.$$

□(Claim 2.1.2)

Following from Claim 2.1.2, we have $BD(B) \geqslant \frac{13}{12}\beta$. This contradicts the supremality of β.

□

REMARK 2.4. (1) For each positive real number $r < 1$, it is not hard to construct a set A such that $BD(A) > r$ and A is not piecewise syndetic.

(2) The reader who is interested in nonstandard analysis should read the original proof in [**J1**]. By translating the nonstandard proof into the standard proof here, certain degree of intuition and motivation are lost. The reader may consider some steps of the proof here to be tedious and awkward. But these steps make perfect sense when one is working in a nonstandard setting.

Theorem 2.1 has two corollaries which are worthy of mentioning. They are proven in [**J1**].

COROLLARY 2.5. *If $BD(A) > 0$ and $\gcd(A - a_0) = 1$, where a_0 is the least element in A, then there is an $h \in \mathbb{N}$ such that hA is thick.*

Given an increasing sequence $\langle x_n : n \in \mathbb{N}\rangle$ of natural numbers, let

$$FS(x_n)_{n=0}^{\infty} = \{\sum_{n \in F} x_n : F \text{ is a non-empty finite subset of } \mathbb{N}\}.$$

COROLLARY 2.6. *If $BD(A) > 0$ and $BD(B) > 0$, then $A + B$ contains a set of form $k + FS(x_n)_{n=0}^{\infty}$.*

REMARK 2.7. It is mentioned in [**B**] that for each positive real number $r < 1$, there is a set A such that $BD(A) > r$ and A does not contain any set of form $k + FS(x_n)_{n=0}^{\infty}$.

3. Buy One Get One Free Scheme

In this section, we show a method that allows one to efficiently derive a theorem about upper Banach density parallel to an existing theorem about Shnirel'man density or lower asymptotic density. Hence, whenever one obtains a result about Shnirel'man density or lower asymptotic density, one can get a result about upper Banach density for free.

Among four densities introduced at the beginning of this article, Shnirel'man density is traditionally the most interesting one to number theorists. There are many classical results about Shnirel'man density such as Mann's theorem [**HR**, page 5], Erdös-Landau's theorem [**HR**, page 10], Plünnecke's theorem [**N2**, page 225], etc. It often happens that after a result about Shnirel'man density is obtained, people like to generalize the result to lower asymptotic density or derive a parallel result about lower asymptotic density. For example, Kneser's theorem [**HR**, page 57] is a parallel theorem to Mann's theorem. Rohrbach's theorem [**HR**, page 45] is parallel to Erdös-Landau's theorem, and Plünnecke's theorem is also true if one replaces σ by $\underline{d}$. From these examples, one can see that the behavior of Shnirel'man density and the behavior of lower asymptotic density are very similar. What about the behavior of upper asymptotic density and the behavior of upper Banach density? Can one derive some parallel results about upper asymptotic density or upper Banach density to the existing results about Shnirel'man density or lower asymptotic density? Since lower asymptotic density and upper asymptotic density are usually introduced as a pair in many books and papers, it is natural for people to consider upper asymptotic density first. However, not many theorems about upper asymptotic density have been proven so far. The reason for that might be because the behavior of upper asymptotic density is quite different from that of lower asymptotic density or Shnirel'man density. For example, one can construct two sets A and B such that $d(A) = d(B) = \overline{d}(A + B + [0, k]) = \frac{1}{2}$ for every $k \in \mathbb{N}$. Hence there is no hope to find a theorem about upper asymptotic density parallel to Mann's theorem. According to the order of these densities, upper Banach density seems farther away from Shnirel'man density or lower asymptotic density than upper asymptotic density does. Therefore, the behavior of upper Banach density might be more different from the behavior of Shnirel'man density or lower asymptotic density. However, this is not the case! In fact, the behavior of upper Banach density bears extreme resemblance with the behavior of Shnirel'man density or lower asymptotic density.

In [**J2**], a general method was developed using nonstandard analysis, which allows us to efficiently derive a theorem about upper Banach density parallel to an existing theorem about Shnirel'man density or lower asymptotic density. The main idea is the following.

A set A has upper Banach density $\geqslant \alpha$ iff there is a copy of the standard set $\mathbb{N}$ inside the nonstandard version of $\mathbb{N}$ such that the nonstandard version of A has lower asymptotic density $\geqslant \alpha$ inside the copy of $\mathbb{N}$. The above statement is also true with lower asymptotic density replaced by Shnirel'man density. So when a set A with $BD(A) = \alpha$ is given, one can apply the existing theorem about Shnirel'man density or lower asymptotic density to that remote copy of $\mathbb{N}$ to obtain a result about the nonstandard version of A. By pushing down the nonstandard result to the standard world, one can obtain a parallel theorem about upper Banach density.

By following the same idea in the first section, one can find that it is no harder than the proof of Theorem 2.1 for translating each nonstandard proof in [**J2**] to a standard proof. One only needs to finitize every hyperfinite argument. But by doing so, one has to dissect the proof and get into those hard-core finite combinatorial arguments of each existing theorem about Shnirel'man density or lower asymptotic density. Those arguments are difficult and very different from theorem to theorem. So this retail style translation would force us to deal with each theorem individually. Therefore, it is not very efficient and usually obscures the general picture.

Recently, I found that the ergodic methods developed in [**F**] can play the same role as the nonstandard methods on this subject. The ergodic methods also offer a wholesale style treatment for deriving parallel theorems about upper Banach density. So I would like to use ergodic methods here as the standard alternatives of the nonstandard proofs in [**J2**].

Theorem 3.20 of [**F**] almost establishes the needed relation between upper Banach density and lower asymptotic density. It says that if $BD(A) > 0$, then there is a set R such that $\bar{d}(R) = \underline{d}(R) > 0$ and for every n, there is an m such that $m + R \cap [0, n] \subseteq A$. In order to use this idea for our purpose, we need to make a little adjustment. We would like to prove the following.

THEOREM 3.1. *Suppose $BD(A) = \alpha$. Then one can find a set R with $\bar{d}(R) = \underline{d}(R) = \alpha$ that for every n, there is an m such that $m + R \cap [0, n] = A \cap [m, m+n]$.*

Let's first prove a corollary of Theorem 3.1, which relates upper Banach density to Shnirel'man density.

COROLLARY 3.2. *Suppose $BD(A) = \alpha$ and $\epsilon > 0$. Then one can find a set Q with $\sigma(Q) \geqslant \alpha - \epsilon$ that for every n, there is an m such that $m + Q \cap [1, n] = A \cap [m+1, m+n]$.*

PROOF. Let R be the set produced in Theorem 3.1. Similar to the proof of Lemma 2.2, we construct an increasing sequence $\langle a_n : n = 0, 1, \ldots, k\rangle$.

Let the sequence to be an empty sequence and $Q = R$ if

$$\inf_{n \geq 1} \frac{R(n)}{n} \geqslant \alpha - \epsilon.$$

Then the proof is finished because $\sigma(Q) \geqslant \alpha - \epsilon$.

Otherwise, we construct an increasing sequence $a_0 < a_1 < \cdots < a_k$ inductively with $a_0 = 0$. Suppose a_i is defined for $i \geqslant 0$. If the set

$$S = \{h > a_i : \frac{R(a_i + 1, h)}{h - a_i} \leqslant \alpha - \epsilon\}$$

is not empty, then let $a_{i+1} = \min S$. Otherwise the construction is terminated.

Claim 3.2.1: The sequence has finite length.

Proof Suppose the sequence is never terminated. Then for each n one has

$$\frac{R(a_n)}{a_n} \leqslant \alpha - \epsilon,$$

which contradicts the fact that $\underline{d}(R) = \alpha$. □(Claim 3.2.1)

Let a_k be the last element of the sequence and let

$$Q = \{m - a_k : m \in R \text{ and } m \geqslant a_k\}.$$

Clearly, $\sigma(Q) \geqslant \alpha - \epsilon$. For each n, there exists an m' such that

$$m' + R \cap [0, a_k + n] = A \cap [m', m' + a_k + n].$$

Hence,

$$\begin{aligned} m' + a_k + Q \cap [1, n] &= m' + R \cap [a_k + 1, a_k + n] \\ &= A \cap [m' + a_k + 1, m' + a_k + n]. \end{aligned}$$

Now let $m = m' + a_k$ and this ends the proof.

□

Now we prove Theorem 3.1

PROOF. All ideas of the proof of Theorem 3.1 are from **[F]** except Claim 3.1.1 below.

Let A be a set with $BD(A) = \alpha$. Let $\Omega = \{0,1\}^{\mathbb{Z}}$ be the set of all functions from $\mathbb{Z}$ to $\{0,1\}$ with the product topology (consider $\{0,1\}$ as a two-point discrete space). Consider A as a subset of $\mathbb{Z}$ and let ϕ_A be the characteristic function of A. Then $\phi_A \in \Omega$. Let T be a bijection from Ω to Ω such that for every $\omega \in \Omega$, $T(\omega)(x) = \omega(x+1)$ for every $x \in \mathbb{Z}$. Let T^0 be the identity map and for every positive integer k, let $T^k = T \circ T^{k-1}$ and let T^{-k} be the inverse function of T^k. Let

$$X = \overline{\{T^k(\phi_A) : k \in \mathbb{Z}\}},$$

the orbit closure of ϕ_A under T. Then X is a separable compact topological subspace of Ω. Let $C(X)$ be the set of all real-valued continuous functions on X. Then $C(X)$ is a separable Banach space under the supreme norm $\|\cdot\|$, *i.e.* $\|f\| = \sup\{|f(x)| : x \in X\}$.

Now we choose a sequence of intervals $\langle [a_n, b_n] : n \in \mathbb{N}\rangle$ such that $\lim_{n\to\infty}(b_n - a_n) = \infty$ and $\lim_{n\to\infty} \frac{A(a_n,b_n)}{b_n - a_n + 1} = \alpha$. Without loss of generality, we can assume that for every $f \in C(X)$, the limit

$$\lim_{n\to\infty} \frac{1}{b_n - a_n + 1} \sum_{a_n \le k \le b_n} f(T^k(\phi_A)) \tag{3.1}$$

exists. The way to do this is the following. Let C_0 be a countable dense subset of $C(X)$. Using Cantor's diagonal argument, one can replace the sequence of the intervals mentioned above by a subsequence so that the limit in 3.1 exists for every $f \in C_0$. Since every function in $C(X)$ is a uniform limit of a function sequence in C_0, then the limit in 3.1 also exists for every $f \in C(X)$.

Now we define a positive linear functional L on $C(X)$. For each $f \in C(X)$, let

$$L(f) = \lim_{n\to\infty} \frac{1}{b_n - a_n + 1} \sum_{a_n \le k \le b_n} f(T^k(\phi_A)).$$

It is not hard to verify that L is a well-defined linear functional, $|L(f)| \leqslant \|f\|$, and $L(f) \geqslant 0$ whenever $f \geqslant 0$. By Riesz Representation Theorem, there is a finite Borel measure μ on X such that for every $f \in C(X)$,

$$L(f) = \int_X f d\mu.$$

Since

$$\mu(X) = \int_X 1 d\mu = L(1)$$
$$= \lim_{n\to\infty} \frac{1}{b_n - a_n + 1} \sum_{a_n \le k \le b_n} 1(T^k(\phi_A)) = 1,$$

then μ is a probability measure. It is also easy to verify that $L(f \circ T) = L(f)$ for every $f \in C(X)$. So T is a μ–measure preserving transformation from X to X. Let

$$E = \{\omega \in X : \omega(0) = 1\}.$$

Then E is a clopen subset of X. Hence the characteristic function Φ_E of E is an element in $C(X)$. Since for each $k \in \mathbb{N}$

$$\begin{aligned}\Phi_E(T^k(\phi_A)) = 1 &\Longleftrightarrow T^k(\phi_A) \in E \\ &\Longleftrightarrow T^k(\phi_A)(0) = 1 \\ &\Longleftrightarrow \phi_A(k) = 1 \\ &\Longleftrightarrow k \in A,\end{aligned}$$

then

$$\begin{aligned}\mu(E) &= \int_X \Phi_E d\mu = L(\Phi_E) \\ &= \lim_{n\to\infty} \frac{1}{b_n - a_n + 1} \sum_{a_n \le k \le b_n} \Phi_E(T^k(\phi_A)) \\ &= \lim_{n\to\infty} \frac{A(a_n, b_n)}{b_n - a_n + 1} = \alpha.\end{aligned}$$

Since Φ_E is a μ–measurable function and T is a μ–measure–preserving transformation, by Birkhoff ergodic theorem [**F**, page 59], there exists a measurable function $\bar{f} \in L^1(X)$ such that for μ–almost all $\omega \in X$

$$\lim_{n\to\infty} \frac{1}{n+1} \sum_{k=0}^{n} \Phi_E(T^k(\omega)) = \bar{f}(\omega).$$

Claim 3.1.1: For μ–almost all $\omega \in X$, $\bar{f}(\omega) = \alpha$.

Proof: Suppose that there is an $\omega \in X$ such that

$$\lim_{n\to\infty} \frac{1}{n+1} \sum_{k=0}^{n} \Phi_E(T^k(\omega)) = \beta > \alpha.$$

Let $R = \{m \in \mathbb{N} : \omega(m) = 1\}$. Then

$$\begin{aligned}k \in R &\Longleftrightarrow \omega(k) = 1 \\ &\Longleftrightarrow T^k(\omega)(0) = 1 \\ &\Longleftrightarrow T^k(\omega) \in E \\ &\Longleftrightarrow \Phi_E(T^k(\omega)) = 1.\end{aligned}$$

Hence,

$$\lim_{n\to\infty} \frac{1}{n+1} \sum_{k=0}^{n} \Phi_E(T^k(\omega)) = \lim_{n\to\infty} \frac{R(0,n)}{n+1} = \beta.$$

So $\underline{d}(R) = \beta$. Since $\omega \in X$, then for every n, there is a k_n such that $\omega(i) = T^{k_n}(\phi_A)(i)$ for $i = 0, 1, \ldots, n$. So

$$k_n + R \cap [0, n] = A \cap [k_n, k_n + n].$$

Since

$$\lim_{n \to \infty} \frac{R(0, n)}{n + 1} = \beta,$$

then

$$\lim_{n \to \infty} \frac{A(k_n, k_n + n)}{n + 1} = \beta.$$

This shows that $BD(A) \geqslant \beta > \alpha = BD(A)$, a contradiction. So $\bar{f}(\omega) \leqslant \alpha$ is true for μ–almost all $\omega \in X$.

On the other hand, since T is μ–invariant, then

$$\begin{aligned}
\int_X \bar{f} d\mu &= \int_X (\lim_{n \to \infty} \frac{1}{n+1} \sum_{k=0}^{n} \Phi_E(T^k(\omega))) d\mu \\
&= \lim_{n \to \infty} \frac{1}{n+1} \sum_{k=0}^{n} \int_X \Phi_E(T^k(\omega)) d\mu \\
&= \lim_{n \to \infty} \frac{1}{n+1} \sum_{k=0}^{n} \int_X \Phi_E(\omega) d\mu \\
&= \lim_{n \to \infty} \frac{1}{n+1} \sum_{k=0}^{n} \mu(E) \\
&= \mu(E) = \alpha.
\end{aligned}$$

Hence $\int_X (\alpha - \bar{f}) d\mu = 0$. Since $\alpha - \bar{f}$ is μ–almost surely non–negative, then we conclude that $\bar{f}(\omega) = \alpha$ for μ–almost all $\omega \in X$. □(Claim 3.1.1)

Now let's choose an $\omega \in X$ such that

$$\lim_{n \to \infty} \frac{1}{n+1} \sum_{k=0}^{n} \Phi_E(T^k(\omega)) = \alpha$$

and let $R = \{m \in \mathbb{N} : \omega(m) = 1\}$. Then $\underline{d}(R) = \alpha$. For every $n \in \mathbb{N}$, since $\omega \in X$, there is a $k_n \in \mathbb{Z}$ such that $\omega(i) = T^{k_n}(\phi_A)(i) = \phi_A(k_n + i)$ for $i = 0, 1, \ldots, n$. So

$$k_n + R \cap [0, n] = A \cap [k_n, k_n + n].$$

Without loss of generality, we can assume that $0 \in R$ (or simply replace R by $R - r_0$, where r_0 is the least element in R). Then it is easy to see that $m = k_n \geqslant 0$ for every n. This ends the proof. □

With Theorem 3.1 and Corollary 3.2 in our hands, we can now derive parallel theorems about upper Banach density. We derive four theorems in this section for a demonstration of the general method of the applications of Theorem 3.1 and Corollary 3.2.

We first derive a theorem parallel to Plünnecke's Theorem. A set B is called a basis of order h if $hB = \mathbb{N}$. Plünnecke's Theorem says that if B is a basis of order h, then for every set A, $\sigma(A + B) \geqslant \sigma(A)^{1 - \frac{1}{h}}$.

THEOREM 3.3. *If B is a basis of order h, then for every set A,*

$$BD(A+B) \geqslant BD(A)^{1-\frac{1}{h}}. \tag{3.2}$$

PROOF. Let $BD(A) = \alpha$. For each $\epsilon > 0$, let Q be the set obtained in Corollary 3.2 such that $\sigma(Q) \geqslant \alpha - \epsilon$. By Plünnecke's Theorem, one has

$$\sigma(Q+B) \geqslant \sigma(Q)^{1-\frac{1}{h}} = (\alpha - \epsilon)^{1-\frac{1}{h}}.$$

For every n, since there is a k such that $k + Q \cap [1,n] = A \cap [k+1, k+n]$, then $k + (Q+B) \cap [1,n] \subseteq (A+B) \cap [k+1, k+n]$. This shows that

$$\begin{aligned} BD(A+B) &\geqslant \lim_{n\to\infty} \sup_{k\geq 0} \frac{(A+B)(k+1,k+n)}{n} \\ &\geqslant \inf_{n\geq 1} \frac{(Q+B)(1,n)}{n} \\ &= \sigma(Q+B) \geqslant (\alpha-\epsilon)^{1-\frac{1}{h}}. \end{aligned}$$

Let $\epsilon \to 0$ and we have 3.2.

□

Next we want to derive a theorem parallel to Erdös–Landau's Theorem and to Rohrbach's Theorem. Let B be a basis of order h. The real number h^* is called the average order of B if

$$h^* = \sup_{n\geq 1} \frac{1}{n} \sum_{k=1}^{n} h(k),$$

where $h(k) = \min\{h' : k \in h'B\}$. Clearly, $h^* \leqslant h$. Erdös-Landau's Theorem says that if B is a basis of average order h^*, then for every set A,

$$\sigma(A+B) \geqslant \sigma(A) + \frac{1}{2h^*}\sigma(A)(1-\sigma(A)).$$

A set B is called an asymptotic basis of order h if there is an $m \in \mathbb{N}$ such that $hB \cup [0,m] = \mathbb{N}$. Let B be an asymptotic basis of order h and m be the number mentioned above. The real number h^* is called an average order of an asymptotic basis B if

$$h^* = \limsup_{n\to\infty} \frac{1}{n} \sum_{k=m+1}^{n} h(k).$$

Again $h^* \leqslant h$. Rohrbach's Theorem says that if B is an asymptotic basis of average order h^*, then

$$\underline{d}(A+B) \geqslant \underline{d}(A) + \frac{1}{2h^*}\underline{d}(A)(1-\underline{d}(A)).$$

Our parallel theorem is the following.

THEOREM 3.4. *If B is an asymptotic basis of average order h^*, then for every set A,*

$$BD(A+B) \geqslant BD(A) + \frac{1}{2h^*}BD(A)(1-BD(A)). \tag{3.3}$$

PROOF. Let $BD(A) = \alpha$. and let R be the set obtained in Theorem 3.1 with $\underline{d}(R) = \alpha$. By Rohrbach's Theorem, we have

$$\begin{aligned}\underline{d}(R+B) &\geqslant \underline{d}(R) + \frac{1}{2h^*}\underline{d}(R)(1-\underline{d}(R)) \\ &= \alpha + \frac{1}{2h^*}\alpha(1-\alpha).\end{aligned}$$

For every n, since there is a k such that

$$k + R \cap [0, n] = A \cap [k, k+n],$$

then there is a k such that

$$k + (R+B) \cap [0,n] \subseteq (A+B) \cap [k, k+n].$$

This shows that

$$\begin{aligned}BD(A+B) &\geqslant \liminf_{n\to\infty} \frac{(A+B)(k,k+n)}{n+1} \\ &\geqslant \underline{d}(R+B) \\ &\geqslant \alpha + \frac{1}{2h^*}\alpha(1-\alpha).\end{aligned}$$

Hence, 3.3 is true. □

REMARK 3.5. In Theorem 3.3, the basis B of order h can be replaced by a piecewise basis of order h. Also in Theorem 3.4, the asymptotic basis of average order h^* can be replaced by a piecewise asymptotic basis of average order h^*. The definitions of these concepts can be found in [**J2**]. The reason why we don't want to make these replacements here is that if we do, then we have to, without nonstandard analysis, dissect the proofs of Plünnecke's Theorem, Erdös–Landau's Theorem, and Rohrbach's Theorem and get into the finite combinatorial arguments.

Next we derive a parallel theorem to Mann's Theorem. Mann's Theorem says that if $0 \in A \cap B$, then $\sigma(A+B) \geqslant \min\{1, \sigma(A) + \sigma(b)\}$.

THEOREM 3.6. *For any two sets A and B,*

$$BD(A + B + \{0,1\}) \geqslant \min\{1, BD(A) + BD(B)\}.$$

The proof of Theorem 3.6 in [**J2**] uses Besicovitch's Theorem [**HR**, page 6]. Here we would like to prove Theorem 3.6 using next theorem parallel to Kneser's Theorem.

REMARK 3.7. (1) The term $\{0,1\}$ can be replaced by any two consecutive positive integers $\{a, a+1\}$. However, it can't be removed. For example, if one let A and B both be the set of all even numbers, then $BD(A+B) = BD(A) = BD(B) = \frac{1}{2}$.

(2) For each real number α, let $\lceil \alpha \rceil$ be the least integer greater than or equal to α. By Mann's Theorem, a set A with $\sigma(A) > 0$ and containing 0 is a basis of order at most $\lceil \frac{1}{\sigma(A)} \rceil$. Let's call a set B a Banach basis of order h if hB is thick. By Theorem 3.6, a set A with $BD(A) > 0$ and containing two consecutive numbers is a Banach basis of order at most $2\lceil \frac{1}{BD(A)} \rceil - 1$. This result is also optimal. For example, let

$$A = \{10n : n \in \mathbb{N}\} \cup \{10n+1 : n \in \mathbb{N}\}.$$

Then $BD(A) = \frac{1}{5}$. So A is a Banach basis of order 9. Since for every n, $10n + 9$ is not in $8A$, then A is not a Banach basis of order 8.

(3) Let

$$A = \bigcup_{n=1}^{\infty} [2^{2^{2n}}, 2^{2^{2n}+1} - 1] \text{ and} \tag{3.4}$$

$$B = \bigcup_{n=1}^{\infty} [2^{2^{2n+1}}, 2^{2^{2n+1}+1} - 1]. \tag{3.5}$$

Then $\bar{d}(A) = \bar{d}(B) = \bar{d}(A + B + [0, k]) = \frac{1}{2}$ for any $k \in \mathbb{N}$. So Mann's Theorem has no nice parallel theorem about upper asymptotic density. But if one considers $2A$ instead of $A + B$, then there is a result by G. A. Frieman, which says that if A is a set such that $\gcd(A - a_0) = 1$, where a_0 is the least element in A, then $\bar{d}(A) < \frac{1}{2}$ implies $\bar{d}(2A) \geqslant \frac{3}{2}\bar{d}(A)$ and $\bar{d}(A) \geqslant \frac{1}{2}$ implies $\bar{d}(2A) \geqslant \frac{1+\bar{d}(A)}{2}$. This result is a corollary of [**N2**, Theorem 1.15], one of Frieman's theorems in [**N2**]. On the other hand, for any α and β with $0 \leqslant \alpha \leqslant \frac{1}{2}$ and $\frac{1}{2} < \beta \leqslant 1$, it is not hard to construct a set A and a set B similar to 3.4 or 3.5 such that $\bar{d}(A) = \alpha$, $\bar{d}(2A) = \frac{3}{2}\alpha$ and $\bar{d}(B) = \beta$, $\bar{d}(2B) = \frac{1+\beta}{2}$.

Next we derive a theorem, which is not in [**J2**], parallel to Kneser's Theorem. Kneser's Theorem says that for any $k + 1$ sets $A_0, A_1, \ldots, A_k$, if

$$\underline{d}(\sum_{i=0}^{k} A_i) < \liminf_{n \to \infty} \sum_{i=0}^{k} \frac{A_i(n)}{n},$$

then there is a positive integer g and a set $E \subseteq \sum_{i=0}^{k} A_i$ such that E is essentially the union of congruence classes modulo g, *i.e.* there is a positive integer m such that

$$\{e + ng : e \in E \text{ and } n \in \mathbb{Z}\} \cap \mathbb{N} \setminus [0, m] = E \setminus [0, m],$$

and

$$\underline{d}(E) = \bar{d}(E) \geqslant (\liminf_{n \to \infty} \sum_{i=0}^{k} \frac{A_i(n)}{n}) - \frac{k}{g}.$$

Note that since

$$\sum_{i=0}^{k} \underline{d}(A_i) \leqslant \liminf_{n \to \infty} \sum_{i=0}^{k} \frac{A_i(n)}{n},$$

then Kneser's Theorem is also true if one replaces

$$\liminf_{n \to \infty} \sum_{i=0}^{k} \frac{A_i(n)}{n}$$

by $\sum_{i=0}^{k} \underline{d}(A_i)$. We would like to use this replacement instead of the original form of Kneser's Theorem in the next theorem because we want to avoid introducing too much notation to obscure the main idea.

THEOREM 3.8. *Let $A_0, A_1, \ldots, A_k$ be $k + 1$ sets. If*

$$BD(\sum_{i=0}^{k} A_i) < \sum_{i=0}^{k} BD(A_i), \tag{3.6}$$

then there are $g, l \in \mathbb{N}$, a sequence of intervals $\langle [a_n, b_n] : n = 1, 2, \ldots \rangle$ with $b_n - a_n + 1 = gn$, and a set $E \subseteq \sum_{i=0}^k A_i$ such that $E \cap [a_n, b_n]$ is the union of l arithmetic sequences of length n with common difference g and

$$BD(\sum_{i=0}^{k} A_i) \geqslant BD(E) \geqslant \frac{l}{g} \geqslant \sum_{i=0}^{k} BD(A_i) - \frac{k}{g}.$$

PROOF. Let $BD(A_i) = \alpha_i$ and $\sum_{i=0}^k \alpha_i = \alpha$. Suppose $BD(\sum_{i=0}^k A_i) < \alpha$. For each i let R_i be the set obtained in Theorem 3.1 corresponding to A_i. Then $\underline{d}(R_i) = \alpha_i$. Without loss of generality, let's assume $0 \in R_i$.

Claim 3.8.1 $\underline{d}(\sum_{i=0}^k R_i) < \alpha$.

Proof: Suppose $\underline{d}(\sum_{i=0}^k R_i) \geqslant \alpha$. For each n, let m_i be such that

$$m_i + R_i \cap [0, n] = A_i \cap [m_i, m_i + n]$$

and let $M = \sum_{i=0}^k m_i$. Then

$$M + (\sum_{i=0}^{k} R_i) \cap [0, n] \subseteq (\sum_{i=0}^{k} A_i) \cap [M, M + n].$$

Hence

$$\frac{(\sum_{i=0}^k A_i)(M, M+n)}{n+1} \geqslant \frac{(\sum_{i=0}^k R_i)(0, n)}{n+1}.$$

This shows

$$BD(\sum_{i=0}^{k} A_i) \geqslant \underline{d}(\sum_{i=0}^{k} R_i) \geqslant \alpha,$$

a contradiction to 3.6. □(Claim 3.8.1)

By Kneser's Theorem, there exist $g, l, m \in \mathbb{N}$ and a set $R_E \subseteq \sum_{i=0}^k R_i$ such that

$$R_E \setminus [0, m] = \{e + gn : e \in R_E \text{ and } n \in \mathbb{Z}\} \cap \mathbb{N} \setminus [0, m]$$

and

$$\underline{d}(R_E) = \frac{l}{g} \geqslant \alpha - \frac{k}{g}.$$

Now we define $[a_n, b_n]$ and E_n by induction on n. Let $[c_1, d_1]$ be such that $d_1 - c_1 + 1 = g$ and $c_1 > m$. Then there exists an $m_{1,i}$ for each $i = 0, 1, \ldots, k$ such that

$$m_{1,i} + R_i \cap [0, d_1] - A_i \cap [m_{1,i}, m_{1,i} + d_1].$$

So

$$\sum_{i=0}^{k} m_{1,i} + R_E \cap [0, d_1] \subseteq (\sum_{i=0}^{k} A_i) \cap [\sum_{i=0}^{k} m_{1,i}, \sum_{i=0}^{k} m_{1,i} + d_1].$$

Let

$$[a_1, b_1] = [\sum_{i=0}^{k} m_{1,i} + c_1, \sum_{i=0}^{k} m_{1,i} + d_1]$$

and let

$$E_1 = \sum_{i=0}^{k} m_{1,i} + R_E \cap [c_1, d_1].$$

Suppose we have found $[a_{n-1}, b_{n-1}]$ and E_{n-1}. Let $c_n > b_{n-1}$ and $d_n = c_n + gn - 1$. Then we do the exactly same as above with $a_1, b_1, c_1, b_1, m_{1,i}$ being replaced by $a_n, b_n, c_n, d_n, m_{n,i}$, and define

$$E_n = \sum_{i=0}^{k} m_{n,i} + R_E \cap [c_n, d_n].$$

Finally, we define $E = \bigcup_{n=1}^{\infty} E_n$. Note that the intervals $[a_n, b_n]$'s are pairwise disjoint. Now it is easy to check that the numbers g, l, the set E, and the sequence $\langle [a_n, b_n] : n = 1, 2, \ldots \rangle$ satisfy all the requirements.

□

Next we use Theorem 3.8 to prove Theorem 3.6.

PROOF. Let $BD(A) = \alpha$ and $BD(B) = \beta$. The theorem is obviously true if $BD(A + B) \geqslant \min\{1, \alpha + \beta\}$. Suppose $BD(A + B) < \min\{1, \alpha + \beta\}$. By Theorem 3.8, there are g, l, the set E, and the sequence $\langle [a_n, b_n] : n = 1, 2, \ldots \rangle$ such that $b_n - a_n + 1 = gn$, $E \cap [a_n, b_n]$ is the union of l arithmetic sequences of length n with common difference g, and $\frac{l}{g} \geqslant \alpha + \beta - \frac{1}{g}$. Clearly, $l < g$ because $BD(A + B) < 1$. Hence $(E + \{0, 1\}) \cap [a_n, b_n]$ contains one more arithmetic sequence of length n or $n - 1$ with common difference g, which is not in E_n. This shows that

$$\begin{aligned} BD(A + B + \{0, 1\}) &\geqslant BD(E + \{0, 1\}) \\ &= \frac{l+1}{g} = \frac{l}{g} + \frac{1}{g} \geqslant \alpha + \beta \\ &= BD(A) + BD(B). \end{aligned}$$

□

References

[B] V. Bergelson, *Ergodic Ramsey Theory–an Update, Ergodic theory of $\mathbb{Z}^d$ actions.* (Warwick, 1993-1994) 1-61, London Mathematical Society Lecture Note Ser. 228, Cambridge University Press, Cambridge 1996.

[F] H. Furstenberg, *Recurrence in Ergodic Theory and Combinatorial Number Theory.* Princeton University Press, 1981.

[HR] H. Halberstam and K. F. Roth, *Sequences.* Oxford University Press, 1966

[J1] R. Jin,*Sumset phenomenon.* The Proceedings of American Mathematical Society, *130*, No.3 (2002), 855—861.

[J2] R. Jin, *Nonstandard methods for upper Banach density problems.* The Journal of Number Theory, *91* (2001), 20—38.

[KL] H. J. Keisler and S. Leth, *Meager Sets on the Hyperfinite Time Line.* The Journal of Symbolic Logic, *56* (1991), 71–102.

[N1] M. B. Nathanson, *Additive Number Theory–the Classical Bases.* Springer, 1996.

[N2] M. B. Nathanson, *Additive Number Theory–Inverse Problems and the Geometry of Sumsets.* Springer, 1996.

DEPARTMENT OF MATHEMATICS, COLLEGE OF CHARLESTON, CHARLESTON, SOUTH CAROLINA, 29424

E-mail address: jinr@cofc.edu

DIMACS Series in Discrete Mathematics
and Theoretical Computer Science
Volume **64**, 2004

On Homogeneous Linear Recurrence Relations and Approximation of Zeros of Complex Polynomials

Bahman Kalantari

ABSTRACT. Let $p(z)$ be a complex polynomial of degree n. To each complex number a we associate a sequence called the *Basic Sequence* $\{B_m(a) = a - p(a)D_{m-2}(a)/D_{m-1}(a)\}$, where $D_m(a)$ is defined via a homogeneous linear recurrence relation and depends only on the normalized derivatives $p^{(i)}(a)/i!$. Each $D_m(a)$ is also representable as a Toeplitz determinant. Except possibly for the locus of points equidistant to two distinct roots, given any input a, the Basic Sequence converges to a root of p. The roots of p partition the Euclidean plane into Voronoi regions. Under some regularity assumptions (e.g. simplicity of the roots), for almost all inputs within the Voronoi polygon of a root, the corresponding Basic Sequence converges to that root. The discovery of the Basic Sequence, its error estimates, and several of its properties are motivated by our previous analysis of a fundamental family of iteration functions $\{B_m(z)\}$, called the Basic Family. Given any fixed $m \geq 2$ and an appropriate input a_0, the fixed-point iterates $a_{k+1} = B_m(a_k)$ converge to a simple root having order m. The Basic Sequence corresponds to the pointwise evaluation of the Basic Family. We present several fractal images that confirm the theoretical convergence results: as m increases, the basins of attractions to the roots, as computed with respect to the iteration function $B_m(z)$, rapidly converge to the Voronoi regions. Thus the regions with chaotic behavior rapidly shrink to the boundaries of the Voronoi regions. The interplay between the Basic Sequence and the Basic Family result in new algorithms for approximation of polynomial roots. Also as a byproduct of the analysis we give determinantal representation of solution of homogeneous linear recurrence relations, easily computable bounds on the modulus of terms of such recurrence relations, and fast computation of determinant of certain Toeplitz matrices. Finally, we prove a relationship between the Basic Sequence and the iterates of Bernoulli Method for approximation of extreme roots.

1. Introduction.

The problem of approximating roots of a real or complex polynomial $p(z)$ is a fundamental and classic problem. A comprehensive bibliography is given by McNamee [**18**]. Some aspects of the history, applications, and algorithms,

2000 *Mathematics Subject Classification.* Primary 12D10, 65Y20, 41A20, 11B39; Secondary 28A80, 30C10.

Key words and phrases. Zeros of Complex Polynomials, Basic Family of Iteration Functions, Recurrence Relations, Convergence.

are described in Pan [**20**]. For a complexity result see Smale [**22**]. New results on this problem continue to be discovered. In [**6**], Kalantari et al. study a fundamental family of iteration functions, called the Basic Family, $\{B_m(z) = z - p(z)D_{m-2}(z)/D_{m-1}(z)\}_{m=2}^{\infty}$, where $D_m(z)$ is a Toeplitz determinant defined with respect to the normalized derivatives of $p(z)$, but is also derivable via a homogeneous linear recurrence relation. Many results on the properties of the members of this family including their close tie with a determinantal generalization of Taylor's theorem can be found in [**5**], [**6**], [**7**], [**9**], [**10**] [**8**], [**11**], and [**13**]. The first member of the sequence, $B_2(z)$, is Newton's iteration function, and $B_3(z)$ is Halley's iteration function. For the history of these two iteration functions see Ypma [**25**] and Traub [**23**].

In this paper we will prove several new properties of the Basic Family. On the one hand these properties are derived by studying general homogeneous linear recurrence relations that are defined via a special set of initial conditions. On the other hand the properties of the Basic Family reveal new results on such recurrence relations. In tracing a circle of relations, we are led to the study of a classical method for approximation of dominant roots of polynomial, called the Bernoulli Method. As a byproduct we even reveal new properties on this method. Thus, not only in this paper we give new results on approximation of polynomial roots, but on homogeneous linear recurrence relations as well.

Let $p(z)$ be a given complex polynomial of degree n. To each point a in the complex plane we associate a sequence $\{d_m(a)\}$ defined via a homogeneous linear recurrence relation, derived from a special representation of the Taylor's expansion of $p(z)$ at a. The *Basic Sequence* at a is defined to be $\{b_m(a) = a - p(a)d_{m-2}(a)/d_{m-1}(a)\}$. We will show that $d_m(a) = D_m(a)$. Thus the basic sequence corresponds to the pointwise evaluation of the Basic Family. The Basic Sequence has important properties with respect to the approximation of roots of $p(z)$. One of these properties is as follows: Except possibly for the locus of points that are equidistant from two distinct roots (a set of measure zero consisting of the union of a finite set of lines) for any input a, the corresponding Basic Sequence is well-defined and it converges to a root of $p(z)$.

Under some regularity assumptions, e.g. the simplicity of all the roots, the Basic Sequence at a converges to the root of $p(z)$ that is nearest to a. By virtue of this, regions of convergence in the complex plane can best be described by *Voronoi regions*. The roots partition the plane into Voronoi regions. Voronoi region of a root is the set of points that are closer to this root than to any other root. The Voronoi regions consist of convex polygons and their boundaries (see e.g. [**21**]). Thus, under some regularity assumptions, the Voronoi region of each root is the set of points whose corresponding Basic Sequence converges to that root.

As an example of the Basic Sequence consider the complex polynomial $p(z) = 1 + p_1 z + \cdots + p_n z^n$, and let $a = 0$. The associated sequence becomes

$$d_m(0) = \sum_{i=1}^{n} (-1)^{i-1} p_i d_{m-i}(0),$$

where $d_0(0) = 1$, and $d_j(0) = 0$ for $j < 0$. The evaluation of the corresponding Basic Sequence $\{b_m(0)\}$ in many cases provides a very simple scheme for the approximation of a root of $p(z)$ (often the root nearest to the origin). This approach is very different than the classic iterative method of Newton which requires repeated polynomial evaluations.

In order to estimate the error in the approximation of a root θ via the individual member of the Basic Sequence, i.e. the quantity $|b_m(a) - \theta|$, as well as the discovery of further properties of the Basic Sequence, we will make use a nontrivial determinantal generalization of Taylor's theorem derived in **[9]**. We will prove that in fact the Basic Sequence corresponds to the pointwise evaluation of the Basic Family, $\{B_m(z)\}$ (i.e. $B_m(a) = b_m(a)$). In particular it follows that the Basic Sequence has the following important property: given a fixed $m \geq 2$, for any a_0 within a certain radius of a simple root, the iterates defined by $a_{k+1} = b_m(a_k)$ converge to that root with order m.

The discovery of the Basic Sequence is motivated by our previous analysis of the Basic Family. The results proved in this paper are consequences of the interplay between properties of the Basic Sequence proved here and those of the Basic Family. Not only this interplay gives new algorithms for approximation of polynomial roots, it results in determinantal representation of solution of homogeneous linear recurrence relations defined via a single initial input. This representation allows an easily computable upper bound on the modulus of the terms via Hadamard inequality. Also in some cases it is possible to state lower bounds using a recent determinantal lower bound, as well as the use of Gerschgorin's theorem.

This paper is organized as follows: In Section 2 we derive properties of general homogeneous recurrence relations defined via a single nonzero initial condition. In Section 3, we define the Basic Sequence and state several of its properties. In Section 4, we give the relationship between the Basic Sequence and the Basic Family of iteration functions, thereby state additional properties of the Basic Sequence. In Section 5, we give the connection between the Basic Sequence and the Bernoulli Method. In Section 6, we give some properties on the solution of general homogeneous linear recurrence relations that are defined via a single initial condition. In Section 7, we present some experimental results via a graphics package that produces fractal images. Section 8 consists of our concluding remarks.

2. Properties of Homogeneous Linear Recurrence Relations Having a Single Nonzero Initial Condition.

Let $\mathbb{C}$ be the field of complex numbers. For a complex number $c = a + ib$ ($i = \sqrt{-1}$) its modulus is $|c| = \sqrt{a^2 + b^2}$. Consider a general homogeneous linear recurrence relation of degree n, i.e.,

$$a_m = c_1 a_{m-1} + c_2 a_{m-2} + \cdots + c_n a_{m-n}, \tag{2.1}$$

where c_i's are given numbers in $\mathbb{C}$, and $c_n \neq 0$. We assume that $a_0 \neq 0$ and for all $j < 0$, $a_j = 0$. Without loss of generality we assume that the initial conditions are

$$a_0 = 1, \ , a_j = 0, \ \forall \ j < 0. \tag{2.2}$$

We shall refer to (2.1) and (2.2) as *homogeneous linear recurrence relation with a single initial condition.* The *characteristic polynomial* of the sequence $\{a_m\}$ is the polynomial

$$Q(z) = z^n - (c_1 z^{n-1} + c_2 z^{n-2} + \cdots + c_{n-1} z + c_n). \tag{2.3}$$

The *characteristic equation* refers to the equation

$$Q(z) = 0. \tag{2.4}$$

It is well known that a_m can be represented in terms of the roots of $Q(z)$. More precisely, let $\eta_1, \ldots, \eta_t$ be the set of distinct roots of $Q(z)$ having multiplicities $n_1, \ldots, n_t$. Then for each $i = 1, \ldots, t$ there exist a polynomial $q_i(z)$, such that either $q_i(z)$ is identically zero, or it is a polynomial of degree $\leq n_i - 1$, such that

$$a_m = q_1(m)\eta_1^m + \cdots + q_t(m)\eta_t^m, \ \forall\ m \geq -n+1. \tag{2.5}$$

PROPOSITION 2.1. *Consider the homogeneous linear recurrence relation with a single initial initial condition defined by (2.1) and (2.2). Assume that $|\eta_i| \neq |\eta_j|$, if $i \neq j$. Let r be the index satisfying*

$$|\eta_r| = \max\{|\eta_i| : q_i(z) \not\equiv 0,\ i = 1, \ldots, t\}$$

(i.e. the largest modulus root whose corresponding coefficient polynomial in (2.5) is not the zero polynomial). Then r is well-defined and

$$\lim_{m \to \infty} \frac{a_m}{a_{m-1}} = \eta_r.$$

Moreover, if all the roots of $Q(z)$ are simple, then $q_i(z)$ is a nonzero constant for all $i = 1, \ldots, n$. In particular, η_r is the root of $Q(z)$ with the largest modulus.

PROOF. Since $a_0 = 1$ we conclude that not all $q_i(z)$'s are zero. Hence r is well-defined. From the assumption that all distinct roots of $Q(z)$ have different modulus and the definition of η_r, for any $i \neq r$ such that $q_i(z) \not\equiv 0$, we have $|\eta_i/\eta_r| < 1$. Since for any such index $q_i(z)$ is a polynomial it follows that

$$\lim_{m \to \infty} q_i(m)\left(\frac{\eta_i}{\eta_r}\right)^m = 0. \tag{2.6}$$

First we claim that the sequence $\{a_m\}$ cannot have a subsequence $\{a_{m_k}\}$ where $a_{m_k} = 0$ for all k. Suppose such a subsequence exists. If all but one of the $q_i(z)$'s are identically zero, then from (2.1) we have $a_m = q_l(m)\eta_l^m$, for some l and $q_l(m) \not\equiv 0$. Thus, $a_m \neq 0$ for all $m \geq 0$. So under the assumption that $\{a_{m_k} = 0\}$ there must exist at least two indices for which $q_i(z)$ is not identically zero. Since for any $a, b \in \mathbb{C}$, $|a - b| \geq |a| - |b|$, we have

$$\left|\frac{a_{m_k}}{\eta_r^{m_k}}\right| = \left|\sum_{i=1}^{t} q_i(m_k)\left(\frac{\eta_i}{\eta_r}\right)^{m_k}\right| \geq \left|q_r(m_k)\right| - \sum_{i=1,\ i\neq r,\ q_i(z)\not\equiv 0}^{t} \left|q_i(m_k)\left(\frac{\eta_i}{\eta_r}\right)^{m_k}\right|. \tag{2.7}$$

Taking the limit in (2.7) and using (2.6), we get

$$0 \geq \lim_{k \to \infty} |q_r(m_k)|. \tag{2.8}$$

But the limit in (2.8) is either a nonzero number, or infinity. This is a contradiction. Thus a_m can be zero at most for a finite set of indices m.

Since $q_i(z)$ is a polynomial, if it is not identically zero then we have

$$\lim_{m \to \infty} \frac{q_i(m)}{q_i(m-1)} = 1. \tag{2.9}$$

Thus for m large enough the ratio a_m/a_{m-1} is well-defined and is given by:

$$\frac{a_m}{a_{m-1}} = \frac{\eta_r^m \sum_{i=1}^t q_i(m)\left(\frac{\eta_i}{\eta_r}\right)^m}{\eta_r^{m-1} \sum_{i=1}^t q_i(m-1)\left(\frac{\eta_i}{\eta_r}\right)^{m-1}}. \tag{2.10}$$

Now using (2.6) and (2.9) it follows that the ratio in (2.10) converges to η_r.

Now assume that all the roots of $Q(z)$ are simple. In this case $t = n$, and $q_i(z)$ is a constant, say $q_i \in \mathbb{C}$, for all $i = 1, \dots, n$. We wish to show that $q_i \neq 0$ for all $i = 1, \dots, n$.

From the initial condition (2.2), while substituted into (2.5), we get the following system of linear equations in q_i's

$$\begin{pmatrix} \eta_1^{-n+1} & \eta_2^{-n+1} & \cdots & \eta_n^{-n+1} \\ \eta_1^{-n+2} & \eta_2^{-n+2} & \cdots & \eta_n^{-n+2} \\ \vdots & \vdots & \cdots & \vdots \\ \eta_1^{-1} & \eta_2^{-1} & \cdots & \eta_n^{-1} \\ 1 & 1 & \cdots & 1 \end{pmatrix} \begin{pmatrix} q_1 \\ q_2 \\ \vdots \\ q_{n-1} \\ q_n \end{pmatrix} = \begin{pmatrix} 0 \\ 0 \\ \vdots \\ 0 \\ 1 \end{pmatrix}$$

Suppose that one of the q_i's is zero. Without loss of generality we may assume that $q_1 = 0$. Substituting $q_1 = 0$ in the above system of linear equations gives:

$$\begin{pmatrix} \eta_2^{-n+1} & \eta_3^{-n+1} & \cdots & \eta_n^{-n+1} \\ \eta_2^{-n+2} & \eta_3^{-n+2} & \cdots & \eta_n^{-n+2} \\ \vdots & \vdots & \cdots & \vdots \\ \eta_2^{-1} & \eta_3^{-1} & \cdots & \eta_n^{-1} \end{pmatrix} \begin{pmatrix} q_2 \\ q_3 \\ \vdots \\ q_n \end{pmatrix} = \begin{pmatrix} 0 \\ 0 \\ \vdots \\ 0 \end{pmatrix}$$

But the coefficient matrix is a product of a Vandermonde matrix in $\eta_2, \dots, \eta_n$ and the diagonal matrix $\operatorname{diag}(\eta_1^{-2}, \cdots, \eta_n^{n-1})$. Since the η_i's are distinct the Vandermonde matrix is invertible. Hence if $q_1 = 0$, then $q_2 = \cdots = q_n = 0$. But this contradicts that $a_0 \neq 0$. □

We now review some properties of a_m which are important from the computational point of view. From (2.1) and the identity equations $a_i = a_i$, $i = 1, \dots, m$, we get

$$\begin{pmatrix} a_m \\ a_{m-1} \\ \vdots \\ a_{m-n} \end{pmatrix} = \begin{pmatrix} c_1 & c_2 & \cdots & c_{n-1} & c_n \\ 1 & 0 & \cdots & 0 & 0 \\ \vdots & \vdots & \cdots & \vdots & \vdots \\ 0 & 0 & \cdots & 1 & 0 \end{pmatrix} \begin{pmatrix} a_{m-1} \\ a_{m-2} \\ \vdots \\ a_{m-n-1} \end{pmatrix}$$

Thus, if for each $k \geq 0$ we let $v_k = (a_k, \cdots, a_{k-n})^T$, and let A be the coefficient matrix above, then we have

$$v_m = Av_{m-1} = \cdots = A^m v_0, \ v_0 = (1, 0, \dots, 0)^T. \tag{2.11}$$

The matrix A is the companion matrix of the characteristic polynomial $Q(z)$. Using (2.11) it can be shown that v_m and hence a_m can be computed in $O(n^2 \log m)$ arithmetic operations, see Gries and Levin [**1**]. It is also possible to compute a_m in $O(n \log n \log m)$ arithmetic operations, see Fiduccia [**4**].

3. The Basic Sequence: Approximation of Polynomial Roots Using a Homogeneous Linear Recurrence Relation.

Let

$$p(z) = p_0 + p_1 z + \cdots + p_n z^n, \tag{3.1}$$

be a given polynomial of degree n with coefficients $p_i \in \mathbb{C}$, $i = 0, \dots, n$. We wish to approximate roots of $p(z)$. For each point $a \in \mathbb{C}$ we associate a sequence $\{d_m(a)\}$ defined via the following homogeneous linear recurrence relation of degree n, with a single nonzero initial condition:

$$d_m(a) = \sum_{i=1}^{n} (-1)^{i-1} p^{i-1}(a) \frac{p^{(i)}(a)}{i!} d_{m-i}(a), \tag{3.2}$$

where

$$d_0(a) = 1, \; d_j(a) = 0, \; \forall \, j < 0. \tag{3.3}$$

Define the *characteristic polynomial of* a, denoted by $Q_a(z)$, to be the characteristic polynomial corresponding to the above recurrence relation, i.e.

$$Q_a(z) = z^n - \sum_{i=1}^{n} (-1)^{i-1} p^{i-1}(a) \frac{p^{(i)}(a)}{i!} z^{n-i}.$$

From the relationship between a recurrence relation and its characteristic polynomial it follows that if $\eta_1, \dots, \eta_t$ is the set of distinct roots of $Q_a(z)$ having multiplicities $n_1, \dots, n_t$, then there exist polynomials $q_1(z), \dots, q_t(z)$ such that for each $i = 1, \dots, t$, $q_i(z)$ is either identically zero or a polynomial of degree $\leq n_i - 1$, such that

$$d_m(a) = q_1(m)\eta_1^m + \cdots + q_t(m)\eta_t^m, \; \forall \, m \geq -n+1. \tag{3.4}$$

THEOREM 3.1. *If η is a root of $Q_a(z)$, then $\theta = a - p(a)/\eta$ is a root of $p(z)$. Conversely, if θ is a root of $p(z)$, then $\eta = -p(a)/(\theta - a)$ is a root of $Q_a(z)$.*

PROOF. Suppose that η is a root of $Q_a(z)$, i.e.

$$\eta^n - \sum_{i=1}^{n} (-1)^{i-1} p^{i-1}(a) \frac{p^{(i)}(a)}{i!} \eta^{n-i} = 0. \tag{3.5}$$

Multiplying both sides by $p(a)/\eta^n$ we get

$$p(a) - \sum_{i=1}^{n} (-1)^{i-1} p^{i}(a) \frac{p^{(i)}(a)}{i!} \eta^{n-i} = 0. \tag{3.6}$$

Let $\theta = a - p(a)/\eta$. Then

$$\eta = -\frac{p(a)}{(\theta - a)}. \tag{3.7}$$

Substituting (3.7) in (3.6) gives

$$\sum_{i=0}^{n} \frac{p^{(i)}(a)}{i!} (\theta - a)^i = 0. \tag{3.8}$$

But from Taylor's theorem the above coincides with $p(\theta)$. Hence θ is a root of $p(z)$. The converse follows simply by reversing the steps. □

Let

$$R_p = \{\theta_1, \dots, \theta_t\} \tag{3.9}$$

be the set of distinct roots of $p(z)$. The elements of R_p partition the Euclidean plane into Voronoi regions and their boundaries. The Voronoi region of a root θ is a convex polygon defined by the locus of points which are closer to this root than to any other root. More precisely, the Voronoi region of a root θ is

$$V(\theta) = \{z \in \mathbb{C} : |z - \theta| < |z - \theta'|,\ \theta' \in R_p,\ \theta' \neq \theta\}. \tag{3.10}$$

Let S_p be the locus of points that are equidistant from two distinct roots, i.e.

$$S_p = \{z \in \mathbb{C} : |z - \theta| = |z - \theta'|, \text{ where } \theta, \theta' \in R_p,\ \theta \neq \theta'\}. \tag{3.11}$$

This is a set of measure zero consisting of the union of finite number of lines.

DEFINITION 3.2. Given $a \in \mathbb{C}$ the *Basic Sequence* at a is defined as

$$b_m(a) = a - p(a)\frac{d_{m-2}(a)}{d_{m-1}(a)}, \quad m = 2, 3, \ldots.$$

THEOREM 3.3. *Given $p(z)$, for any input $a \notin S_p$, the Basic Sequence is well-defined satisfying*

$$\lim_{m\to\infty} b_m(a) = \theta,$$

for some $\theta \in R_p$. Moreover, for each root θ of $p(z)$, given any $a \in V(\theta)$, the dominant root of $Q_a(z)$ is $\eta = -p(a)/(\theta - a)$. Thus if η^m appears with a nonzero coefficient polynomial in the representation of $d_m(a)$ (see (3.4)), then $b_m(a)$ converges to θ. In particular, if all roots of $p(z)$ are simple, then for all $a \in V(\theta)$, $\lim_{m\to\infty} b_m(a) = \theta$.

PROOF. From Theorem 3.1 for any $a \in \mathbb{C}$, different than a root of $p(z)$, the set of distinct roots of $Q_a(z)$ is

$$T_p(a) = \{\eta_1 = -\frac{p(a)}{\theta_1 - a}, \ldots, \eta_t = -\frac{p(a)}{\theta_t - a}\}. \tag{3.12}$$

If $a \notin S_p$, then $|\eta_i| \neq |\eta_j|$, for $i \neq j$. Now from Proposition 2.1 it follows that

$$\lim_{m\to\infty} \frac{d_m(a)}{d_{m-1}(a)} = \eta,$$

for some $\eta \in T_p(a)$. From (3.12) for each $a \in \mathbb{C}$ the dominant root of $Q_a(z)$ is $-p(a)/(\theta - a)$, where θ is the closest root to a. From Proposition 2.1 it then follows that if η^m appears with a nonzero coefficient in the representation of $d_m(a)$, then $b_m(a)$ converges to θ. If all roots of $p(z)$ are simple, then so are the roots of $Q_a(z)$. Thus from Proposition 2.1 for any $a \in V(\theta)$, $b_m(a)$ converges to θ. □

REMARK 3.4. For points that belong to S_p the convergence of $b_m(a)$ to a root may not occur. For instance if $p(z) = z^2 - 2$, then the origin is a point that belongs to S_p since it is equidistant to $\sqrt{2}$ and $-\sqrt{2}$. It is easy to see that for this choice of a the corresponding sequence $\{d_m(a)\}$ has a subsequence that is zero. Thus, the Basic Sequence is undefined.

4. The Basic Sequence and its Connection to the Basic Family

In this and the next sections we prove several important properties of the sequences $\{d_m(a)\}$ and $\{b_m(a)\}$. In doing so we first need to relate them to some previously known results. We will next describe a family of iteration functions for polynomial root-finding and state some of its known properties.

As before let $p(z)$ be a polynomial of degree $n \geq 2$ with complex coefficients. Set $D_0(z) \equiv 1$, and for each natural number $m \geq 1$, define

$$D_m(z) = \det \begin{pmatrix} p'(z) & \frac{p''(z)}{2!} & \dots & \frac{p(m-1)(z)}{(m-1)!} & \frac{p^{(m)}(z)}{(m)!} \\ p(z) & p'(z) & \ddots & \ddots & \frac{p(m-1)(z)}{(m-1)!} \\ 0 & p(z) & \ddots & \ddots & \vdots \\ \vdots & \vdots & \ddots & \ddots & \frac{p''(z)}{2!} \\ 0 & 0 & \dots & p(z) & p'(z) \end{pmatrix}$$

where det denotes the determinant. Also, given any $m \geq 1$, for each $i = m + 1, \dots, n + m - 1$ define

$$\widehat{D}_{m,i}(z) = \det \begin{pmatrix} \frac{p''(z)}{2!} & \frac{p'''(z)}{3!} & \dots & \frac{p^{(m)}(z)}{(m)!} & \frac{p^{(i)}(z)}{i!} \\ p'(z) & \frac{p''(z)}{2!} & \ddots & \frac{p^{(m-1)}(z)}{(m-1)!} & \frac{p^{(i-1)}(z)}{(i-1)!} \\ p(z) & p'(z) & \ddots & \vdots & \vdots \\ \vdots & \vdots & \ddots & \frac{p''(z)}{2!} & \frac{p^{(i-m+2)}(z)}{(i-m+2)!} \\ 0 & 0 & \dots & p'(z) & \frac{p^{(i-m+1)}(z)}{(i-m+1)!} \end{pmatrix}$$

For each $m \geq 2$, define

$$B_m(z) \equiv z - p(z)\,\frac{D_{m-2}(z)}{D_{m-1}(z)}.$$

THEOREM 4.1. (Kalantari [**13**]) *The following conditions hold:*

(1) *For all $m \geq 1$ we have,*

$$D_m(z) = \sum_{i=1}^{n} (-1)^{i-1} \frac{p^{i-1}(z) p^{(i)}(z)}{i!} D_{m-i}(z), \ D_j = 0, \ j < 0.$$

(2) *Let θ be a simple root of $p(z)$. Then,*

$$B_m(z) = \theta + \sum_{i=m}^{m+n-2} (-1)^m \frac{\widehat{D}_{m-1,i}(z)}{D_{m-1}(z)} (z - \theta)^i.$$

(3) *There exists $r > 0$ such that given any $a_0 \in N_r(\theta) = \{z : |z - \theta| \leq r\}$, the fixed-point iteration $a_{k+1} = B_m(a_k)$ is well-defined, it converges to θ having order m. Specifically,*

$$\lim_{k \to \infty} \frac{(\theta - a_{k+1})}{(\theta - a_k)^m} = (-1)^{m-1} \frac{\widehat{D}_{m-1,m}(\theta)}{D_{m-1}(\theta)} = (-1)^{m-1} \frac{\widehat{D}_{m-1,m}(\theta)}{p'(\theta)^{m-1}}.$$

(4) *Let θ be a simple root of $p(z)$. There exists a neighborhood of θ, $N^*(\theta)$, such that for each a within this neighborhood $p'(a) \neq 0$, $|a - \theta| < 1$, and*

$$|p'(a)| - \sum_{i=0, i \neq 1}^{n} |p(a)|^{\frac{i-1}{2}} \frac{|p^{(i)}(a)|}{i!} \geq \frac{1}{2} |p'(a)|.$$

For any $a \in N^*(\theta)$, if we set

$$h(a) = \left[\sum_{i=0}^{n} \left(\frac{|p^{(i)}(a)|}{i!} \right)^2 \right]^{1/2},$$

then we have

$$|B_m(a) - \theta| \le \left|\frac{2h(a)}{p'(a)}(a-\theta)\right|^m \left(\frac{1}{1-|a-\theta|}\right).$$

(5) *In particular, if θ is a simple root of $p(z)$, there exists $r^* \in (0,1)$ such that given any $a \in N_{r^*}(\theta)$, we have*

$$\theta = \lim_{m\to\infty} B_m(a).$$

THEOREM 4.2. *Given $p(z)$, for any input $a \in \mathbb{C}$ the associated sequences $\{d_m(a)\}$ and the corresponding Basic Sequence $\{b_m(a)\}$ satisfy:*

$$d_m(a) = D_m(a), \; b_m(a) = B_m(a), \; \forall\, m \ge 0.$$

In particular, the following hold:

(1) *$D_m(a)$ and hence $B_m(a)$ can be computed in $O(n \log n \log m)$ operations.*
(2) *$d_m(a)$ can be represented as an $m \times m$ Toeplitz matrix.*
(3) *$D_m(a)$ (and $d_m(a)$) is the first entry of the vector $A(a)^m v_0$, where $v_0 = (1, 0, \dots, 0)^T \in \mathbb{R}^n$ and $A(a)$ is the companion matrix of $Q_a(z)$, i.e.*

$$\begin{pmatrix} (-1)^{n-1}p^{n-1}(a)\frac{p^{(n)}(a)}{n!} & (-1)^{n-2}p^{n-2}(a)\frac{p^{(n-1)}(a)}{n-1!} & \cdots & -p(a)\frac{p''(a)}{2!} & p'(a) \\ 1 & 0 & \cdots & 0 & 0 \\ \vdots & \vdots & \cdots & \vdots & \vdots \\ 0 & 0 & \cdots & 1 & 0 \end{pmatrix}$$

(4) *For each simple root θ of $p(z)$ there exists an open neighborhood of the root such given any fixed $m \ge 2$, and any a_0 within this neighborhood, the sequence $\{a_k\}$ defined by*

$$a_{k+1} = b_m(a_k),$$

converges to θ having order of convergence equal to m.

PROOF. The equality $d_m(a) = D_m(a)$ is immediate since they correspond to the same homogeneous recurrence relations satisfying the same initial conditions. The complexity of computing $D_m(a)$ and $B_m(a)$ follows from two known results. Firstly, given any input a the evaluation of $p^{(j)}(a)/j!$, $j = 0, \dots, n$, called the *normalized derivatives*, can be established in $O(n \log^2 n)$ arithmetic operations, see Kung [**16**]. Secondly, the m-th term of a sequence defined via a homogeneous linear recurrence relation can be computed in $O(n \log n \log m)$ arithmetic operations, see Fiduccia [**4**]. Combing these two results the claimed complexity follows. The remaining parts of the theorem follow from Theorems 3.3 and 4.1. □

5. The Basic Sequence and the Bernoulli Iteration

Here we make a connection between the Basic Sequence and a classic method known as the *Bernoulli Iteration* or the Bernoulli Method due to Daniel Bernoulli, see Hildebrand [**2**], Householder [**3**].

Consider again a complex polynomial $p(z) = p_n z^n + \cdots + p_1 z + p_0$ of degree n, with $p_0 \ne 0$, having t distinct roots $R_p = \{\theta_1, \dots, \theta_t\}$, of multiplicity $n_1, \dots, n_t$, respectively. We can consider $p(z)$ as the characteristic polynomial of the recurrence relation

$$p_n u_m + p_{n-1} u_{m-1} + \cdots + p_0 u_{m-n} = 0. \tag{5.1}$$

For any set of initial conditions we get

$$u_m = \beta_1(m)\theta_1 + \cdots + \beta_t(m)\theta_t, \tag{5.2}$$

where for each $i = 1, \ldots, t$, $\beta_i(z)$ is either identically zero or a polynomial of degree $\leq n_i - 1$. It thus follows, as in the proof of Proposition 2.1, that if the coefficient polynomial corresponding to the dominant root is nonzero, then the ratio u_m/u_{m-1} converges to the dominant root of $p(z)$. Different schemes for selecting the initial conditions so that under the assumption of simplicity of the roots convergence of ratio u_m/u_{m-1} to the dominant root is guaranteed, are discussed in several numerical analysis books, see e.g. Hildebrand [**2**].

In order to construct sequences convergent to the least modulus root, one can consider the polynomial

$$g(y) = p_0 y^n + p_1 y^{n-1} + \cdots + p_n. \tag{5.3}$$

Clearly, θ is a root of $p(z)$ if and only if $1/\theta$ is a root of $g(y)$. Thus if the sequence $\{v_m\}$ is defined via the homogeneous recurrence relation

$$p_0 v_m + p_1 v_{m-1} + \cdots + p_n v_{m-n} = 0, \tag{5.4}$$

then, under some regularity conditions, the ratio v_{m-1}/v_m converges to the least root of $p(z)$. The approximation of the extreme roots via the ratios u_m/u_{m-1} and v_{m-1}/v_m is what is known as the Bernoulli Method, see Householder [**3**] and Hildebrand [**2**].

In what follows we will show that through the use of Taylor's theorem the notion of least modulus and largest modulus is only relative to the input with respect to which the polynomial $p(z)$ is being represented. Thus we may define the ratios u_m/u_{m-1} and v_{m-1}/v_m in more generality and produce convergent sequences to any root. Finally, we establish a relationship between the latter generalization and the Basic Sequence. This results in a deeper understanding of the Bernoulli Method as well as the existence of a precise error formula for its iterates.

Let $a \in \mathbb{C}$ be any complex number that is not a root of $p(z)$, and θ a root of $p(z)$. From Taylor's theorem we have

$$\sum_{i=0}^{n} \frac{p^{(n-i)}(a)}{(n-i)!}(\theta - a)^{n-i} = 0 = \sum_{i=0}^{n} \frac{p^{(i)}(a)}{i!}(\theta - a)^i. \tag{5.5}$$

Letting $z = \theta - a$, from the left-hand-side of (5.5) we get

$$\sum_{i=0}^{n} \frac{p^{(n-i)}(a)}{(n-i)!} z^{n-i} = 0, \tag{5.6}$$

which can be viewed as the characteristic equation of the homogeneous recurrence relation

$$\sum_{i=0}^{n} \frac{p^{(n-i)}(a)}{(n-i)!} u_{m-i}(a) = 0. \tag{5.7}$$

Also by dividing the right-hand-side of (5.5) by $(\theta - a)^n$ and letting $w = 1/(\theta - a)$, we get

$$\sum_{i=0}^{n} \frac{p^{(i)}(a)}{i!} w^{n-i} = 0. \tag{5.8}$$

The above can be viewed as the characteristic equation of the homogeneous recurrence relation

$$\sum_{i=0}^{n} \frac{p^{(i)}(a)}{i!} v_{m-i}(a) = 0. \tag{5.9}$$

DEFINITION 5.1. Let the Bernoulli Sequence associated with the farthest root from a be defined as

$$\alpha_m(a) = a + \frac{u_{m-1}(a)}{u_{m-2}(a)},$$

and the Bernoulli Sequence associated with the closet root to a be defined as

$$\beta_m(a) = a + \frac{v_{m-2}(a)}{v_{m-1}(a)}.$$

In particular, if $a = 0$ the corresponding Bernoulli Sequences coincides with the usual iterates of the Bernoulli Method.

THEOREM 5.2. *Given $p(z)$, let a be any complex number not in S_p (see (3.11)). Consider the Bernoulli Sequences with the initial conditions $u_0(a) = 1$, $u_j(a) = 0$, for all $j < 0$; and $v_0(a) = 1$, $v_j(a) = 0$, for all $j < 0$. We have*

(1)
$$\lim_{m\to\infty} \alpha_m(a) = \theta, \quad \lim_{m\to\infty} \beta_m(a) = \theta',$$
where θ and θ' are root of $p(z)$.

(2) *If all the roots of $p(z)$ are distinct, then θ is the farthest root from a, and θ' the nearest root to a.*

(3)
$$v_m(a) = (-1)^m \frac{d_m(a)}{p(a)^m}.$$

(4)
$$\beta_m(a) = b_m(a) = B_m(a),$$
(i.e. the Bernoulli Sequence associated with the nearest root to a coincides with the Basic Sequence associated with a, and hence the pointwise evaluation of the Basic Family at a).

PROOF. Define

$$\overline{Q}_a(z) = \sum_{i=0}^{n} \frac{p^{(i)}(a)}{i!} z^{n-i}, \quad \widehat{Q}_a(z) = \sum_{i=0}^{n} \frac{p^{(n-i)}(a)}{(n-i)!} z^{n-i}. \tag{5.10}$$

The set of distinct roots of $\overline{Q}_a(z)$ and $\widehat{Q}_a(z)$ are, respectively,

$$\overline{R} = \{\overline{\eta}_1 = \theta_1 - a, \ldots, \overline{\eta}_t = \theta_t - a\}, \quad \widehat{R} = \{\widehat{\eta}_1 = \frac{1}{\theta_1 - a}, \ldots, \widehat{\eta}_t = \frac{1}{\theta_t - a}\}. \tag{5.11}$$

If $a \notin S_p$, then $|\overline{\eta}_i| \neq |\overline{\eta}_j|$, for $i \neq j$. From Proposition 2.1 it follows that $\alpha_m(a)$ converges to a root of $p(z)$. Similarly, it follows that $\beta_m(a)$ converges to a root of $p(z)$. From Proposition 2.1 it also follows that if all the roots are simple, then $\alpha_m(a)$ converges to the farthest root from a, and $\beta_m(a)$ converges to the nearest root to a.

Next we establish the equivalence of $\beta_m(a)$, and $b_m(a)$. First we prove by induction that

$$v_m(a) = (-1)^m \frac{d_m(a)}{p(a)^m}.$$

The above is true for $m \leq 0$. Assume that it is true for all $k = 0, \dots, m-1$. Substituting $v_k(a) = (-1)^k d_k(a)/p(a)^k$ for $k = 0, \dots, m-1$ in (5.9), we get

$$p(a)v_m(a) = -\sum_{i=1}^{n} \frac{p^{(i)}(a)}{i!}(-1)^{m-i}\frac{d_{m-i}(a)}{p(a)^{m-i}}. \tag{5.12}$$

Multiplying the above by $p(a)^{m-1}$ and since $(-1)^{-i+1} = (-1)^{i-1}$ we get

$$p(a)^m v_m(a) = (-1)^m \sum_{i=1}^{n} (-1)^{i-1}\frac{p^{(i)}(a)}{i!} d_{m-i}(a)p(a)^{i-1}. \tag{5.13}$$

But from (3.2) the right-hand-side of the above coincides with $(-1)^m d_m(a)$. From this it easily follows that $\beta_m(a) = b_m(a)$. The equivalence of $\beta_m(a)$ and $B_m(a)$ was already established in Theorem 4.2

□

6. Determinantal Representation of Solution of Homogeneous Recurrence Relations and its Applications

In this section we consider the general homogeneous linear recurrence relation with a singe initial condition, i.e.,

$$a_m = c_1 a_{m-1} + c_2 a_{m-2} + \cdots + c_n a_{m-n}, \ a_0 = 1, \ , a_j = 0, \ \forall \ j < 0. \tag{6.1}$$

We first show that a_m can be represented as the determinant of a certain Toeplitz matrix. Next we show how this representation is useful in obtaining upper bounds on the modulus of a_m, and in some cases lower bounds. We will demonstrate the application of these results to the generalized Fibonacci numbers of order n.

In the following theorems we assume that $\{a_m\}$ satisfies the recurrence relation (6.1).

THEOREM 6.1. *For each $m \geq 1$ we have*

$$a_m = \det \begin{pmatrix} c_1 & -c_2 & \dots & (-1)^{m-2}c_{m-1} & (-1)^{m-1}c_m \\ 1 & c_1 & \ddots & \ddots & (-1)^{m-2}c_{m-1} \\ 0 & 1 & \ddots & \ddots & \vdots \\ \vdots & \vdots & \ddots & \ddots & -c_2 \\ 0 & 0 & \dots & 1 & c_1 \end{pmatrix}$$

where

$$c_i = 0, \ \forall \ i > n.$$

PROOF. Given any set of numbers $c_1, \dots, c_n$ we can always define a polynomial $p(z)$ of degree n and a complex number a such that

$$c_i = (-1)^{i-1}p^{i-1}(a)\frac{p^{(i)}(a)}{i!}, \ i = 1, \dots, n \tag{6.2}$$

From Theorem 4.1 it follows that a_m coincides with the quantity $D_m(a)$ corresponding to this polynomial (see definition of $D_m(a)$). Hence we obtain the claimed determinantal representation of a_m. □

THEOREM 6.2. *For each $m \geq 1$, we have*

$$|a_m| \leq \left(1 + \sum_{i=1}^{n} |c_i|\right)^{m/2}.$$

PROOF. From Hadamard inequality (see e.g. Marcus and Minc [**17**]), given an $m \times m$ matrix $A = (a_{ij})$, we have

$$\det(A) \leq \left(\prod_{i=1}^{m} \sum_{i=1}^{m} |a_{ij}|^2\right)^{1/2}.$$

But this implies the claimed bound. □

The following is an immediate corollary which applies to a sequence known as the the generalized Fibonacci numbers (see Miles [**19**]):

COROLLARY 6.3. *Consider the Fibonacci sequence of order n defined via a single nonzero initial condition, i.e.,*

$$F_m = F_{m-1} + F_{m-2} + \cdots + F_{m-n}, \ F_0 = 1, \ , F_j = 0, \ \forall\, j < 0. \tag{6.3}$$

Then, we have

$$F_m \leq (n+1)^{m/2}.$$

For $n = 2$ it is well-known that F_m is bounded above by $[(1+\sqrt{5})/2]^m$. Note that our bound is $\sqrt{3}^m$ and is comparable to the known bound. The problem of computing the determinant of a tridiagonal matrix that corresponds to the determinantal representation of the ordinary Fibonacci numbers (i.e. $n = 2$) appears as an exercise in Knuth [**15**] (see page 84). However, other than the determinantal representation given in the above corollary we are not aware of any other determinantal representation of the generalized Fibonacci numbers.

THEOREM 6.4. *Suppose that there exists a positive number $\delta \in (0,1)$ such that*

$$\sum_{i=0}^{n} |c_i| = (1+\delta)|c_1|.$$

Then we have

$$|a_m| \geq (1-\delta)^{\lceil \frac{m}{2} \rceil}(1+\delta)^{\lfloor \frac{m}{2} \rfloor}|c_1|^m.$$

The proof of Theorem 6.4 appears in Kalantari [**13**]. For the sake of completeness we reproduce this proof which in particular makes use of the following result:

THEOREM 6.5. (Kalantari [**10**]) *Let A be an $m \times m$ complex matrix. Assume we are given positive numbers l and u such that if λ is an eigenvalue of A, then $l \leq |\lambda| \leq u$. Let*

$$\kappa = \kappa(l,u) \equiv \left\lceil \frac{mu - |trace(A)|}{u - l} \right\rceil.$$

If $|trace(A)| \geq ml$, then

$$|\det(A)| \geq l^{\kappa} u^{m-\kappa}.$$

PROOF OF Theorem 6.4. Let A_m be the matrix corresponding to a_m (see Theorem 6.1). The matrix A_m is diagonally dominant. Since

$$\delta = \frac{1}{|c_1|} \sum_{i=0, i\neq 1}^{m} |c_i|,$$

from Gerschgorin's theorem it follows that if λ is an eigenvalue of A_m, we have

$$l = (1-\delta)|c_1| \leq \lambda \leq (1+\delta)|c_1| = u.$$

Since $|trace(\Delta_m)| = m|c_1| \geq m(1-\delta)|c_1|$, Theorem 6.4 applies to A_m, where

$$\kappa = \lceil \frac{mu - |trace(\Delta_m)|}{u-l} \rceil = \lceil \frac{m(1+\delta)|a_1| - m|a_1|}{(1+\delta)|a_1| - (1-\delta)|a_1|} \rceil = \lceil \frac{m}{2} \rceil,$$

and we get

$$|a_m| \geq \Big((1-\delta)|c_1|\Big)^{\lceil \frac{m}{2} \rceil} \Big((1+\delta)|c_1|\Big)^{\lfloor \frac{m}{2} \rfloor},$$

which coincides with the claimed lower bound.

7. Experimental Results Via Fractal Images

In this section we present some of our experimentation with a preliminary graphic package designed to produce fractal images that depict the convergence behavior of the Basic Family $B_m(z)$, as applied to a given polynomial, for a fixed m, as well as the convergence behavior of the Basic Sequence $\{B_m(a)\}_{m=2}^{\infty}$.

Consider a polynomial $p(z)$ and a fixed natural number $m \geq 2$. The basins of attraction of a root of $p(z)$ with respect to the iteration function $B_m(z)$ are regions in the complex plane such that given an initial point within them the corresponding sequence $a_{k+1} = B_m(a_k)$ will converge to that root. The fractal nature and the images of the basins of attractions of Newton's method are now quite familiar for some special polynomials such as $p(z) = z^n - 1$. However, this is not the case as far as the behavior of other members of the Basic Family is concerned.

In Figure 1 we consider a polynomial with a random set of roots, depicted as dots. The figure shows the evolution of the basins of attraction of the roots to the Voronoi regions as m takes the values $2, 4, 10$, and 50. In Figure 2-4 we show the basin of attractions for the polynomials $p(z) = z^3 - 1$, $P(z) = z^4 - 1$, and $p(z) = z^9 - 1$; corresponding to different values of m. Note that in all the figures as m increases the regions with chaotic behavior dramatically decrease to merely the boundaries of the Voronoi regions. Note also that, as predicted by the theory, for each root the basins of attractions converge to the Voronoi region of that root. The roots of $p(z) = z^n - 1$ are the roots of unity and hence the Voronoi regions are completely symmetric. In all the figures in the case of $m = 2$, i.e. Newton's method, the basins of attractions are chaotic. However, these regions rapidly improve by increasing m.

More vivid and color pictures of the above images as well as many more images will become available at www.cs.rutgers.edu/~kalantar. Moreover, in a forthcoming paper [**14**] we use a graphic approach to examine the computational advantages of the higher order members of the Basic Family, as well as the use of the Basic Sequence in computing a single root or all the roots of a given polynomial.

8. Concluding Remarks

We have shown that given a complex polynomial $p(z)$, to each complex number a one can associate a Basic Sequence, $\{B_m(a) = a - p(a)D_{m-2}(a)/D_{m-1}(a)\}$, where $D_m(a)$ is dependent only on the normalized derivatives at a, and is also definable via a homogeneous linear recurrence relation of degree n. Except possibly for a set of measure zero (union of finite number of lines) for any other input a the

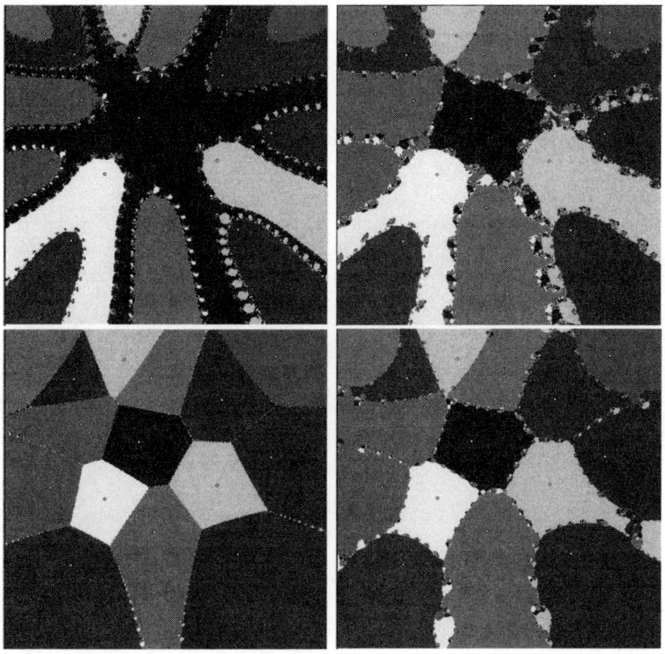

FIGURE 1. Evolution of basins of attraction to Voronoi regions via $B_m(z)$: random points, $m = 2, 4, 10, 50$ clockwise from upper left corner.

corresponding Basic Sequence converges to a root of $p(z)$. Under some regularity assumption (e.g. simplicity of all the roots) the Basic Sequence will necessarily converge to the nearest root of $p(z)$.

The evaluation of the Basic Sequence provides a very simple scheme for the approximation of a root of $p(z)$. As an example given the complex polynomial $p(z) = 1 + p_1 z + \cdots + p_n z^n$, and $a = 0$, the corresponding Basic Sequence $\{B_m(0)\}$ in many cases provides a very simple method for the approximation of a root of $p(z)$. Even if the sequence $\{D_m(0)\}$ has an infinite subsequence of zeros so that the Basic Sequence is not well-defined for all m, the Basic Sequence could still have subsequences that converge to a root of $p(z)$. For instance consider $p(z) = (z^2 + 2)(z - \theta)$, say for $\theta = 1$ and $\theta = 4$. For $\theta = 1$ the Basic Sequence $\{B_m(0)\}$ converges to θ even-though the origin is equidistant to the complex roots. Although the root $\theta = 4$ is not the nearest root to the origin, the Basic Sequence at $a = 0$ still has subsequences convergent to θ. If we now set $p(z)$ to be $p(z) = (z^2 + 2)(z - \theta)(z - \beta)^2$, where β is a real number, again at $a = 0$ we observe the convergence of the corresponding Basic Sequence, or a subsequence of it, to a root of $p(z)$.

The Basic Sequence can be efficiently computed. Its m-th term can be evaluated in $O(n \log n \log m)$ arithmetic operations. For small m it can be computed in $O(mn)$ operations. For each m, $D_m(a)$ can also be represented as a special Toeplitz determinant.

When the given input a is close enough to a simple root, for any $m \geq 2$ the Basic Sequence can be turned into an iterative method of order m, by replacing the given input a with $B_m(a)$, and repeating. There are many computational

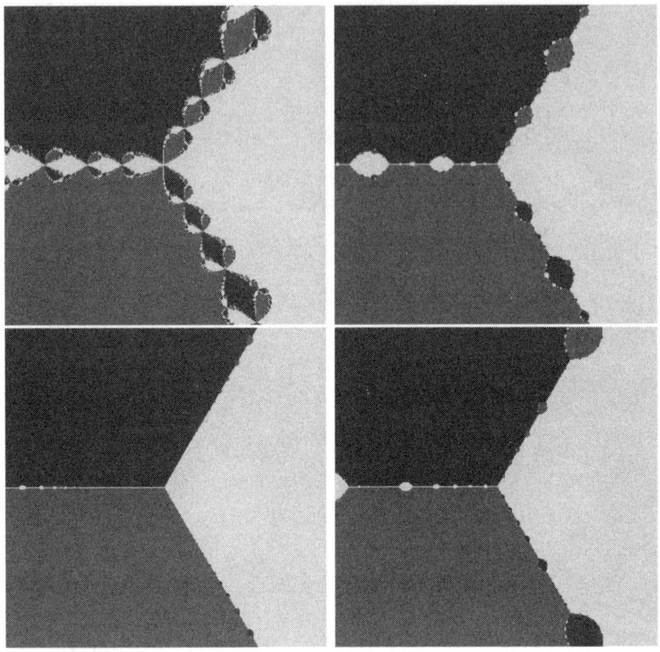

FIGURE 2. Evolution of basins of attraction to Voronoi regions via $B_m(z)$: $p(z) = z^3 - 1$, $m = 2, 3, 10, 50$ clockwise from upper left corner.

implication of these results. As an example one possible algorithm for polynomial root-finding is the following: for a given input a continue producing $B_m(a)$ until the difference between $B_m(a)$ and $B_{m+1}(a)$ is small. Then for a desirable m replace a with $B_m(a)$ and repeat in order to obtain higher and higher accuracy. It is also possible to switch back and forth between the two schemes.

An approach that is suggested by the properties of the Basic Sequence in order to compute all the roots of a given polynomial is as follows: first we obtain a rectangle that contains all the roots. Then by selecting a sparse number of gird point we can generate the corresponding Basic Sequence and thereby obtain good approximation to a subset of the roots. Then after deflation the same approach can be applied to find additional roots. This method could provide an excellent alternative to the method of Weyl [**24**] (see also Pan [**20**]). Experimentation with this method will be reported in [**14**].

The determinantal generalization of Taylor's theorem (see Theorem 4.1) gives a mechanism for estimating the error $|B_m(a) - \theta|$. The Basic Sequence can be even defined for functions that are not polynomial. In [**12**] we have used the pointwise evaluation of the Basic Family (the Basic Sequence in the terminology of this paper) to give new formulas for approximation of π and e. Although the definition of the Basic Sequence was motivated by the properties of the Basic Family of iteration functions, our new results have implied deeper properties of the Basic Family. Furthermore, the relationship between the Basic Sequence and the Basic Family has motivated new results on general homogeneous linear recurrence relations that

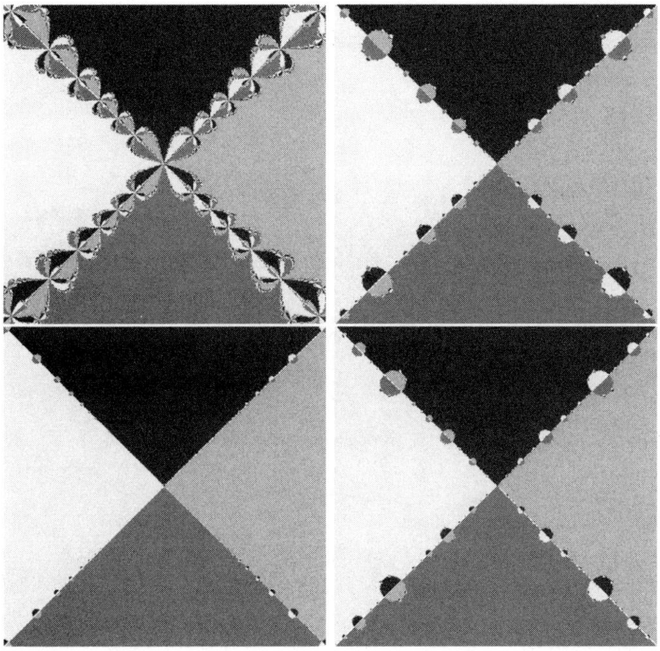

FIGURE 3. Evolution of basins of attraction to Voronoi regions via $B_m(z)$: $p(z) = z^4 - 1$, $m = 2, 3, 4, 50$ clockwise from upper left corner.

are defined via a single nonzero initial condition. Finally, we have shown the equivalence of the Basic Sequence and what we called here "the Bernoulli Sequence." This connection gives a deeper understanding of the Bernoulli Method. In particular, using our determinantal generalization of Taylor's theorem one can actually estimate the error in approximation of the extreme roots via the Bernoulli Method. In general our results should find many theoretical and practical applications in approximation of roots of polynomials and more general functions. In a forthcoming paper, [**14**], we will make use of fractal images to study the computational advantages of members of the Basic Family over Newton's method, as well as the advantages in using the Basic Sequence in computing a single root or all the roots of a given polynomial.

Acknowledgments. I would like to thank Mike Fredman for a communication. I specially would like to thank Mr. Zhengfang Zhu for having produced a beautiful graphic package that was used to produce the fractal figures given in this paper.

References

[1] D. Gries and G. Levin, *Computing Fibonacci numbers (and similarly defined functions) in log time*, IPL 11 (1980), 68-69.

[2] F.B. Hildebrand, "Introduction to Numerical Analysis," 2nd, ed., McGraw-Hill, New York, 1974.

[3] A.S. Householder, "The Numerical Treatment of a Single Nonlinear Equation," McGraw-Hill, New York, 1970.

[4] C.M. Fiduccia, *An efficient formula for linear recurrences*, SIAM J. Comput 14 1985, 106-112.

FIGURE 4. Evolution of basins of attraction to Voronoi regions via $B_m(z)$: $p(z) = z^9 - 1$, $m = 2, 3, 10, 50$ clockwise from upper left corner.

[5] B. Kalantari, and I. Kalantari, *High order iterative methods for approximating square roots*, BIT 36 (1996), 395-399.

[6] B. Kalantari, I. Kalantari, and R. Zaare-Nahandi, *A basic family of iteration functions for polynomial root finding and its characterizations*, J. of Comput. Appl. Math., 80 (1997), 209-226.

[7] B. Kalantari, *On the order of convergence of a determinantal family of root-finding methods*, BIT 39 (1999), 96-109.

[8] B. Kalantari and J. Gerlach, *Newton's method and generation of a determinantal family of iteration functions*, J. of Comp. and Appl. Math., 116 (2000), 195-200.

[9] B. Kalantari, *Generalization of Taylor's theorem and Newton's method via a new family of determinantal interpolation formulas and its applications*, J. of Comp. and Appl. Math., 126 (2000), 287-318.

[10] B. Kalantari, *A lower bound on determinants from linear programming*, Technical Report DCS-TR 330, Department of Computer Science, Rutgers University, New Brunswick, New Jersey, 1997.

[11] B. Kalantari and S. Park, *A computational comparison of the first nine members of a determinantal family of rootfinding methods*, DCS-TR-375. To appear in J. of Comp. and Appl. Math.

[12] B. Kalantari, *New formulas for approximations of Pi and other transcendental numbers*, Numerical Algorithms, 24 (2000), 59-81.

[13] B. Kalantari, *Approximation of polynomial root using a single input and the corresponding derivative values*, Technical Report DCS-TR 369, Department of Computer Science, Rutgers University, New Brunswick, New Jersey, 1998.

[14] B. Kalantari, *A computational study of the Basic Family in locating all zeros of a complex polynomial using fractal images*, forthcoming.

[15] D.E. Knuth, "The Art of Computer Programming, Volume I: Fundamental Algorithms," 2nd Edition, Addison-Wesley, Reading, Mass, 1973.

[16] H.T. Kung, *A new upper bound on the complexity of derivative evaluation*, Info. Proc. Lett., 2 (1974), 146-147.

[17] M. Marcus and H. Minc, *A Survey of Matrix Theory and Matrix Inequalities.* Allyn and Bacon, Boston, Mass., 1964.

[18] J.M. McNamee, *A bibliography on root of polynomials*, J. of Comp. and Appl. Math., 47 (1993), 391-394.

[19] E.P. Miles, *Generalized Fibonacci numbers and associated matrices*, Amer. Math. Monthly, 67 (1960), 745-757.

[20] V. Y. Pan, *Solving a polynomial equation: some history and recent progress*, SIAM Review, 39 (1997),187–220.

[21] F.P. Preparata, "Computational Geometry An Introduction," Springer-Verlag, New York, 1985.

[22] S. Smale, *The fundamental theorem of algebra and complexity theory*, Bull. Amer. Math. Soc. 4 (1981), 1-36.

[23] J.F. Traub, "Iterative Methods for the Solution of Equations," Prentice Hall, Englewood Cliffs, New Jersey, 1964.

[24] H. Weyl, *Randbemerkungen zu Hauptproblemen der Mathematik. II. Fundamentalsatz der Algebra und Grundlagen der Mathematik*, Math. Z. 20 (1924), 142–150.

[25] T.J. Ypma, Historical development of Newton-Raphson method, *SIAM Rev.*, 37 (1995), 531-551.

Department of Computer Science, Rutgers University, New Brunswick, New Jersey 08903

E-mail address: kalantari@cs.rutgers.edu

DIMACS Series in Discrete Mathematics
and Theoretical Computer Science
Volume **64**, 2004

Formal power series arising from multiplication of quantum integers

Melvyn B. Nathanson

Abstract. For the quantum integer $[n]_q = 1 + q + q^2 + \cdots + q^{n-1}$ there is a natural polynomial multiplication such that $[mn]_q = [m]_q \otimes_q [n]_q$. This multiplication is described by the functional equation $f_{mn}(q) = f_m(q)f_n(q^m)$, defined on a given sequence $\mathcal{F} = \{f_n(q)\}_{n=1}^{\infty}$ of polynomials such that $f_n(0) = 1$ for all n. The principal result of this paper is the following: If $\mathcal{F} = \{f_n(q)\}_{n=1}^{\infty}$ is a solution of the functional equation, then there exists a formal power series $F(q)$ such that the sequence $\{f_n(q)\}_{n=1}^{\infty}$ converges to $F(q)$.

Quantum mulitplication also suggests the functional equation

$$f(q)F(q^m) = F(q),$$

where $f(q)$ is a fixed polynomial or formal power series with constant term $f(0) = 1$, and $F(q) = 1 + \sum_{k=1}^{\infty} b_k q^k$ is a formal power series. It is proved that this functional equation has a unique solution $F(q)$ for every polynomial or formal power series $f(q)$. If the degree of $f(q)$ is at most $m - 1$, then there is an explicit formula for the coefficients b_k of $F(q)$ in terms of the coefficients of $f(q)$ and the m-adic representation of k.

The paper also contains a review of convergence properties of formal power series with coefficients in an arbitrary field or integral domain.

1. Quantum multiplication

Nathanson [**1, 2**] introduced the functional equation for multiplication of quantum integers as follows.

Let $\mathbf{N} = \{1, 2, 3, \ldots\}$ denote the set of positive integers and $\mathbf{N}_0 = \mathbf{N} \cup \{0\}$ the set of nonnegative integers. For $n \in \mathbf{N}$, the polynomial

$$[n]_q = 1 + q + q^2 + \cdots + q^{n-1}$$

is called the *quantum integer* n. With the usual multiplication of polynomials, however, we observe that $[m]_q[n]_q \neq [mn]_q$ for all $m \geq 2$ and $n \geq 2$. We would like to define a polynomial multiplication such that the product of the quantum integers $[m]_q$ and $[n]_q$ is $[mn]_q$.

2000 *Mathematics Subject Classification.* Primary 30B12, 81R50. Secondary 11B13.

Key words and phrases. Quantum integers, quantum polynomial, polynomial functional equation, q-series, formal power series.

This work was supported in part by grants from the NSA Mathematical Sciences Program and the PSC-CUNY Research Award Program.

Let $\mathcal{F} = \{f_n(q)\}_{n=1}^{\infty}$ be a sequence of polynomials with coefficients in a field. We define a multiplication operation on the polynomials in $\mathcal{F}$ by

$$f_m(q) \otimes_q f_n(q) = f_m(q) f_n(q^m).$$

If $f_n(q) = [n]_q$ is the n-th quantum integer, then

$$\begin{aligned} [m]_q \otimes_q [n]_q &= [m]_q [n]_{q^m} \\ &= \left(1 + q + q^2 + \cdots + q^{m-1}\right)\left(1 + q^m + q^{2m} + \cdots + q^{m(n-1)}\right) \\ &= 1 + q + q^2 + \cdots + q^{mn-1} \\ &= [mn]_q \end{aligned}$$

for all positive integers m and n, and so the sequence $\mathcal{F} = \{[n]_q\}_{n=1}^{\infty}$ satisfies the functional equation

$$f_{mn}(q) = f_m(q) \otimes_q f_n(q). \tag{1}$$

This is called the *functional equation for quantum multiplication.*

Many other sequences of polynomials also satisfy this functional equation. For example, for every positive integer t, the sequence of polynomials $\{q^{t(n-1)}\}_{n=1}^{\infty}$ satisfies (1).

PROBLEM 1. *Determine all sequences of polynomials $\mathcal{F} = \{f_n(q)\}_{n=1}^{\infty}$ that satisfy the functional equation*

$$f_{mn}(q) = f_m(q) \otimes_q f_n(q).$$

The sequence of $\mathcal{F} = \{f_n(q)\}_{n=1}^{\infty}$ is called *nonzero* if $f_n(q) \neq 0$ for some integer n. The functional equation implies that

$$f_1(q) = f_1(q)^2,$$

and so $f_1(q) = 0$ or 1. If $f_1(q) = 0$, then

$$f_n(q) = f_{1 \cdot n}(q) = f_1(q) f_n(q) = 0$$

for all $n \in \mathbf{N}$. Therefore, a solution $\mathcal{F} = \{f_n(q)\}_{n=1}^{\infty}$ of (1) is nonzero if and only if $f_1(q) = 1$.

The *support* of $\mathcal{F}$ is the set

$$\operatorname{supp}(\mathcal{F}) = \{n \in \mathbf{N} : f_n(q) \neq 0\}.$$

If a nonzero sequence of polynomials $\mathcal{F} = \{f_n(q)\}_{n=1}^{\infty}$ is a solution of the functional equation (1), then $\operatorname{supp}(\mathcal{F})$ is a multiplicative subsemigroup of the positive integers. Nathanson [**1**] proved that $\operatorname{supp}(\mathcal{F})$ is a semigroup of the form $S(P)$, where P is a set of prime numbers and $S(P)$ is the semigroup of positive integers generated by P.

Every nonzero polynomial $f(q)$ with coefficients in a field can be written uniquely in the form

$$f(q) = a q^{v(f)} g(q), \tag{2}$$

where $a \neq 0$, $v(f) \in \mathbf{N}_0$, and $g(q)$ is a polynomial with constant term $g(0) = 1$. Let $\mathcal{F} = \{f_n(q)\}_{n=1}^{\infty}$ be a nonzero sequence of polynomials that is a solution of the functional equation (1). We represent each nonzero polynomial $f_n(q)$ in the form

$$f_n(q) = a(n) q^{v(f_n)} g_n(q),$$

where $a(n) \neq 0$, $v(f_n) \in \mathbf{N}_0$, and $g_n(0) = 1$. We define $g_n(q) = 0$ and $a(n) = 0$ for all $n \in \mathbf{N}_0 \setminus \operatorname{supp}(\mathcal{F})$.

For all $m, n \in \operatorname{supp}(\mathcal{F})$, we have

$$\begin{aligned} a(mn)q^{v(f_{mn})}g_{mn}(q) &= f_{mn}(q) \\ &= f_m(q)f_n(q^m) \\ &= a(m)q^{v(f_m)}g_m(q)a(n)q^{mv(f_n)}g_n(q^m) \\ &= a(m)a(n)q^{v(f_m)+mv(f_n)}g_m(q)g_n(q^m). \end{aligned}$$

Since

$$g_{mn}(0) = g_m(0)g_n(0) = 1,$$

it follows that

$$a(mn) = a(m)a(n)$$

for all $m, n \in \operatorname{supp}(\mathcal{F})$, and so a is a completely multiplicative arithmetic function on the semigroup $\operatorname{supp}(\mathcal{F})$ with $a(n) \neq 0$ for all $n \in \operatorname{supp}(\mathcal{F})$. Similarly,

$$q^{v(f_{mn})} = q^{v(f_m)+mv(f_n)},$$

and so

$$v(f_m) + mv(f_n) = v(f_{mn}) = v(f_{nm}) = v(f_n) + nv(f_m)$$

for all integers m and n in $\operatorname{supp}(\mathcal{F})$. This implies that there exists a nonnegative rational number t such that

$$\frac{v(f_m)}{m-1} = \frac{v(f_n)}{n-1} = t$$

for all $m, n \in \operatorname{supp}(\mathcal{F}) \setminus \{1\}$. Moreover,

$$t(n-1) \in \mathbf{N}_0$$

and

$$q^{v(f_n)} = q^{t(n-1)}$$

for all $n \in \operatorname{supp}(\mathcal{F})$.

Finally, we see that $\mathcal{G} = \{g_n(q)\}_{n=1}^{\infty}$ is also a solution of the functional equation for quantum multiplication with $\operatorname{supp}(\mathcal{G}) = \operatorname{supp}(\mathcal{F})$, and $g_n(0) = 1$ for all $n \in \operatorname{supp}(\mathcal{G})$. It follows that to solve Problem 1 it suffices to classify solutions $\mathcal{F} = \{f_n(q)\}_{n=1}^{\infty}$ of (1) such that $f_n(0) = 1$ for all $n \in \operatorname{supp}(\mathcal{F})$.

Let $\mathcal{F} = \{f_n(q)\}_{n=1}^{\infty}$ be a solution of (1). The operation $\otimes_q$ is commutative on $\mathcal{F}$ since

$$f_m(q) \otimes_q f_n(q) = f_{mn}(q) = f_{nm}(q) = f_n(q) \otimes_q f_m(q).$$

Equivalently,

$$f_m(q)f_n(q^m) = f_n(q)f_m(q^n) \tag{3}$$

for all positive integers m and n. Moreover,

$$\deg(f_m) + m\deg(f_n) = \deg(f_n) + n\deg(f_m),$$

and so $\deg(f_m) = t(m-1)$ and $\deg(f_n) = t(n-1)$ for some nonnegative rational number t. This suggests the following problem.

PROBLEM 2. *Let m and n be positive integers. For a fixed polynomial $f_m(q)$, determine all polynomials $f_n(q)$ that satisfy the functional equation*

$$f_m(q)f_n(q^m) = f_n(q)f_m(q^n).$$

Let $f_m(q) = \sum_{i=0}^{t(m-1)} a_i q^i$ and $f_n(x) = \sum_{j=0}^{t(n-1)} b_j q^j$ be polynomials of degrees $t(m-1)$ and $t(n-1)$, respectively, with constant terms $a_0 = b_0 = 1$. If $f_m(q)$ and $f_n(q)$ satisfy equation (3), then

$$f_m(q) \sum_{j=0}^{t(n-1)} b_j q^{mj} = \left(\sum_{j=0}^{t(n-1)} b_j q^j \right) \left(1 + a_1 q^n + \cdots + a_{t(m-1)} q^{nt(m-1)} \right).$$

Letting n tend to infinity, we obtain the functional equation

$$f_m(q) \sum_{j=0}^{\infty} b_j q^{mj} = \sum_{j=0}^{\infty} b_j q^j,$$

or, equivalently,

$$f_m(q) F(q^m) = F(q),$$

where

$$F(q) = \sum_{k=0}^{\infty} b_k q^k$$

is a formal power series with constant term $b_0 = 1$.

PROBLEM 3. *Let m be positive integer and $f(q)$ a polynomial or formal power series with constant term $f(0) = 1$. Determine all formal power series $F(q)$ that satisfy the functional equation*

$$f(q) F(q^m) = F(q). \tag{4}$$

Problems 1 and 2 are unsolved. Problem 3, however, will be solved in Section 4, where we prove that for every polynomial $f(q)$, and, more generally, for every formal power series $f(q)$ with constant term $f(0) = 1$, there is a unique formal power series $F(q)$ that is a solution of the functional equation

$$f(q) F(q^m) = F(q).$$

Moreover, we shall explicitly construct the coefficients of the formal power series $F(q)$ when $f(q)$ is a polynomial of degree at most $m - 1$.

2. Convergence of formal power series

We review here some elementary convergence properties of formal power series with coefficients in a ring.

2.1. The ring of formal powers series. A *formal power series* in the variable q with coefficients in a ring R is an expression of the form

$$\sum_{n=0}^{\infty} a_n q^n = a_0 + a_1 q = a_2 q^2 + \cdots,$$

where the coefficients a_n are elements of R. The coefficient a_0 is called the *constant term* of the power series. We denote by $R[[q]]$ the set of all formal power series with coefficients in R. The sum and product of formal power series are defined, as usual, by

$$\sum_{n=0}^{\infty} a_n q^n + \sum_{n=0}^{\infty} b_n q^n = \sum_{n=0}^{\infty} (a_n + b_n) q^n$$

and

$$\sum_{n=0}^{\infty} a_n q^n \cdot \sum_{n=0}^{\infty} b_n q^n = \sum_{n=0}^{\infty} c_n q^n,$$

where

$$c_n = \sum_{i=0}^{n} a_i b_{n-i}.$$

With these operations of addition and multiplication, $R[[q]]$ is a ring.

In this paper we shall always assume that R is an integral domain. Then the ring of formal power series $R[[q]]$ is also an integral domain.

A formal power series $f(q)$ is *invertible* if there exists a formal power series $g(q)$ such that $f(q)g(q) = 1$. If $f(q) \in R[[q]]$ is invertible, then there is a unique formal power series $g(q)$ such that $f(q)g(q) = 1$. The inverse $g(q)$ is denoted $f(q)^{-1}$. For example, $(1-q)^{-1} = \sum_{n=0}^{\infty} q^n$.

THEOREM 1. *The formal power series $f(q) = \sum_{n=0}^{\infty} a_n q^n$ is invertible in $R[[q]]$ if and only if the coefficient a_0 is invertible in the ring R.*

PROOF. If the power series $\sum_{n=0}^{\infty} b_n q^n$ is the multiplicative inverse of $f(q)$, then $a_0 b_0 = 1$ and so a_0 is invertible in R.

Conversely, suppose that a_0 is invertible in R. The formal power series $g(q) = \sum_{n=0}^{\infty} b_n q^n$ is a solution of the equation $f(q)g(q) = 1$ if and only if the coefficients b_n satisfy the identities

$$a_0 b_0 = 1$$

and

$$\sum_{i=0}^{n} a_i b_{n-i} = a_0 b_n + a_1 b_{n-1} + \cdots + a_n b_0 = 0 \qquad \text{for all } n \geq 1.$$

These conditions are equivalent to

$$b_0 = a_0^{-1}$$

and

$$b_n = -a_0^{-1} \sum_{i=1}^{n} a_i b_{n-i} \qquad \text{for all } n \geq 1.$$

We can now construct inductively the unique sequence $\{b_n\}_{n=1}^{\infty}$ of elements of the ring R such that $g(q) = \sum_{n=0}^{\infty} b_n q^n = f(q)^{-1}$. This completes the proof. □

Let $f(q) = \sum_{n=0}^{\infty} a_n q^n$ be a formal power series. We define the *valuation*

$$v : R[[q]] \to \mathbf{N}_0 \cup \{\infty\}$$

as follows: If $f(q) \neq 0$, then $v(f)$ is the smallest integer n such that $a_n \neq 0$. If $f(q) = 0$, then we set $v(f) = \infty$.

Let $f(q) = \sum_{n=0}^{\infty} a_n q^n$ and $g(q) = \sum_{n=0}^{\infty} b_n q^n$ be formal power series. We write

$$f(q) \equiv g(q) \pmod{q^N}$$

if $a_n = b_n$ for $n = 0, 1, \ldots, N-1$. This is equivalent to the inequality $v(f-g) \geq N$.

THEOREM 2. *Let R be an integral domain, and let $f, g \in R[[q]]$. The valuation $v : R[[q]] \to \mathbf{N}_0 \cup \{\infty\}$ satisfies the following properties:*

(i)

$$v(-f) = v(f)$$

(ii)
$$v(fg) = v(f) + v(g)$$

(iii)
$$v(f \pm g) \geq \min(v(f), v(g))$$

(iv)
$$v(f \pm g) = \min(v(f), v(g)) \qquad \textit{if } v(f) \neq v(g).$$

(v)
$$v(f(q^k) - f(0)) \geq k.$$

PROOF. Let $v(f) = s$ and $v(g) = t$. Then $f(q) = \sum_{n=s}^{\infty} a_n q^n$ and $g(q) = \sum_{n=t}^{\infty} b_n q^n$, where a_s and b_t are nonzero elements of the ring R. Then $-a_s \neq 0$, and so $v(-f) = s$. This proves (i).

Statement (ii) is a consequence of the identity

$$f(q)g(q) = a_s b_t q^{s+t} + \sum_{n=s+t+1}^{\infty} c_n q^n,$$

where $c_n \in R$ for $n \geq s+t+1$. Since R is an integral domain, it follows that $a_s b_t \neq 0$, and so $v(fg) = s + t = v(f) + v(g)$.

To prove (iii) and (iv), we observe that if $s = t$, then

$$f(q) \pm g(q) = \sum_{n=s}^{\infty} (a_n \pm b_n) q^n$$

and $v(f \pm g) \geq s = \min(v(f), v(g))$. If $s < t$, then

$$f(q) \pm g(q) = \sum_{n=s}^{t-1} a_n q^n + \sum_{n=t}^{\infty} (a_n \pm b_n) q^n$$

and so $v(f \pm g) = s = \min(v(f), v(g))$.

Finally, if $s = 0$, then $f(q) - f(0) = \sum_{n=r}^{\infty} a_n q^n$, where $v(f - f(0)) = r \geq 1$ and $a_r \neq 0$. Then

$$f(q^k) - f(0) = a_r q^{kr} + \sum_{n=r+1}^{\infty} a_n q^{kn},$$

and

$$v(f(q^k) - f(0) = kr \geq k.$$

If $s \geq 1$, then $f(0) = 0$ and

$$f(q^k) - f(0) = f(q^k) = a_s q^{ks} + \sum_{n=s+1}^{\infty} a_n q^{kn},$$

and

$$v(f(q^k) - f(0) = ks \geq k.$$

This proves (v). □

We introduce a topology on the ring of formal power series as follows. Let $\{f_k(q)\}_{k=1}^{\infty}$ be a sequence of formal power series in $R[[q]]$, where

$$f_k(q) = \sum_{n=0}^{\infty} a_{k,n} q^n.$$

Let

$$f(q) = \sum_{n=0}^{\infty} a_n q^n \in R[[q]].$$

The sequence $\{f_k(q)\}_{k=1}^{\infty}$ *converges* to $f(q)$, that is,

$$\lim_{k\to\infty} f_k(q) = f(q),$$

if for every $n \geq 0$ there exists an integer $k_0(n)$ such that $a_{k,n} = a_n$ for all $k \geq k_0(n)$. For example, the sequence $\{[k]_q\}_{k=1}^{\infty}$ of quantum integers converges to $(1-q)^{-1}$.

Similarly, if

$$f_{\mathbf{k}}(q) = \sum_{n=0}^{\infty} a_{\mathbf{k},n} q^n.$$

is a formal power series for all

$$\mathbf{k} = (k_1, \ldots, k_d) \in \mathbf{N}_0^d,$$

where $\mathbf{N}_0^d$ is the set of all d-tuples of nonnegative integers, then

$$\lim_{\mathbf{k}\to\infty} f_{\mathbf{k}}(q) = \lim_{k_1,\ldots,k_d\to\infty} f_{k_1,\ldots,k_d}(q) = f(q),$$

if for every $n \geq 0$ there exists an integer $k_0(n)$ such that $a_{\mathbf{k},n} = a_n$ for all $\mathbf{k} \in \mathbf{N}_0^d$ with $k_i \geq k_0(n)$ for all $i = 1, \ldots, d$.

We consider convergence in the case $d = 2$ in Theorem 9.

THEOREM 3. *Let $\{f_k(q)\}_{k=1}^{\infty}$ be a sequence of formal power series in $R[[q]]$. The sequence $\{f_k(q)\}_{k=1}^{\infty}$ converges if and only if*

$$\lim_{k,\ell\to\infty} v(f_k - f_\ell) = \infty.$$

Moreover, if $f(q) \in R[[q]]$, then

$$\lim_{k\to\infty} f_k(q) = f(q)$$

if and only if

$$\lim_{k\to\infty} v(f - f_k) = \infty.$$

PROOF. This follows immediately from the definitions of the valuation v and convergence in the ring of formal power series. □

THEOREM 4. *If*

$$\lim_{k\to\infty} f_k(q) = f(q) \qquad \text{and} \qquad \lim_{k\to\infty} g_k(q) = g(q),$$

then

$$\lim_{k\to\infty} (f_k(q) + g_k(q)) = f(q) + g(q) \tag{5}$$

and

$$\lim_{k\to\infty} f_k(q) g_k(q) = f(q) g(q). \tag{6}$$

If $f_k(q)$ is invertible for all $k \geq 1$, then $f(q)$ is invertible and

$$\lim_{k\to\infty} f_k(q)^{-1} = f(q)^{-1}. \tag{7}$$

PROOF. Statements (5) and (6) are straightforward verifications.

To prove (7), let $a_{k,0}$ and a_0 be the constant terms of the formal power series $f_k(q)$ and $f(q)$, respectively, for all $k \geq 1$. Since $f_k(q)$ is invertible, it follows that $a_{k,0}$ is a unit in R. Since $a_0 = a_{k,0}$ for all sufficiently large k, it follows that a_0 is a unit, and so $f(q)$ is invertible by Theorem 1. Applying Theorem 2 to the identity

$$f(q)^{-1} - f_k(q)^{-1} = (f_k(q) - f(q))f(q)^{-1}f_k(q)^{-1},$$

we obtain

$$v(f^{-1} - f_k^{-1}) = v(f_k - f) + v(f^{-1}) + v(f_k^{-1}) \geq v(f_k - f).$$

Since $f_k(q)$ converges to $f(q)$, it follows that $v(f_k - f)$ tends to infinity, hence $v(f^{-1} - f_k^{-1})$ also tends to infinity and $f_k(q)^{-1}$ converges to $f(q)^{-1}$. This completes the proof. □

2.2. Infinite series and infinite products. We define convergence of the *infinite series*

$$\sum_{k=0}^{\infty} f_k(q)$$

of formal power series as follows. The *m-th partial sum*

$$s_m(q) = \sum_{k=0}^{m-1} f_k(q)$$

is simply a finite sum of formal power series. If the sequence $\{s_m(q)\}_{m=0}^{\infty}$ converges to $f(q) \in R[[q]]$, then we write that the infinite series converges to $f(q)$, that is,

$$\sum_{k=0}^{\infty} f_k(q) = \lim_{m\to\infty} s_m(q) = f(q).$$

For example, if $f_k(q) = \sum_{n=k}^{\infty} q^n$ for $k = 0, 1, 2, \ldots$, then

$$\sum_{k=0}^{\infty} f_k(q) = \sum_{n=0}^{\infty}(n+1)q^n.$$

THEOREM 5. *Let $\{f_k(q)\}_{k=0}^{\infty}$ be a sequence of formal power series. The infinite series $\sum_{k=0}^{\infty} f_k(q)$ converges if and only if $\lim_{k\to\infty} v(f_k) = \infty$.*

PROOF. Let $f_k(q) = \sum_{n=0}^{\infty} a_{k,n}q^n$. If $\lim_{k\to\infty} v(f_k) = \infty$, then for every nonnegative integer n we have $a_{k,n} = 0$ for all sufficiently large k. The coefficient of q^n in the partial sum $s_m(q)$ is

$$\sum_{k=0}^{m} a_{k,n}. \tag{8}$$

Since this sum is constant for all sufficiently large m, it follows that the sequence $\{s_m(q)\}_{m=0}^{\infty}$ converges.

Conversely, if $\lim_{k\to\infty} v(f_k) \neq \infty$, then there exists a nonnegative integer n such that $a_{k,n} \neq 0$ for infinitely many integers k. It follows that the sum (8) is *not* constant for sufficiently large m, and so the sequence $\{s_m(q)\}_{m=0}^{\infty}$ does not converge. This completes the proof. □

Theorem 5 allows us to substitute one formal power series into another. Let $f(q)$ and $h(q)$ be formal power series with $v(h) \geq 1$. If

$$f(q) = \sum_{n=0}^{\infty} a_n q^n,$$

then we define the composite function

$$(f \circ h)(q) = f(h(q)) = \sum_{n=0}^{\infty} a_n h(q)^n. \tag{9}$$

Since

$$v(a_n h(q)^n) = v(a_n) + nv(h) \geq n,$$

it follows that the infinite series (9) converges. Moreover,

$$v(f \circ h) = v(f)v(h). \tag{10}$$

THEOREM 6. *Let $a_0 \in R$. If $\{f_k(q)\}_{k=1}^{\infty}$ is a sequence of formal power series such that $f_k(q)$ has constant term a_0 for all $k \geq 1$, and if $\{m_k\}_{k=1}^{\infty}$ is a sequence of nonnegative integers such that*

$$\lim_{k\to\infty} m_k = \infty,$$

then

$$\lim_{k\to\infty} f_k(q^{m_k}) = a_0. \tag{11}$$

In particular, for any formal power series $f(q)$,

$$\lim_{k\to\infty} f(q^{m_k}) = f(0). \tag{12}$$

PROOF. We have $v(f_k - a_0) \geq 1$ for all $k \geq 1$. The convergence of (11) follows from (10) and the observation that

$$v\left(f_k(q^{m^k}) - a_0\right) = v\left((f_k - a_0)(q^{m_k})\right) = v\left(f_k - a_0\right)m_k \geq m_k,$$

hence

$$\lim_{k\to\infty} v\left(f_k(q^{m^k}) - a_0\right) = \infty.$$

The convergence of (12) follows from Theorem 3. □

THEOREM 7. *Let $(k_i, \ell_i)_{i=0}^{\infty}$ be a sequence of ordered pairs of nonnegative integers such that every ordered pair (k, ℓ) occurs at most once in the sequence. If the infinite series*

$$f(q) = \sum_{k=0}^{\infty} f_k(q)$$

and

$$g(q) = \sum_{\ell=0}^{\infty} g_\ell(q)$$

converge in $R[[q]]$, then the infinite series

$$\sum_{i=0}^{\infty} f_{k_i}(q) g_{\ell_i}(q) \tag{13}$$

converges, and

$$f(q)g(q) = \sum_{i=0}^{\infty} f_{k_i}(q)g_{\ell_i}(q).$$

In particular,

$$f(q)g(q) = \sum_{n=0}^{\infty} \sum_{k+\ell=n} f_k(q)g_\ell(q) \tag{14}$$

PROOF. By Theorem 5,

$$\lim_{k\to\infty} v(f_k) = \lim_{\ell\to\infty} v(g_\ell) = \infty,$$

and so, for every N, there exists an integer $j_0 = j_0(N)$ such that $v(f_k) \geq N$ and $v(g_\ell) \geq N$ for all $k > j_0$ and $\ell > j_0$. There are only finitely many ordered pairs (k_i, ℓ_i) with both $k_i \leq j_0$ and $\ell_i \leq j_0$, and so there is an integer $i_0 = i_0(N)$ such that, for all $i > i_0$, we have $k_i > j_0$ or $\ell_i > j_0$, hence

$$v(f_{k_i}g_{\ell_i}) = v(f_{k_i}) + v(g_{\ell_i}) \geq N$$

and

$$\lim_{i\to\infty} v(f_{k_i}g_{\ell_i}) = \infty.$$

By Theorem 5, the series (13) converges. Moreover, for $m \geq i_0$,

$$\begin{aligned} s_m(q) &= \sum_{i=0}^{m} f_{k_i}(q)g_{\ell_i}(q) \\ &\equiv \sum_{i=0}^{i_0} f_{k_i}(q)g_{\ell_i}(q) \pmod{q^N} \\ &\equiv \sum_{k=0}^{j_0} f_k(q) \sum_{\ell=0}^{j_0} g_\ell(q) \pmod{q^N} \\ &\equiv f(q)g(q) \pmod{q^N}, \end{aligned}$$

and so

$$v(fg - s_m) \geq N$$

and

$$\lim_{m\to\infty} v(fg - s_m) = \infty.$$

By Theorem 3,

$$\sum_{i=0}^{\infty} f_{k_i}(q)g_{\ell_i}(q) = \lim_{m\to\infty} s_m(q) = f(q)g(q).$$

Identity (14) is a special case of this result. □

We define convergence of the *infinite product*

$$\prod_{k=1}^{\infty} f_k(q)$$

of formal power series as follows. The *m–th partial product*

$$p_m(q) = \prod_{k=1}^{m} f_k(q)$$

is simply a finite product of formal power series. If the sequence $\{p_m(q)\}_{m=1}^{\infty}$ converges to $f(q) \in R[[q]]$, then we write that the infinite product converges to $f(q)$, that is,

$$\prod_{k=1}^{\infty} f_k(q) = \lim_{m\to\infty} p_m(q) = f(q).$$

For example, from the unique representation of an integer as the sum of distinct powers of 2, we have

$$\prod_{k=1}^{m}(1+q^{2^{k-1}}) = (1+q)(1+q^2)(1+q^4)\cdots(1+q^{2^{m-1}}) = \sum_{n=0}^{2^m-1} q^n,$$

and so

$$\prod_{m=1}^{\infty}(1+q^{2^{m-1}}) = \sum_{n=0}^{\infty} q^n = (1-q)^{-1}.$$

THEOREM 8. *Let $\{f_k(q)\}_{k=1}^{\infty}$ be a sequence of formal power series such that $f_k(q)$ has constant term 1 for all $k \geq 1$. The infinite product $\prod_{k=1}^{\infty} f_k(q)$ converges if and only if $\lim_{k\to\infty} v(f_k(q)-1) = \infty$ if and only if the infinite series $\sum_{k=1}^{\infty}(f_k(q)-1)$ converges.*

PROOF. This follows immediately from the definition of convergence. □

2.3. Change of variables in formal power series. The next result allows us to substitute variables in identities of formal power series.

THEOREM 9. *Let $\{g_k(q)\}_{k=1}^{\infty}$ be a sequence of formal power series in $R[[q]]$ that converges to $g(q)$ and let $\{h_\ell(q)\}_{\ell=1}^{\infty}$ be a sequence of formal power series that converges to $h(q)$. If*

$$v(h_\ell) \geq 1 \qquad \text{for all } \ell \geq 1,$$

then

$$\lim_{k,\ell\to\infty} g_k(h_\ell(q)) = g(h(q)).$$

PROOF. For every nonnegative integer N and formal power series

$$f(q) = \sum_{n=0}^{\infty} a_n q^n,$$

let

$$f^{(N)}(q) = \sum_{n=0}^{N-1} a_n q^n.$$

Then

$$f(q) \equiv f^{(N)}(q) \pmod{q^N}.$$

By the remark following Theorem 5, since $v(h_\ell) \geq 1$ for all ℓ, it follows that $v(h) \geq 1$, and so the formal power series $g(h(q))$ and $g_k(h_\ell(q))$ are well-defined for all nonnegative integers k and ℓ. Since $\lim_{k\to\infty} g_k(q) = g(q)$, there exists an integer $k_0(N)$ such that

$$g_k^{(N)}(q) = g^{(N)}(q) \qquad \text{for all } k \geq k_0(N).$$

Similarly, since $\lim_{\ell\to\infty} h_\ell(q) = h(q)$, there exists an integer $\ell_0(N)$ such that

$$h_\ell^{(N)}(q) = h^{(N)}(q) \qquad \text{for all } \ell \geq \ell_0(N).$$

Therefore, for $k \geq k_0(N)$ and $\ell \geq \ell_0(N)$ we have

$$g_k\left(h_\ell(q)\right) \equiv g_k^{(N)}\left(h_\ell^{(N)}(q)\right) = g^{(N)}\left(h^{(N)}(q)\right) \equiv g(h(q)) \pmod{q^N}.$$

This completes the proof. □

THEOREM 10. *Let $\{g_k(q)\}_{k=1}^\infty$ be a sequence of formal power series in $R[[q]]$ that converges to $g(q) \in R[[q]]$, and let $h(q)$ be a formal power series with $v(h) \geq 1$. Then*

$$\lim_{k\to\infty} g_k(h(q)) = g(h(q)).$$

PROOF. This follows from Theorem 9 with $h_\ell(q) = h(q)$ for all $\ell \geq 1$. □

Theorem 10 is often applied in the following form.

THEOREM 11. *Let $\{f_k(q)\}_{k=0}^\infty$ and $\{g_k(q)\}_{k=1}^\infty$ be sequences of formal power series such that the infinite series $\sum_{k=1}^\infty f_k(q)$ converges and the infinite product $\prod_{k=1}^\infty g_k(q)$ converges. Let $h(q)$ be a formal power series with $v(h) \geq 1$. If*

$$\sum_{k=0}^\infty f_k(q) = \prod_{k=1}^\infty g_k(q),$$

then

$$\sum_{k=0}^\infty f_k(h(q)) = \prod_{k=1}^\infty g_k(h(q)).$$

PROOF. Let

$$\sum_{k=0}^\infty f_k(q) = \lim_{m\to\infty} \sum_{k=0}^{m-1} f_k(q) = F(q).$$

Theorem 10 implies that

$$\sum_{k=0}^\infty f_k(h(q)) = \lim_{m\to\infty} \sum_{k=0}^{m-1} f_k((h(q)) = F(h(q)).$$

Similarly,

$$\prod_{k=1}^\infty g_k(q) = \lim_{m\to\infty} \prod_{k=1}^{m} g_k(q) = F(q)$$

implies that

$$\prod_{k=1}^\infty g_k(h(q)) = \lim_{m\to\infty} \prod_{k=1}^{m} g_k(h(q)) = F(h(q)).$$

It follows that

$$\sum_{k=1}^\infty f_k(h(q)) = \prod_{k=1}^\infty g_k(h(q)).$$

This completes the proof. □

THEOREM 12. *Let $\{g_k(q)\}_{k=1}^\infty$ be a sequence of formal power series in $R[[q]]$, and let $h(q)$ be a formal power series with $v(h) \geq 1$. If*

$$\lim_{k\to\infty} g_k(h(q)) = g(h(q)),$$

then

$$\lim_{k\to\infty} g_k(q) = g(q).$$

PROOF. Let

$$g_k(q) = \sum_{n=0}^{\infty} b_{k,n} q^n$$

and

$$g(q) = \sum_{n=0}^{\infty} b_n q^n.$$

We must show that for every nonnegative integer n we have $b_{k,n} = b_n$ for all sufficiently large k. The proof will be by induction on n. Let

$$h(q) = \sum_{m=M}^{\infty} c_m q^m, \qquad \text{where } v(h) = M \geq 1 \text{ and } c_M \neq 0.$$

Then

$$h(q) \equiv c_M q^M \pmod{q^{M+1}}.$$

The congruences

$$g_k(h(q)) \equiv b_{k,0} \pmod{q}$$

and

$$g(h(q)) \equiv b_0 \pmod{q}$$

imply that $b_{k,0} = b_0$ for all sufficiently large k. Let $N \geq 1$. Suppose there exists an integer $k_0(N)$ such that, for $k \geq k_0(N)$ and $0 \leq n \leq N-1$, we have $b_{k,n} = b_n$. Then

$$\sum_{n=N}^{\infty} b_{k,n}(h(q))^n \equiv b_{k,N} \left(c_M q^M\right)^N \equiv b_{k,N} c_M^N q^{MN} \pmod{q^{MN+1}}$$

and

$$\sum_{n=N}^{\infty} b_n(h(q))^n \equiv b_N \left(c_M q^M\right)^N \equiv b_N c_M^N q^{MN} \pmod{q^{MN+1}}.$$

Since

$$\lim_{k\to\infty} \sum_{n=N}^{\infty} b_{k,n}(h(q))^n = \sum_{n=N}^{\infty} b_n(h(q))^n,$$

it follows that for suffficiently large k we have

$$b_{k,N} c_M^{MN} = b_N c_M^{MN},$$

and so

$$b_{k,N} = b_N.$$

This completes the proof. □

THEOREM 13. *Let R and S be rings, and let $\varphi : R \to S[[q]]$ be a ring homomorphism. Let v be the valuation in the formal power series ring $S[[q]]$. Let $h(q) \in S[[q]]$ with $v(h) \geq 1$. The map $\Phi : R[[q]] \to S[[q]]$ defined by*

$$\Phi\left(\sum_{n=0}^{\infty} a_n q^n\right) = \sum_{n=0}^{\infty} \varphi(a_n) h(q)^n \tag{15}$$

is a ring homomorphism.

PROOF. Since

$$v\left(\varphi(a_n)h(q)^n\right) = v\left(\varphi(a_n)\right) + nv\left(h\right) \geq nv(h) \geq n,$$

it follows from Theorem 5 that the infinite series on the right side of (15) converges in the formal power series ring $S[[q]]$. Let $f(q) = \sum_{n=0}^{\infty} a_n q^n$ and $g(q) = \sum_{n=0}^{\infty} b_n q^n$. Then

$$\begin{aligned}
\Phi(f(q)) + \Phi(g(q)) &= \sum_{n=0}^{\infty} \varphi(a_n)h(q)^n + \sum_{n=0}^{\infty} \varphi(b_n)h(q)^n \\
&= \sum_{n=0}^{\infty} \left(\varphi(a_n) + \varphi(b_n)\right) h(q)^n \\
&= \sum_{n=0}^{\infty} \varphi\left(a_n + b_n\right) h(q)^n \\
&= \Phi\left(\sum_{n=0}^{\infty} \left(a_n + b_n\right) q^n\right) \\
&= \Phi\left(f(q) + g(q)\right).
\end{aligned}$$

Similarly,

$$\begin{aligned}
\Phi(f(q))\Phi(g(q)) &= \left(\sum_{n=0}^{\infty} \varphi(a_n)h(q)^n\right)\left(\sum_{n=0}^{\infty} \varphi(b_n)h(q)^n\right) \\
&= \sum_{n=0}^{\infty} \left(\sum_{k+\ell=n} \varphi(a_k)\varphi(b_\ell)\right) h(q)^n \\
&= \sum_{n=0}^{\infty} \varphi\left(\sum_{k+\ell=n} a_k b_\ell\right) h(q)^n \\
&= \Phi\left(\sum_{n=0}^{\infty}\left(\sum_{k+\ell=n} a_k b_\ell\right) q^n\right) \\
&= \Phi\left(\left(\sum_{n=0}^{\infty} a_n q^n\right)\left(\sum_{n=0}^{\infty} b_n q^n\right)\right) \\
&= \Phi\left(f(q)g(q)\right).
\end{aligned}$$

This completes the proof. □

3. Limits of solutions of the functional equation for quantum multiplication

The main result (Theorem 18) in this section states that if $\mathcal{F} = \{f_n(q)\}_{n=1}^{\infty}$ is a solution of the functional equation $f_{mn}(q) = f_m(q)f_n(q^m)$ with $\mathrm{supp}(\mathcal{F}) = S(P)$ for a nonempty set P of prime numbers, and if $f_p(0) = 1$ for all $p \in P$, then there exists a formal power series $F(q)$ such that

$$F(q) = \lim_{\substack{n\to\infty \\ n\in S(P)}} f_n(q).$$

For example, the sequence $\mathcal{F} = \{[n]_q\}_{n=1}^{\infty}$ has support $\mathrm{supp}(\mathcal{F}) = \mathbf{N} = S(P)$, where P is the set of all primes, and $\lim_{n\to\infty}[n]_q = (1-q)^{-1}$.

THEOREM 14. *Let $\mathcal{F} = \{f_n(q)\}_{n=1}^{\infty}$ be a sequence of polynomials that satisfies the functional equation*

$$f_{mn}(q) = f_m(q) f_n(q^m).$$

Let $\{a_i\}_{i=1}^{\infty}$ be a sequence of positive integers such that $a_i \geq 2$ for infinitely many i. Let $n_0 = 1$ and

$$n_k = a_1 a_2 \cdots a_k$$

for $k \geq 1$. If $f_{a_i}(0) = 1$ for all $i \geq 1$, then the infinite product

$$\prod_{i=1}^{\infty} f_{a_i}(q^{n_{i-1}})$$

converges, and

$$\lim_{k \to \infty} f_{n_k}(q) = \prod_{i=1}^{\infty} f_{a_i}(q^{n_{i-1}}). \tag{16}$$

In particular, if $a_i = a \geq 2$ for all i, then there exists a formal power series $F_{a^\infty}(q)$ such that

$$F_{a^\infty}(q) = \lim_{k \to \infty} f_{a^k}(q). \tag{17}$$

PROOF. We shall show by induction that

$$f_{n_k}(q) = \prod_{i=1}^{k} f_{a_i}\left(q^{n_{i-1}}\right) \tag{18}$$

for all $k \geq 1$. The functional equation implies that

$$\begin{aligned} f_{n_1}(q) &= f_{a_1}(q) = f_{a_1}(q^{n_0}) \\ f_{n_2}(q) &= f_{a_1 a_2}(q) = f_{a_1}(q) f_{a_2}(q^{a_1}) = f_{a_1}(q^{n_0}) f_{a_2}(q^{n_1}) \\ f_{n_3}(q) &= f_{a_1 a_2 a_3}(q) = f_{a_1 a_2}(q) f_{a_3}(q^{a_1 a_2}) \\ &= f_{a_1}(q^{n_0}) f_{a_2}(q^{n_1}) f_{a_3}(q^{n_2}), \end{aligned}$$

and so formula (18) holds for $k = 1, 2, 3$. If (18) holds for k, then

$$\begin{aligned} f_{n_{k+1}}(q) &= f_{n_k a_{k+1}}(q) \\ &= f_{n_k}(q) f_{a_{k+1}}(q^{n_k}) \\ &= \left(\prod_{i=1}^{k} f_{a_i}\left(q^{n_{i-1}}\right) \right) f_{u_{k+1}}(q^{n_k}) \\ &= \prod_{i=1}^{k+1} f_{a_i}\left(q^{n_{i-1}}\right). \end{aligned}$$

This completes the induction.

Since $f_{a_i}(0) = 1$ for all $i \geq 1$, it follows that $f_{n_k}(0) = 1$ and $v(f_{n_k}(q) - 1) \geq 1$ for all $k \geq 1$. By Theorem 6,

$$v\left(f_{n_k}\left(q^{n_{k-1}}\right) - 1\right) \geq n_{k-1}.$$

Since $\lim_{k \to \infty} n_k = \infty$, Theorem 8 implies the convergence of the infinite product (16). This immediately implies (17). □

THEOREM 15. *Let* $\mathcal{F} = \{f_n(q)\}_{n=1}^{\infty}$ *be a sequence of polynomials that satisfies the functional equation*

$$f_{mn}(q) = f_m(q)f_n(q^m).$$

If $a \geq 2$, $b \geq 2$, *and* $f_a(0) = f_b(0) = 1$, *then*

$$F_{a^\infty}(q) = \lim_{k\to\infty} f_{a^k}(q) = \lim_{k\to\infty} f_{b^k}(q) = F_{b^\infty}(q).$$

PROOF. The functional equation implies that $f_{a^k}(0) = f_{b^k}(0) = f_{(ab)^k}(0) = 1$ for all $k \geq 1$. By Theorem 6,

$$\lim_{k\to\infty} f_{b^k}(q^{a^k}) = 1.$$

It follows from Theorem 14 that

$$\begin{aligned} F_{(ab)^\infty}(q) &= \lim_{k\to\infty} f_{(ab)^k}(q) \\ &= \lim_{k\to\infty} f_{a^k}(q) f_{b^k}(q^{a^k}) \\ &= \lim_{k\to\infty} f_{a^k}(q) \lim_{k\to\infty} f_{b^k}(q^{a^k}) \\ &= \lim_{k\to\infty} f_{a^k}(q) \\ &= F_{a^\infty}(q). \end{aligned}$$

Similarly, $F_{(ab)^\infty}(q) = F_{b^\infty}(q)$ and so $F_{a^\infty}(q) = F_{b^\infty}(q)$. This completes the proof. □

THEOREM 16. *Let* $\mathcal{F} = \{f_n(q)\}_{n=1}^{\infty}$ *be a sequence of polynomials that satisfies the functional equation*

$$f_{mn}(q) = f_m(q)f_n(q^m).$$

If supp$(\mathcal{F}) = S(P)$ *for some nonempty finite set* P *of prime numbers, and if* $f_p(0) = 1$ *for all* $p \in P$, *then there exists a formal power series* $F(q)$ *such that*

$$F(q) = \lim_{\substack{n\to\infty \\ n\in S(P)}} f_n(q). \tag{19}$$

PROOF. Since $f_p(0) = 1$ for all $p \in P$, it follows that $f_m(0) = 1$, $v(f_m) = 0$, and $v(f_m - 1) \geq 1$ for all $m \in S(P)$.

By Theorem 15, there exists a formal power series $F(q)$ such that

$$F(q) = F_{p^\infty}(q) = \lim_{k\to\infty} f_{p^k}(q) \qquad \text{for all } p \in P.$$

Since the set P is finite, for every integer N there exists an integer

$$k_0 = k_0(N) \geq N$$

such that

$$v(F - f_{p^k}) \geq N \qquad \text{for all } p \in P \text{ and } k \geq k_0.$$

Let $n_0 = n_0(N) = \prod_{p\in P} p^{k_0 - 1}$. If $n > n_0$ and $n \in S(P)$, then $n = p^{k_0}m$ for some $p \in P$ and $m \in S(P)$. Moreover,

$$\begin{aligned} F(q) - f_n(q) &= F(q) - f_{p^{k_0}}(q) f_m(q^{p^{k_0}}) \\ &= F(q) - f_{p^{k_0}}(q) - f_{p^{k_0}}(q)\left(f_m(q^{p^{k_0}}) - 1\right). \end{aligned}$$

We have

$$v(F - f_{p^{k_0}}) \geq N$$

and

$$\begin{aligned} v\left(f_{p^{k_0}}(q)\left(f_m(q^{p^{k_0}})-1\right)\right) &= v\left(f_{p^{k_0}}\right)+v\left(f_m(q^{p^{k_0}})-1\right) \\ &= v\left(f_m(q^{p^{k_0}})-1\right) \\ &= v(f_m-1)p^{k_0} \\ &\geq p^{k_0} \\ &\geq N. \end{aligned}$$

It follows that

$$v(F-f_n) \geq \min\left(v(F-f_{p^{k_0}}), v\left(f_{p^{k_0}}(q)(f_m(q^{p^{k_0}})-1)\right)\right) \geq N$$

for all $n > n_0$. This completes the proof of (19). □

THEOREM 17. *Let $\mathcal{F} = \{f_n(q)\}_{n=1}^{\infty}$ be a sequence of polynomials that satisfies the functional equation*

$$f_{mn}(q) = f_m(q)f_n(q^m).$$

If supp$(\mathcal{F}) = S(P)$, where P is an infinite set of prime numbers, and if $f_p(0) = 1$ for all $p \in P$, then there exists a formal power series $F_P(q)$ such that

$$F_P(q) = \lim_{\substack{p\to\infty \\ p\in P}} f_p(q). \tag{20}$$

PROOF. If p_1 and p_2 are prime numbers in P such that $p_1 \geq N$ and $p_2 \geq N$, then

$$\begin{aligned} f_{p_1}(q) - f_{p_2}(q) &= f_{p_1}(q) - f_{p_1p_2}(q) + f_{p_1p_2}(q) - f_{p_2}(q) \\ &= f_{p_1}(q) - f_{p_1}(q)f_{p_2}(q^{p_1}) + f_{p_2}(q)f_{p_1}(q^{p_2}) - f_{p_2}(q) \\ &= f_{p_1}(q)\left(1 - f_{p_2}(q^{p_1})\right) + f_{p_2}(q)\left(f_{p_1}(q^{p_2}) - 1\right). \end{aligned}$$

Since

$$v(f_{p_1}) = v(f_{p_2}) = 0$$

and, by Theorem 2,

$$v\left(1 - f_{p_2}(q^{p_1})\right) \geq p_1 \geq N,$$

and

$$v\left(f_{p_1}(q^{p_2}) - 1\right) \geq p_2 \geq N,$$

it follows that

$$v(f_{p_1} - f_{p_2}) \geq N.$$

Therefore,

$$\lim_{p_1,p_2\to\infty} v(f_{p_1} - f_{p_2}) = \infty,$$

and Theorem 3 implies the existence of the limit (20). This completes the proof. □

THEOREM 18. *Let $\mathcal{F} = \{f_n(q)\}_{n=1}^{\infty}$ be a sequence of polynomials that satisfies the functional equation*

$$f_{mn}(q) = f_m(q)f_n(q^m).$$

Let supp$(\mathcal{F}) = S(P)$, where P is a nonempty set of prime numbers. If $f_p(0) = 1$ for all $p \in P$, then there exists a formal power series $F(q)$ such that

$$F(q) = \lim_{\substack{n\to\infty \\ n\in S(P)}} f_n(q). \tag{21}$$

PROOF. If P is finite, then the result is Theorem 16.

Suppose that P is an infinite set of prime numbers. Let N be a positive integer. By Theorem 17, there is a formal power series $F_P(q)$ such that

$$F_P(q) = \lim_{\substack{p\to\infty \\ p\in P}} f_p(q),$$

and so there exists a prime number

$$p_0 = p_0(N) \geq N$$

such that

$$v(F_P - f_p) \geq N \qquad \text{for all } p \in P \text{ with } p \geq p_0.$$

If $n = pm \in S(P)$, where $p \in P$ and $p \geq p_0$, then

$$\begin{aligned} F_P(q) - f_n(q) &= F_P(q) - f_p(q) f_m(q^p) \\ &= (F_P(q) - f_p(q)) + f_p(q)\left(1 - f_m(q^p)\right). \end{aligned}$$

Since $v\left(1 - f_m(q^p)\right) \geq p \geq p_0 \geq N$, it follows that

$$v(F_P - f_n) \geq N.$$

By Theorem 15, there is a formal power series $F(q)$ such that

$$F(q) = \lim_{k\to\infty} f_{p^k}(q)$$

for all $p \in P$. There exists an integer

$$k_0 = k_0(N) \geq N$$

such that

$$v\left(F - f_{p^k}\right) \geq N \qquad \text{for all } k \geq k_0 \text{ and } p \in P \text{ with } p < p_0.$$

If $n = p^k m \in S(P)$, where $k \geq k_0$ and $p \in P$ with $p < p_0$, then

$$\begin{aligned} F(q) - f_n(q) &= F(q) - f_{p^k}(q) f_m(q^{p^k}) \\ &= \left(F(q) - f_{p^k}(q)\right) + f_{p^k}(q)\left(1 - f_m(q^{p^k})\right). \end{aligned}$$

Since $v\left(1 - f_m(q^{p^k})\right) \geq p^k \geq k \geq k_0 \geq N$, it follows that

$$v(F - f_n) \geq N.$$

We shall show that the formal power series $F(q)$ and $F_P(q)$ are identical. Let p_1 and p_2 be prime numbers in P such that $p_1 < p_0 \leq p_2$, and let $k \geq k_0$. Let $n = p_1^k p_2$. Then

$$\begin{aligned} v(F_P - f_n) &\geq N, \\ v(F - f_n) &\geq N, \end{aligned}$$

and so

$$v(F - F_P) = v\left((F - f_n) + (f_n - F_P)\right) \geq N.$$

This holds for all N, and so

$$F(q) = F_P(q).$$

Let $n_0 = \prod_{\substack{p<p_0 \\ p\in P}} p^{k_0 - 1} = n_0(N)$ and let $n \in S(P)$ with $n \geq n_0$. If p divides n for some prime $p \in P$, $p \geq p_0$, then $v(F - f_n) = v(F_P - f_n) \geq N$. If every prime factor of n is less than p_0, then n is an integer of the form $n = \prod_{\substack{p<p_0 \\ p\in P}} p^{a_p}$, where

$a_p \in \mathbf{N}_0$ for all $p \in P$. Since $n \geq n_0$, it follows that $a_p \geq k_0$ for some p, and so $v(F - f_n) \geq N$. Therefore,

$$\lim_{\substack{n \to \infty \\ n \in S(P)}} v(F - f_n) = \infty.$$

This completes the proof. □

Let $\mathcal{F} = \{f_n(q)\}_{n=1}^{\infty}$ be a sequence of polynomials satisfying the functional equation (1). We assume that $\operatorname{supp}(\mathcal{F}) = S(P)$, where P is a nonempty set of prime numbers, and $f_p(0) = 1$ for all $p \in P$. By Theorem 18, there exists a formal power series $F(q)$ such that

$$F(q) = \lim_{\substack{n \to \infty \\ n \in S(P)}} f_n(q).$$

PROBLEM 4. *Which formal power series are limits of solutions of the functional equation (1)?*

PROBLEM 5. *To what extent is it possible to reconstruct the sequence $\mathcal{F}$ from the formal power series $F(q)$? If $F(q)$ is the limit of a solution of the functional equation (1), determine all solutions $\mathcal{F}$ of (1) whose limit is $F(q)$.*

Let $\mathcal{G} = \{g_n(q)\}_{n=1}^{\infty}$ be a solution of the functional equation (1) with support $S(P_G)$ and with limit

$$F(q) = \lim_{\substack{n \to \infty \\ n \in S(P_G)}} g_n(q).$$

We observe that if P is a nonempty subset of P_G, then we can construct another solution $\mathcal{F} = \{f_n(q)\}_{n=1}^{\infty}$ ofthe functional equation (1) by *restriction* of $\mathcal{G}$ to the subsemigroup $S(P)$ of $S(P_G)$ as follows:

$$f_n(q) = \begin{cases} g_n(q) & \text{if } n \in S(P), \\ 0 & \text{if } n \notin S(P). \end{cases}$$

Since

$$F(q) = \lim_{\substack{n \to \infty \\ n \in S(P)}} f_n(q),$$

it follows that $F(q)$ does not uniquely determine $\mathcal{F}$.

The suggests the following definition. A solution $\mathcal{F} = \{f_n(q)\}_{n=1}^{\infty}$ of the functional equation (1) will be called *maximal* if $\mathcal{F}$ cannot be constructed from another solution $\mathcal{G}$ of the functional equation by restriction to a subsemigroup of $\operatorname{supp}(\mathcal{G})$.

THEOREM 19. *Every solution of the functional equation (1) is contained in at least one maximal solution.*

PROOF. Let $\mathcal{F} = \{f_n(q)\}_{n=1}^{\infty}$ be a solution of (1) with $\operatorname{supp}(\mathcal{F}) = S(P)$. Let X be the set of all pairs $(\mathcal{G}, P_G)$ such that

(i) P_G is a set of prime numbers and $P \subseteq P_G$,
(ii) $\mathcal{G}$ is a solution of the functional equation (1) with support $S(P_G)$,
(iii) The restriction of $\mathcal{G}$ to $S(P)$ is $\mathcal{F}$.

The set X is partially ordered as follows: $(\mathcal{G}, P_G) \preceq (\mathcal{H}, P_H)$ if $P_G \subseteq P_H$ and the restriction of $\mathcal{H}$ to $S(P_G)$ is $\mathcal{G}$. Every chain in X has an upper bound, and so, by Zorn's Lemma, the set X contains a maximal element $(\mathcal{H}, P_H)$. Then $\mathcal{H}$ is a maximal solution of the functional equation (1) that restricts to $\mathcal{F}$. □

PROBLEM 6. *Describe the maximal solutions of the functional equation (1). For which sets P of prime numbers does there exist a maximal solution $\mathcal{F}$ of (1) with $supp(\mathcal{F}) = S(P)$?*

4. A functional equation for formal power series

We shall prove that the functional equation $f(q)F(q^m) = F(q)$ has a unique solution not only for polynomials $f(q)$ but also for formal power series $f(q)$ with constant term 1.

THEOREM 20. *Let m be a positive integer. For every formal power series $f(q)$ with constant term 1, there exists a unique formal power series*

$$F(q) = \prod_{i=0}^{\infty} f\left(q^{m^i}\right) = f(q)f(q^m)f\left(q^{m^2}\right)\cdots \tag{22}$$

that satisfies the functional equation

$$f(q)F(q^m) = F(q). \tag{23}$$

PROOF. The infinite product (22) converges in the ring of formal power series, since

$$v\left(f\left(q^{m^i}\right) - 1\right) = v(f(q)-1)m^i \geq m^i$$

and so

$$\lim_{i\to\infty} v\left(f\left(q^{m^i}\right) - 1\right) = \infty.$$

Then

$$\begin{aligned} f(q)F(q^m) &= f(q)\prod_{i=0}^{\infty} f\left((q^m)^{m^i}\right) \\ &= f(q)\prod_{i=0}^{\infty} f\left(q^{m^{i+1}}\right) \\ &= f(q)\prod_{i=1}^{\infty} f\left(q^{m^i}\right) \\ &= \prod_{i=0}^{\infty} f\left(q^{m^i}\right) \\ &= F(q). \end{aligned}$$

This proves that the formal power series (22) is a solution of (23).

Conversely, let $F(q)$ be a formal power series that is a solution of (23). We shall prove that

$$F(q) = \prod_{i=0}^{k-1} f\left(q^{m^i}\right) F\left(q^{m^k}\right) \tag{24}$$

for every positive integer k. From the functional equation (23), we obtain

$$F(q) = f(q)F(q^m) = f(q)\left(f(q^m)F((q^m)^m)\right) = f(q)f(q^m)F(q^{m^2}).$$

This proves (24) for $k = 1$ and 2. If (24) holds for some integer $k \geq 2$, then

$$\begin{aligned} F(q) &= f(q)F(q^m) \\ &= f(q)\prod_{i=0}^{k-1} f\left(q^{m^{i+1}}\right) F\left(q^{m^{k+1}}\right) \\ &= \prod_{i=0}^{k} f\left(q^{m^i}\right) F\left(q^{m^{k+1}}\right), \end{aligned}$$

and so (24) holds for $k+1$. This completes the induction. Since $F(0) = 1$, it follows from Theorem 6 that

$$\begin{aligned} F(q) &= \lim_{k\to\infty} \prod_{i=0}^{k-1} f\left(q^{m^i}\right) \lim_{k\to\infty} F\left(q^{m^k}\right) \\ &= \lim_{k\to\infty} \prod_{i=0}^{k-1} f\left(q^{m^i}\right) \\ &= \prod_{i=0}^{\infty} f\left(q^{m^i}\right). \end{aligned}$$

This completes the proof. □

If m is a positive integer and $f(q)$ a polynomial of degree less than m with constant term $f(0) = 1$, then we can explicitly construct the coefficients of the unique formal power series that satisfies the equation $f(q)F(q^m) = F(q)$. We need the following notation. Let m be an integer, $m \geq 2$. Every integer k has a unique m–adic representation

$$k = \sum_{t=0}^{\infty} e_{t,k} m^t,$$

where

$$e_{t,k} \in \{0, 1, 2, \ldots, m-1\}$$

for all t, and $e_{t,k} = 0$ for all but finitely many t. For $i = 1, 2, \ldots, m-1$, let

$$d_i(k) = \sum_{e_{t,k}=i} 1.$$

Then $d_i(k)$ is the arithmetic function that counts the number of digits equal to i in the m–adic representation of k. For example, if $m = 3$, then $71 = 2 \cdot 3^0 + 2 \cdot 3^1 + 1 \cdot 3^2 + 2 \cdot 3^3$, and so $d_1(71) = 1$ and $d_2(71) = 3$.

We adopt the convention that $0^0 = 1$.

THEOREM 21. *Let $f(q) = 1 + \sum_{i=1}^{m-1} a_i q^i$ be a polynomial of degree at most $m-1$. Let $F(q) = 1 + \sum_{k=1}^{\infty} b_k q^k$ be the unique formal power series that satisfies the functional equation*

(25) $$f(q)F(q^m) = F(q).$$

Then the coefficients of $F(q)$ are given by the formula

(26) $$b_k = \prod_{i=1}^{m-1} a_i^{d_i(k)} \qquad \text{for all } k = 1, 2, 3, \ldots.$$

PROOF. Let $a_0 = b_0 = 1$. From the functional equation (25) we have

$$\left(\sum_{i=0}^{m-1} a_i q^i\right)\left(\sum_{j=0}^{\infty} b_j q^{mj}\right) = \sum_{i=0}^{m-1}\sum_{j=0}^{\infty} a_i b_j q^{i+mj} = \sum_{k=0}^{\infty} b_k q^k,$$

if and only if

$$a_i b_j = b_{i+mj} \tag{27}$$

for all $i = 0, 1, \ldots, m-1$ and $j \in \mathbf{N}_0$.

We shall prove that the the unique solution of (25) is the formal power series $\sum_{k=1}^{\infty} b_k q^k$ with coefficients (26).

For $j = 0$, we have

$$b_k = a_k = \prod_{i=0}^{m-1} a_i^{d_i(k)}$$

for $k = 0, 1, \ldots, m-1$. This proves (26) for $0 \leq k < m$.

Let $r \geq 1$, and assume that identity (26) holds for all nonnegative integers $k < m^r$. Let k be an integer in the interval $m^r \leq k < m^{r+1}$. We can write

$$k = i + mj, \qquad \text{where } 0 \leq i \leq m-1 \text{ and } m^{r-1} \leq j < m^r.$$

If

$$j = \sum_{t=0}^{\infty} e_{t,j} m^t,$$

then

$$mj = \sum_{t=0}^{\infty} e_{t,j} m^{t+1} = \sum_{t=1}^{\infty} e_{t-1,j} m^t, = \sum_{t=1}^{\infty} e_{t,mj} m^t,$$

and

$$e_{t,mj} = e_{t-1,j} \qquad \text{for all } t \in \mathbf{N}.$$

Then

$$k = i + mj = i + \sum_{t=1}^{\infty} e_{t,mj} m^t,$$

and so

$$d_t(k) = \begin{cases} d_t(j) & \text{for } t \neq i \\ d_t(j) + 1 & \text{for } t = i. \end{cases}$$

Therefore,

$$\begin{aligned}
\prod_{t=0}^{m-1} a_t^{d_t(k)} &= a_i^{d_i(k)} \prod_{\substack{t=0 \\ t\neq i}}^{m-1} a_t^{d_t(k)} \\
&= a_i^{d_i(j)+1} \prod_{\substack{t=0 \\ t\neq i}}^{m-1} a_t^{d_t(j)} \\
&= a_i \prod_{t=0}^{m-1} a_t^{d_t(j)} \\
&= a_i b_j \\
&= b_{i+mj} \\
&= b_k
\end{aligned}$$

by (27). This completes the proof. □

Acknowledgement. I wish to thank Yang Wang for a helpful discussion about self-similar sets.

References

[1] M. B. Nathanson, *A functional equation arising from multiplication of quantum integers*, J. Number Theory, to appear.

[2] ———, *Additive number theory and the ring of quantum integers*, arXiv: math.NT/0204006.

Department of Mathematics, Lehman College (CUNY), Bronx, New York 10468
E-mail address: nathansn@alpha.lehman.cuny.edu

DIMACS Series in Discrete Mathematics
and Theoretical Computer Science
Volume **64**, 2004

Formal Languages and Number Theory

Jeffrey Shallit

ABSTRACT. I illustrate the connection between formal language theory and number theory. I give three examples (dealing with unary regular languages, unary context-free languages, and primitive words) where number theory may be applied to solve problems in formal language theory. I also give one example (transcendence in finite characteristic) where formal language theory may be applied to solve a problem in number theory.

1. Introduction.

An *alphabet* is a (usually) finite set of symbols Σ. A *word* or *string* — the terms are synonymous — is a (usually) finite sequence of symbols chosen from Σ. By Σ^* we mean the free monoid over Σ (the set of all finite strings with symbols chosen from Σ). A *language* is a (finite or infinite) subset of Σ^*. For example, if x^R denotes the reversal of the string x, then the language PAL of palindromes over $\{a, b\}$ can be defined as follows:

$$\begin{aligned} \texttt{PAL} &= \{x \in \{a,b\}^* \ : \ x = x^R\} \\ &= \{\epsilon, a, b, aa, bb, aaa, aba, bab, bbb, \ldots\}. \end{aligned}$$

Note that ϵ denotes the empty string.

The basic operations on languages include union, intersection, complementation, concatenation (defined by $L_1 L_2 := \{xy \ : x \in L_1, \ y \in L_2\}$) and Kleene closure (defined by $L^* := \bigcup_{i \geq 0} L^i$.)

Formal language theory is the study of the properties of languages. By contrast, number theory is the study of the properties of integers. Nevertheless, these two areas have many interesting intersections. For a survey, see a previous paper of mine [**23**] and the new book [**7**].

In this paper, I illustrate the connections between formal language theory and number theory by discussing four examples. In the first three examples, given in sections 2, 3, and 4, I show how number theory may be used to solve problems in formal language theory. These examples are surely in the spirit of the theme of the conference, entitled "Unusual Applications of Number Theory". In the last

---

2000 *Mathematics Subject Classification.* Primary 11B85; Secondary 68Q42, 68Q45, 11T99.

Key words and phrases. Finite automata; context-free grammar; transcendence in finite characteristic.

Research supported in part by a grant from NSERC.

example, given in section 5, I show how formal language theory may be applied to solve a problem in number theory. (This example would be suitable for a "co-conference".) I do not make any attempt to be comprehensive or provide all details. Further, while the sections 2 and 3 discuss original results, sections 4 and 5 are based on results of Jean-Paul Allouche.

2. State complexity of the intersection of unary languages

A *deterministic finite automaton* (DFA) is a simple model of a computer. It consists of a finite nonzero number of states. One state, called the *start state*, is drawn with a single arrow entering. Other states, called *accepting* or *final* states, are drawn with two concentric circles. A string is said to be *accepted* if it is the label of a path beginning with the start state and ending at some final state. The *language accepted by a finite automaton* M, written $L(M)$, is the set of all strings accepted by M. For example, in Figure 1 below, the given automaton accepts the base-2 representations of the prime numbers ≤ 11.

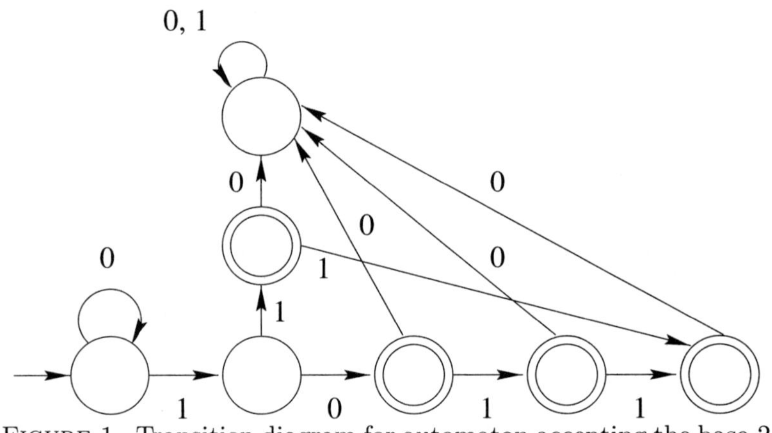

FIGURE 1. Transition diagram for automaton accepting the base-2 representations of the primes $p \leq 11$

More formally, a DFA is a quintuple: $M = (Q, \Sigma, \delta, q_0, F)$ where

- Q is a finite set of states;
- the *size* of M is $|M| := |Q|$, the number of states;
- Σ is the input alphabet;
- $q_0 \in Q$ is the start state;
- $F \subseteq Q$ is the set of final states;
- $\delta : Q \times \Sigma \to Q$ is the *transition function*, which is extended to $\delta : Q \times \Sigma^* \to Q$ in the obvious way.

The *language accepted by* M is denoted by $L(M)$ and is given by $\{w \in \Sigma^* : \delta(q_0, w) \in F\}$. A language L is said to be *regular* if it is accepted by some DFA M.

The *state complexity* of a regular language L, $\mathrm{sc}(L)$, is the minimum number of states required by any DFA which accepts it; see, for example, [**28**].

The state complexity problem is the following: given regular languages L, L' with state complexity n, n' respectively, what are good bounds on the state complexity of $L \cap L'$, $L \cup L'$, LL', L^*, etc.?

For the state complexity of intersection, we have the following upper bound, which is well-known.

PROPOSITION 1. *We have*

$$\mathrm{sc}(L \cap L') \leq \mathrm{sc}(L)\mathrm{sc}(L').$$

PROOF. Let L be accepted by the DFA $(Q, \Sigma, \delta, q_0, F)$ and L' be accepted by the DFA $(Q', \Sigma, \delta', q_0', F')$. Then $L \cap L'$ can be accepted by a DFA $(Q'', \Sigma, \delta'', q_0'', F'')$ where

- $Q'' := Q \times Q'$;
- $q_0'' := [q_0, q_0']$;
- $F'' := F \times F'$; and
- $\delta''([p, q], a) = [\delta(p, a), \delta(q, a)]$.

□

As Yu and Zhuang observed [**27**], the upper bound of $\mathrm{sc}(L)\mathrm{sc}(L')$ can be achieved if L, L' are over an alphabet of size at least 2:

PROPOSITION 2. *Define*

$$L := \{x \in \{a, b\}^* \ : \ |x|_a \equiv 0 \pmod{n}\};$$
$$L' := \{y \in \{a, b\}^* \ : \ |y|_b \equiv 0 \pmod{n'}\}.$$

Then

$$\mathrm{sc}(L \cap L') = nn'.$$

But what if L, L' are unary, that is, defined over an alphabet of one symbol? Clearly if $\gcd(n, n') = 1$ then the bound nn' can again be achieved, by taking $L = (a^n)^*$ and $L' = (a^{n'})^*$. But what if $\gcd(n, n') > 1$? This problem was stated as an unsolved problem by Yu [**26**].

To obtain a lower bound on the state complexity of intersection of unary regular languages, we examine the topology of unary DFA's. A connected unary DFA has the property that its transition diagram consists of

- a *tail* of $t \geq 0$ states and
- a *cycle* of $c \geq 1$ states.

It is then not hard to prove the following:

THEOREM 3. *Let M, M' be unary DFA's with tails of size t, t' and cycles of size c, c', respectively. If L, L' are the corresponding languages, we have*

$$\mathrm{sc}(L \cap L') \leq \max(t, t') + \mathrm{lcm}(c, c'). \tag{1}$$

Furthermore, for all $t, t' \geq 0$ and $c, c' \geq 1$ there exist unary languages for which the bound (1) is achieved.

(This theorem was obtained independently by the author [**24**] and G. Pighizzini [**20**]; also see [**22**].)

Thus, to estimate the worst-case behavior for the state complexity of intersection of unary languages with n and n' states, respectively, we must estimate the function

$$F(n, n') = \max_{\substack{1 \leq c \leq n \\ 1 \leq c' \leq n'}} \left(\max(n - c, n' - c') + \mathrm{lcm}(c, c') \right).$$

This in turn suggests studying the somewhat simpler and more natural function

$$G(n, n') = \max_{\substack{1 \le c \le n \\ 1 \le c' \le n'}} \operatorname{lcm}(c, c').$$

Although the asymptotic behavior of F and G is still not known precisely, there is a relation to Jacobsthal's function $g(n)$, which is defined to be the least integer r such that every set of r consecutive integers contains at least one integer relatively prime to n [**14**].

Iwaniec [**13**] proved using the linear sieve that $g(n) = O((\log n)^2)$. The following lower bound for our problem then follows:

THEOREM 4. *If $n \le n'$, we have $F(n, n') \ge G(n, n') \ge nn' - c_1(\log n)^2 n$ for some constant c_1.*

Note: results on the average state complexity of operations on unary DFA's were obtained by Nicaud [**18**].

3. Grammatical complexity of unary context-free grammars

Context-free grammars are a method for generating languages. The modern mathematical formulation is due to Chomsky [**8**], although the basic idea goes back to Indian philologist Panini, c. 400 B.C.E.

A context-free grammar consists of a start symbol and *rewriting* or *production rules*, e.g.:

$$\begin{aligned} S &\rightarrow aSa \\ S &\rightarrow bSb \\ S &\rightarrow a \\ S &\rightarrow b \\ S &\rightarrow \epsilon \end{aligned}$$

which generates the palindromes over $\{a, b\}$. Often multiple productions are abbreviated using $|$, e.g.,

$$S \rightarrow aSa \mid bSb \mid a \mid b \mid \epsilon.$$

To derive a word, one starts with the start symbol S and then successively replaces an occurrence of a variable with a right-hand-side of a production, until a string of symbols (without variables) is obtained.

More formally, a context-free grammar (CFG) is a 4-tuple $G = (V, \Sigma, P, S)$ where

- V is a finite set of *variables*;
- Σ is a finite alphabet;
- P is a set of *production rules* of the form $A \rightarrow \gamma$, where $A \in V$ and $\gamma \in (V \cup \Sigma)^*$; and
- S is the *start symbol*.

We write $\alpha \Longrightarrow \beta$ if β can be obtained from α by the use of one production rule. We write $\Longrightarrow^*$ for the reflexive, transitive closure of $\Longrightarrow$. Then $L(G)$, the language generated by G is formally defined as

$$L(G) := \{x \in \Sigma^* \ : \ S \Longrightarrow^* x\}.$$

Context-free grammars generate a class of languages, the context-free languages, which are a strict superset of the class of regular languages.

A *leftmost derivation* in a grammar is a series of steps deriving a terminal string, in which the leftmost variable is replaced at each step. A grammar G is said to be *ambiguous* if there exists at least one word in $L(G)$ possessing at least two distinct leftmost derivations.

We can measure the size of a context-free grammar as the number of symbols needed to write down its description. Suppose a CFG G generates a regular language. How big can the corresponding DFA be, in terms of the size of G?

If the CFG is over an alphabet with at least 2 symbols, the answer is, there is no recursive bound. More precisely, Meyer and Fischer [**17**] proved that given any recursive function f, for arbitrarily large integers n there exists a CFG of size n describing a regular language L such that any DFA accepting L has at least $f(n)$ states.

But how about the unary case? As Ginsburg and Rice [**12**] proved, any unary CFG generates a regular language. Further, the author together with M.-w. Wang (and independently, G. Pighizzini [**21**]) have shown there exists a constant such that any unary CFG of size n describing a regular language can be accepted by a DFA with at most $O(2^{cn^2})$ states.

But is this bound achievable? We give an example based on number theory that achieves this bound. Consider the following productions:

$$
\begin{aligned}
A_0 &\rightarrow a \\
A_{i+1} &\rightarrow A_i A_i \quad (i \geq 0) \\
B_i &\rightarrow a A_i \\
C_0 &\rightarrow a \\
C_{i+1} &\rightarrow a \mid C_i C_i \quad (i \geq 0) \\
D_i &\rightarrow D_i B_i \mid C_i \quad (i \geq 0) \\
S_i &\rightarrow \epsilon \mid D_0 \mid D_1 \mid D_2 \mid \cdots \mid D_i \quad (i \geq 0)
\end{aligned}
$$

It follows that, for $i \geq 0$,

$$
\begin{aligned}
A_i &\Longrightarrow^* \{a^{2^i}\} \\
B_i &\Longrightarrow^* \{a^{2^i+1}\} \\
C_i &\Longrightarrow^* \{a, a^2, a^3, \ldots, a^{2^i}\} \\
D_i &\Longrightarrow^* \{a, a^2, a^3, \ldots, a^{2^i}\}\{a^{2^i+1}\}^* - \{a^j \;:\; j \not\equiv 0 \pmod{2^i+1}\}, \\
S_i &\Longrightarrow^* \{\epsilon\} \cup \{a^k \;:\; k \not\equiv 0 \pmod{\operatorname{lcm}(2^0+1, 2^1+1, \ldots, 2^i+1)}\}.
\end{aligned}
$$

Now let $G_n = (V_n, \{a\}, P_n, S_n)$, where

$$V_n = \{A_i, B_i, C_i, D_i, S_i \;:\; 0 \leq i \leq n\}$$

and P_n is the set of $O(n)$ productions given above involving these variables. It is clear that $L(G_n)$ is regular. The shortest string not generated by G_n is of length

$$\operatorname{lcm}(2^0+1, 2^1+1, \ldots, 2^n+1)$$

and so any DFA accepting $L(G_n)$ must have at least this many states.

It remains to estimate

$$\operatorname{lcm}(2^0+1, 2^1+1, \ldots, 2^n+1).$$

We use the following theorem of Bézivin [1989]:

THEOREM 5. *Let a, b be integers with $b \neq 0$ and $\gcd(a, b) = 1$. Let α, β be zeroes of the polynomial $X^2 - aX - b$. For $m \geq 2$ define*

$$u_m(n) = \frac{\alpha^{mn} - \beta^{mn}}{\alpha^n - \beta^n}.$$

Then

$$\lim_{n \to \infty} \frac{\log(u_m(1)u_m(2) \cdots u_m(n))}{\log \operatorname{lcm}(u_m(1), u_m(2), \ldots, u_m(n))} = \frac{(m-1)L(m)\pi^2}{6H(m)},$$

where

$$L(m) = \prod_{p \mid m} \left(1 - \frac{1}{p^2}\right)$$

and

$$H(m) = \sum_{\substack{d \mid m \\ d > 1}} \frac{\varphi(d)\varphi(m/d)d}{m}.$$

Now take $a = 3, b = -2, m = 2$ in Bézivin's theorem. Then $\alpha = 2$ and $\beta = 1$, and we obtain

$$\lim_{n \to \infty} \frac{\log((2^0 + 1)(2^1 + 1) \cdots (2^n + 1))}{\log \operatorname{lcm}(2^0 + 1, 2^1 + 1, \ldots, 2^n + 1))} = \frac{\pi^2}{8}.$$

On the other hand, it is easy to see that

$$\lim_{n \to \infty} \frac{(2^0 + 1)(2^1 + 1) \cdots (2^n + 1)}{2^0 \cdot 2^1 \cdots 2^n} = c_2,$$

where $c_2 \doteq 4.768$, so it follows that

$$\log((2^0 + 1)(2^1 + 1) \cdots (2^n + 1)) \sim \log c_2 + \frac{n(n+1)}{2} \log 2.$$

Putting this together with the Bézivin result, we get

$$\log \operatorname{lcm}(2^0 + 1, 2^1 + 1, \ldots, 2^n + 1) \sim \frac{4 \log 2}{\pi^2} n^2.$$

We remark that other constructions are possible which achieve the 2^{cn^2} bound. For example, instead of $\operatorname{lcm}(2^0+1, 2^1+1, \ldots, 2^n+1)$, we could instead use a result of Szymiczek [**25**] and consider $\operatorname{lcm}(2^1 - 1, 2^2 - 1, \ldots, 2^n - 1)$. Or we could use a result of Matiyasevich and Guy [**16**] and consider $\operatorname{lcm}(F_1, F_2, \ldots, F_n)$, where F_i is the ith Fibonacci number. For related results, see [**15, 1, 2**].

4. The primitive words problem

Let Σ be a finite alphabet with at least two letters. A word $w \in \Sigma^*$ is said to be *primitive* if it cannot be expressed in the form x^k with $k \geq 2$. For example, *abaab* is primitive, but *abaaba* $= (aba)^2$ is not. A major open problem in formal languages is the following: is the language P of primitive words over $\{a, b\}$ context-free? The answer is almost certainly no, but nobody currently knows how to prove this. Petersen [**19**] proved the following weaker result:

THEOREM 6. *P is not unambiguously context-free (i.e., if P is a CFL, then any grammar for it is ambiguous).*

Petersen used the Chomsky-Schützenberger theorem [**9**], which states that if L is a context-free language having an unambiguous grammar, and $a_n := |L \cap \Sigma^n|$, then $\sum_{n \geq 0} a_n X^n$ is a formal power series in $\mathbb{Z}[[X]]$ which is algebraic over $\mathbb{Q}(X)$.

Recently a remarkably simple proof of Petersen's result was found by Allouche [**5, 6**] using the theory of automatic sequences. In this section we explain Allouche's proof.

First, we give an example of Chomsky-Schützenberger theorem. Consider the unambiguous grammar

$$\begin{aligned} S &\rightarrow M \mid U \\ M &\rightarrow 0M1M \mid \epsilon \\ U &\rightarrow 0S \mid 0M1U \end{aligned}$$

which represents strings of "if-then-else" clauses. Then this grammar has the following commutative image:

$$\begin{aligned} S &= M + U \\ M &= x^2M^2 + 1 \\ U &= Sx + x^2MU \end{aligned}$$

This system of equations has the following power series solutions:

$$\begin{aligned} M &= 1 + x^2 + 2x^4 + 5x^6 + 14x^8 + 42x^10 + \cdots \\ U &= x + x^2 + 3x^3 + 4x^4 + 10x^5 + 15x^6 + 35x^7 + 56x^8 + \cdots \\ S &= 1 + x + 2x^2 + 3x^3 + 6x^4 + 10x^5 + 20x^6 + 35x^7 + \cdots \end{aligned}$$

By the Chomsky-Schützenberger theorem, each variable satisfies an algebraic equation over $\mathbb{Q}(x)$. For example, we have

$$x(2x-1)S^2 + (2x-1)S + 1 = 0$$

Now we digress a moment to discuss automata as computers of sequences.

We can generalize our notion of automaton to provide an output, not simply accept/reject. Formally, we define a *deterministic finite automaton with output* (DFAO) as a sextuple: $(Q, \Sigma, \delta, q_0, \Delta, \tau)$, where Δ is the finite *output alphabet* and $\tau : Q \rightarrow \Delta$ is the *output mapping*.

Next, we decide on a integer base $k \geq 2$ and represent n as a string of symbols over the alphabet $\Sigma = \{0, 1, 2, \ldots, k-1\}$. To compute f_n, given an automaton M, express n in base-k, say, $a_r a_{r-1} \cdots a_1 a_0$, and compute $f_n = \tau(\delta(q_0, a_r a_{r-1} \cdots a_1 a_0))$. Any sequence that can be computed in this way is said to be *k-automatic* [**11**].

The Thue-Morse sequence $(t_n)_{n \geq 0}$ is defined as follows: t_n is the parity of the number of 1's in the binary expansion of n. We have

$$(t_n)_{n \geq 0} = 0\,1\,1\,0\,1\,0\,0\,1 \cdots$$

Note that $t_0 = 0$; $t_{2n} = t_n$, and $t_{2n+1} = 1 - t_n$ for $n \geq 0$.

Axel Thue (c. 1906) studied this sequence because it is *cubefree*: it contains no subword of the form www, where w is a nonempty word. It is computed by the following DFAO:

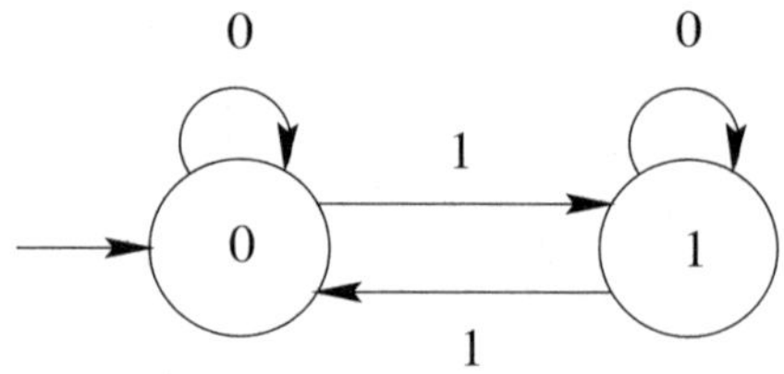

FIGURE 2. Automaton computing the Thue-Morse sequence

We observe that the notion of automatic sequence is *robust*: it does not change under small changes to the computational model. For example, the order in which the base-k digits are fed into the automaton in does not matter (provided it is fixed for all n); other representations also work (such as expansion in base-$(-k)$); automatic sequences are closed under many operations, such as shift, periodic deletion, q-block compression, and q-block substitution; and if a symbol in an automatic sequence occurs with well-defined frequency r, then r is rational [**11**].

The theorem of Christol [**10**] is the most important in the area:

THEOREM 7 (Christol, 1979). *Let $(u_n)_{n\geq 0}$ be a sequence over*

$$\Sigma = \{0, 1, \ldots, p-1\},$$

where p is a prime. Then the formal power series $U(X) = \sum_{n\geq 0} u_n X^n$ is algebraic over $GF(p)[X]$ if and only if $(u_n)_{n\geq 0}$ is p-automatic.

Let us consider an example. Let, as before, $(t_n)_{n\geq 0}$ denote the Thue-Morse sequence, i.e., t_n = sum of the bits in the binary expansion of n, mod 2. Then $t_{2n} \equiv t_n$ and $t_{2n+1} \equiv t_n + 1$. If we set $A(X) = \sum_{n\geq 0} t_n X^n$, then

$$\begin{aligned} A(X) &= \sum_{n\geq 0} t_{2n} X^{2n} + \sum_{n\geq 0} t_{2n+1} X^{2n+1} \\ &= \sum_{n\geq 0} t_n X^{2n} + X \sum_{n\geq 0} t_n X^{2n} + X \sum_{n\geq 0} X^{2n} \\ &= A(X^2) + XA(X^2) + X/(1-X^2) \\ &= A(X)^2(1+X) + X/(1+X)^2. \end{aligned}$$

Hence $(1+X)^3 A^2 + (1+X)^2 A + X = 0$.

We can now return to Allouche's proof of Petersen's result.

PROOF. Let $\psi_k(n)$ be the number of primitive words of length n over a k-letter alphabet. Then it is easy to see (using Möbius inversion) that

$$\psi_k(n) = \sum_{d\,|\,n} \mu(d) k^{n/d}.$$

If P_k were unambiguously context-free then by the Chomsky-Schützenberger theorem

$$R(X) = \sum_{n\geq 1} \psi_k(n) X^n$$

would be algebraic over $\mathbb{Q}(X)$. Then

$$\hat{R}(X) = \sum_{n\geq 1} \frac{\psi_k(n)}{k} X^n$$

would also be algebraic over $\mathbb{Q}(X)$.

Let p be a prime dividing k. Then it is not hard to see that

$$\hat{R}_p(X) = \sum_{n\geq 1} (\frac{\psi_k(n)}{k} \bmod p) X^n$$

would also be algebraic over $GF(p)(X)$. But

$$\frac{\psi_k(n)}{k} = \sum_{d \mid n} \mu(d) k^{n/d-1} = \mu(n) + \sum_{d \mid n d\neq n} \mu(d) k^{n/d-1} \equiv \mu(n) \pmod p.$$

It follows that

$$\hat{R}_p(X) = \sum_{n\geq 1} \mu(n) X^n$$

and so the sequence $(\mu(n) \bmod p)_{n\geq 0}$ must be p-automatic. But then $(\mu(n)^2 \bmod p)_{n\geq 0}$ would be p-automatic.

However, $\mu(n)^2 \equiv 1 \pmod p$ if and only if n is squarefree. By a classical theorem, the density of the squarefree numbers exists and is equal to $6/\pi^2$, an irrational number. But, as remarked earlier, the density of symbols in automatic sequences (if it exists) must be rational, a contradiction. It follows that $R(X)$ is not algebraic over $\mathbb{Q}(X)$ and so P_k is not unambiguously context-free. This completes the proof. □

5. Transcendence in Finite Characteristic

In this section, we turn the tables and illustrate an application of formal language theory to number theory.

Define for $n \geq 1$

$$\zeta_q(n) = \sum_{\substack{P \text{ monic} \\ P\in GF(q)[X]}} \frac{1}{P^n}$$

Thus, for example,

$$\begin{aligned}\zeta_2(1) &= \frac{1}{1} + \frac{1}{X} + \frac{1}{X+1} + \frac{1}{X^2} + \frac{1}{X^2+1} + \frac{1}{X^2+X} + \cdots \\ &= 1 + X^{-2} + X^{-3} + X^{-4} + X^{-5} + X^{-9} + X^{-10} + \cdots \\ &\in GF(2)[[X^{-1}]].\end{aligned}$$

This function ζ_q, now called the Carlitz zeta function, has many properties similar to those of the Riemann zeta function. For example, it admits the following Euler product:

$$\zeta_q(n) = \prod_{\substack{P \text{ irreducible} \\ P\in GF(q)[X]}} \frac{1}{1-\frac{1}{P^n}}.$$

Carlitz also showed that if $q-1 \mid n$, then $\zeta_q(n) = \pi_q^n \cdot r$ where r is a rational function and

$$\pi_q := \prod_{k \geq 1} \left(1 - \frac{X^{q^k} - X}{X^{q^{k+1}} - X} \right).$$

Wade proved the following theorem:

THEOREM 8. *π_q is transcendental.*

Here is another proof of Wade's result, due to Allouche [**4**], using automatic sequences and Christol's theorem (also see [**3**].

PROOF. Taking the logarithmic derivative, we get

$$\begin{aligned}\frac{\pi_q'}{\pi_q} &= \sum_{k \geq 1} \left(\frac{1}{1 - \frac{X^{q^k} - X}{X^{q^{k+1}} - X}} \right) \left(\frac{(X^{q^{k+1}} - X) - (X^{q^k} - X)}{(X^{q^{k+1}} - X)^2} \right) \\ &= \sum_{k \geq 1} \frac{1}{X^{q^{k+1}} - X} \\ &= \left(\sum_{k \geq 1} \frac{1}{X^{q^k} - X} \right) - \frac{1}{X^q - X}.\end{aligned}$$

Now suppose that π_q is algebraic over $GF(q)(X)$. Then so is the formal derivative π_q'. Hence so is π_q'/π_q. But then so is

$$\sum_{k \geq 1} \frac{1}{X^{q^k} - X} = \sum_{k \geq 1} \frac{1}{[k]},$$

the so-called "bracket series" introduced by Wade, who defined $[k] := X^{q^k} - X$.

Thus to prove π_q transcendental, it suffices to show that

$$\sum_{k \geq 1} \frac{1}{X^{q^k} - X}$$

is transcendental. We now have

$$\begin{aligned}\sum_{k\geq 1}\frac{1}{X^{q^k}-X} &= \sum_{k\geq 1}\frac{1}{X^{q^k}\left(1-\left(\frac{1}{X}\right)^{q^k-1}\right)} \\ &= \sum_{k\geq 1}\frac{1}{X^{q^k}}\sum_{n\geq 0}\left(\frac{1}{X}\right)^{n(q^k-1)} \\ &= \frac{1}{X}\sum_{k\geq 1}\frac{1}{X^{q^k-1}}\sum_{n\geq 0}\left(\frac{1}{X}\right)^{n(q^k-1)} \\ &= \frac{1}{X}\sum_{\substack{k\geq 1\\ n\geq 0}}\left(\frac{1}{X}\right)^{(n+1)(q^k-1)} \\ &= \frac{1}{X}\sum_{\substack{k\geq 1\\ n\geq 1}}\left(\frac{1}{X}\right)^{n(q^k-1)}.\end{aligned}$$

Hence

$$\begin{aligned}\sum_{k\geq 1}\frac{1}{X^{q^k}-X} &= \frac{1}{X}\sum_{m\geq 1}\left(\frac{1}{X}\right)^m \sum_{\substack{k,n\geq 1\\ n(q^k-1)=m}} 1 \\ &= \frac{1}{X}\sum_{k\geq 1}\left(\frac{1}{X}\right)^m \sum_{\substack{k\geq 1\\ q^k-1\,|\,m}} 1 \\ &= \frac{1}{X}\sum_{m\geq 1}\left(\frac{1}{X}\right)^m c(m),\end{aligned}$$

where

$$c(m) := \sum_{\substack{k\geq 1\\ q^k-1\,|\,m}} 1.$$

Now, by Christol's theorem, in order to show that $\sum_{k\geq 1}\frac{1}{X^{q^k}-X}$ is transcendental over $GF(q)(X)$, it suffices to prove that $(c(m) \bmod p)_{m\geq 1}$ is not q-automatic, where $q = p^e$ for some e.

If the sequence $(c(m) \bmod p)_{m\geq 1}$ were q-automatic, then the subsequence $(c(q^n - 1) \bmod p)_{n\geq 0}$ would be ultimately periodic. But

$$c(q^n-1) = \sum_{\substack{k\geq 1\\ q^k-1\,|\,q^n-1}} 1 = \sum_{\substack{k\geq 1\\ k\,|\,n}} 1 = d(n),$$

where $d(n)$ is the number of positive integral divisors of n.

It now suffices to show that $(d(n) \bmod p)_{n\geq 1}$ is not ultimately periodic. This can be done by a simple argument using Dirichlet's theorem. This contradiction completes the proof. □

6. Acknowledgments

I thank Gerry Myerson for pointing out the paper of Szymiczek [**25**], and Jean-Eric Pin for pointing out the work of Nicaud [**18**].

References

[1] S. Akiyama. Lehmer numbers and an asymptotic formula for π. *J. Number Theory* **36** (1990), 328–331.

[2] S. Akiyama. A new type of inclusion exclusion principle for sequences and asymptotic formulas for $\zeta(k)$. *J. Number Theory* **45** (1993), 200–214.

[3] J.-P. Allouche. Sur la transcendance de la série formelle π. *Séminaire de Théorie des Nombres de Bordeaux* **2** (1990), 103–117.

[4] J.-P. Allouche. Finite automata and arithmetic. *Sém. Lotharingien de Combinatoire* **B30c** (1993), 1–23.

[5] J.-P. Allouche. Note on the transcendence of a generating function. In A. Laurincikas, E. Manstavicius, and V. Stakenas, editors, *Proceedings of the 2nd International Conference in Honour of J. Kubilius (Palanaga, Lithuania, September 23–27, 1996)*, Vol. 4 of *New Trends in Probability and Statistics*, pp. 461–465. VSP, Utrecht, 1997.

[6] J.-P. Allouche. Transcendence of formal power series with rational coefficients. *Theoret. Comput. Sci.* **218** (1999), 143–160.

[7] J.-P. Allouche and J. O. Shallit. *Automatic Sequences: Theory, Applications, Generalizations.* Cambridge University Press, 2003.

[8] N. Chomsky. Three models for the description of language. *IRE Trans. Info. Theory* **2** (1956), 113–124.

[9] N. Chomsky and M. P. Schützenberger. The algebraic theory of context-free languages. In P. Braffort and D. Hirschberg, editors, *Computer Programming and Formal Systems*, pp. 118–161. North Holland, Amsterdam, 1963.

[10] G. Christol. Ensembles presque périodiques k-reconnaissables. *Theoret. Comput. Sci.* **9** (1979), 141–145.

[11] A. Cobham. Uniform tag sequences. *Math. Systems Theory* **6** (1972), 164–192.

[12] S. Ginsburg and H. G. Rice. Two families of languages related to ALGOL. *J. Assoc. Comput. Mach.* **9** (1962), 350–371.

[13] H. Iwaniec. On the problem of Jacobsthal. *Dem. Math.* **11** (1978), 225–231.

[14] E. Jacobsthal. Über Sequenzen ganzer Zahlen. von denen keine zu n teilerfremd ist. I-III. *Norske Vid. Selsk. Forh. Trondheim* **33** (1960), 117–139.

[15] P. Kiss and F. Mátyás. An asymptotic formula for π. *J. Number Theory* **31** (1989), 255–259.

[16] Yu. V. Matiyasevich and R. K. Guy. A new formula for π. *Amer. Math. Monthly* **93** (1986), 631–635.

[17] A. R. Meyer and M. J. Fischer. Economy of description by automata, grammars, and formal systems. In *Proc. 12th Annual Symposium on Switching and Automata Theory*, pp. 188–191, 1971.

[18] C. Nicaud. Average state complexity of operations on unary automata. In *Proc. 24th Symposium, Mathematical Foundations of Computer Science 1999*, Vol. 1672 of *Lecture Notes in Computer Science*, pp. 231–240. Springer-Verlag, 1999.

[19] H. Petersen. On the language of primitive words. *Theoret. Comput. Sci.* **161** (1996), 141–156.

[20] G. Pighizzini. Unary language concatenation and its state complexity, In S. Yu and A. Păun, eds., *Proc. 5th International Conference on Implementation and Application of Automata (CIAA 2000)*, Lecture Notes in Computer Science Vol. 2088, Springer-Verlag, 2001, pp. 252–262.

[21] G. Pighizzini. In M. Nielsen and B. Rovan, editors, *Proc. 25th International Symposium on Mathematical Foundations of Computer Science (MFCS 2000)*, Vol. 1893 of *Lecture Notes in Computer Science*, pp. 599–608. Springer-Verlag, 2000.

[22] G. Pighizzini and J. Shallit. Unary language operations, state complexity, and Jacobsthal's function. *Int'l. J. Found. Comput. Sci.* **13** (2002), 145–149.

[23] J. Shallit. Number theory and formal languages. In D. A. Hejhal, J. Friedman, M. C. Gutzwiller, and A. M. Odlyzko, editors, *Emerging Applications of Number Theory*, Vol. 109 of *IMA Volumes in Mathematics and Its Applications*, pp. 547–570. Springer-Verlag, 1999.

[24] J. Shallit. State complexity and Jacobsthal's function. In S. Yu and A. Păun, eds., *Proc. 5th International Conference on Implementation and Application of Automata (CIAA 2000)*, Lecture Notes in Computer Science Vol. 2088, Springer-Verlag, 2001, pp. 272–278.

[25] K. Szymiczek. On the distribution of prime factors of Mersenne numbers. *Prace Math.* **13** (1969), 33–49.

[26] S. Yu. State complexity of regular languages. In *International Workshop on Descriptional Complexity of Automata, Grammars and Related Structures, Preproceedings*, pp. 77–88. Department of Computer Science, Otto-von-Guericke University of Magdeburg, July 1999.

[27] S. Yu and Q. Zhuang. On the state complexity of intersection of regular languages. *SIGACT News* **22**(3) (1991), 52–54.

[28] S. Yu, Q. Zhuang, and K. Salomaa. The state complexity of some basic operations on regular languages. *Theoret. Comput. Sci.* **125** (1994), 315–328.

School of Computer Science, University of Waterloo, Waterloo, Ontario N2L 3G1, Canada

E-mail address: `shallit@graceland.uwaterloo.ca`

DIMACS Series in Discrete Mathematics
and Theoretical Computer Science
Volume **64**, 2004

Lattices, Cryptography, and the NTRU Public Key Cryptosystem

Joseph H. Silverman

ABSTRACT. We give a brief introduction to lattices, hard lattice problems, the use of lattices in cryptanalysis, and the construction of lattice-based cryptosystems. Two lattice-based systems are described, the NTRU public key cryptosystem and the PASS authentication and digital signature scheme. Both NTRU and PASS feature highly desirable operating characteristics (high speed, low footprint, fast key generation) compared to other common systems based on factorization (e.g., RSA) or discrete logarithm problems (e.g., Diffie-Hellman, ECC).

Introduction

Modern cryptographic constructions are based on underlying hard mathematical problems. Typical problems that have been used include the integer factorization problem and the discrete logarithm problem, where the latter has been taken in the multiplicative group of a field, in the class group of a number field, and in the group of points on an elliptic curve. For an overview of modern cryptography, see for example [**Sti**].

In this note we describe another hard mathematical problem that can used to construct cryptographic systems, the problem of finding short vectors in lattices of high dimension. The advantage of such systems is that they tend to be extremely fast compared to integer factorization and discrete logarithm systems at equivalent security levels. The problem of finding small vectors in lattices, which also goes by the name "Geometry of Numbers" introduced in Minkowski's influential 1910 monograph [**Min**], has been widely studied from both a theoretical and a computational point of view. For an introduction to the theory, see for example [**Ca, LG, Si**] and the 93 page bibliography in [**LG**]. Computational issues and algorithms are discussed, for example, in [**Coh, dW, LLL, PZ**].

A number of cryptosystems have been proposed that depend, either explicitly or implicitly, on small vectors in lattices. These systems include knapsack systems of Merkle-Hellman [**MH**] and others, the quasi-orthogonal basis systems of Atjai-Dwork [**AD**] and Goldreich-Goldwasser-Halevi [**GGH**], the NTRU system of

2000 *Mathematics Subject Classification.* Primary 94A60; Seconday 11T71.

Key words and phrases. public key cryptography, lattice-based cryptography, lattice, SVP, CVP, NTRU.

Hoffstein-Pipher-Silverman [**HPS**], and the PASS authentication/digital signature scheme of Hoffstein-Lieman-Silverman [**HLS**]. In the first three systems, the key sizes grow like the square of the lattice dimension n, so they become impractical for large values of n, and they are vulnerable to modern lattice search methods for small values of n. The latter two systems, NTRU and PASS, have key sizes that grow (essentially) linearly with the dimension n, so are very practical at lattice dimensions beyond current search techniques.

In this note we will give a brief introduction to lattices from both a theoretical and a computational perspective. We will then describe the GGH, NTRU, and PASS cryptographic systems and their underlying lattices. We will not have space for detailed security analyses of these systems, but will provide the reader with detailed references.

1. Lattices

We start with the formal definition.

DEFINITION 1.1. A *lattice* L of dimension n is a maximal discrete subgroup of $\mathbb{R}^n$.

A more down-to-earth description of a lattice of dimension n is the set of linear combinations

$$L = \{a_1\mathbf{v}_1 + a_2\mathbf{v}_2 + \cdots + a_n\mathbf{v}_n \;:\; a_1, \ldots, a_n \in \mathbb{Z}\},$$

where $\mathbf{v}_1, \ldots, \mathbf{v}_n \in \mathbb{R}^n$ are a collection of linearly independent vectors. The vectors $\mathbf{v}_1, \ldots, \mathbf{v}_n$ are called a *basis for* L.

A fundamental problem is that of finding the shortest nonzero vector in a given lattice. Closely related is the problem of finding the vector in a lattice that is closest to a given vector. We state these formally as "computationally hard problems," of a similar nature to the integer factorization problem and the discrete logarithm problem.

Smallest Vector Problem (SVP):
Find a smallest nonzero vector in a given lattice L.

Closest Vector Problem (CVP):
Given a vector $\mathbf{w} \in \mathbb{R}^n$, find a vector $\mathbf{v} \in L$ (with $\mathbf{v} \neq \mathbf{w}$) that is closest to $\mathbf{w}$.

Clearly SVP is a special case of CVP, simply take $\mathbf{w} = \mathbf{0}$. Interestingly, both in theory and in practice, one can (usually) solve CVP by solving SVP in a lattice of one higher dimension. (See, e.g., [**GMSS**].)

Clearly a lattice has many different bases. In a vector space, the best sort of basis is an orthogonal basis, since computations using an orthogonal basis are especially easy. Lattices do not have orthogonal bases, but one might ask for bases that are as close to orthogonal as possible. Alternatively, one might ask for a basis whose elements are as small as possible, where smallness might be measured in any one of a number of different ways. In order to quantify this idea, we need a couple of definitions.

Let $\mathbf{v}_1, \ldots, \mathbf{v}_n$ be any basis for L. They span a parallelepiped

$$\mathcal{F}(\mathbf{v}_1, \ldots, \mathbf{v}_n) = \{t_1\mathbf{v}_1 + t_2\mathbf{v}_2 + \cdots + t_n\mathbf{v}_n : 0 \leq t_i < 1\}$$

that is a fundamental domain for the quotient $\mathbb{R}^n/L$. The *volume of* L (really its covolume) is defined to be the volume of this fundamental domain. It may be computed as the determinant of the matrix whose columns are the basis vectors:

$$\operatorname{Vol}(L) = \operatorname{Vol}\big(\mathcal{F}(\mathbf{v}_1, \dots, \mathbf{v}_n)\big) = \det\big(\mathbf{v}_1 \mid \mathbf{v}_2 \mid \cdots \mid \mathbf{v}_n\big).$$

(This is also sometimes called the determinant of L and denoted $\det(L)$.)

Hadamard's elementary inequality says that

$$\operatorname{Vol}(L) \le \|\mathbf{v}_1\|\,\|\mathbf{v}_2\| \cdots \|\mathbf{v}_n\|,$$

since the volume of a parallelepiped can never be greater than the product of the lengths of its sides. Notice that Hadamard's inequality is an equality if and only if the basis vectors are pairwise orthogonal, so the extent to which it is an inequality is a measure of the extent to which the basis is nonorthogonal. Thus the following fundamental result says that every lattice has a basis that is moderately orthogonal, where the amount of non-orthogonality is bounded solely in terms of the dimension.

THEOREM 1.2 (Hermite). *Let L be a lattice of dimension n.*

(a) *There is a number γ_n depending only on n so that the shortest non-zero vector $\mathbf{v} \in L$ satisfies*

$$\|\mathbf{v}\| \le \gamma_n \operatorname{Vol}(L)^{1/n}.$$

(Hermite gave the value $\gamma_n = (2/\sqrt{3})^{(n-1)/2}$.)

(b) *There is a number α_n, depending only on n, such that the lattice L has a basis $\mathbf{v}_1, \dots, \mathbf{v}_n$ satisfying*

$$\|\mathbf{v}_1\|\,\|\mathbf{v}_2\| \cdots \|\mathbf{v}_n\| \le \alpha_n \operatorname{Vol}(L).$$

One may take $\alpha_n = \gamma_n^{n/2}$.

The smallest possible value for γ_n that works for all lattices of dimension n is called Hermite's constant.[1] Its exact value is only known for $n \ge 8$. However, it is known that for large values of n,

$$\sqrt{\frac{n}{2\pi e}} \;\lesssim\; \gamma_n^{\text{Hermite}} \;\lesssim\; \sqrt{\frac{n}{\pi e}}.$$

Suppose that we take

$$\begin{aligned}
\mathbf{v}_1 &= \text{shortest non-zero vector in } L,\\
\mathbf{v}_2 &= \text{shortest vector in } L \text{ linearly independent of } \mathbf{v}_1,\\
\mathbf{v}_3 &= \text{shortest vector in } L \text{ linearly independent of } \mathbf{v}_1, \mathbf{v}_2,\\
&\;\;\vdots \qquad\qquad\qquad\qquad \vdots\\
\mathbf{v}_n &= \text{shortest vector in } L \text{ linearly independent of } \mathbf{v}_1, \mathbf{v}_2, \dots, \mathbf{v}_{n-1}.
\end{aligned}$$

Then one can show that the vectors $\mathbf{v}_1, \dots, \mathbf{v}_n$ satisfy Hermite's theorem.[2] The lengths

$$\|\mathbf{v}_1\|,\ \|\mathbf{v}_2\|, \dots, \|\mathbf{v}_n\|$$

are called the *successive minima of the lattice L*. They are often denoted by $\lambda_1(L), \dots, \lambda_n(L)$.

---

[1] More precisely, Hermite's constant is the square of our γ_n, since people generally work with the quadratic form $\|\mathbf{v}\|^2$, rather than the length $\|\mathbf{v}\|$.

[2] The set of vectors constructed in this way need not be a basis for L. However, it generates a sublattice of L whose index can be bounded.

2. Lattice Reduction and the LLL Algorithm

Lattice reduction is the process of finding small vectors and/or small quasi-orthogonal bases for a lattice. In principal one can find the smallest vector in a lattice by exhaustive search, since Hermite's theorem gives an upper bound for its length; but in practice an exhaustive search takes exponential time (in the dimension of the lattice).

It is known that the general problem of finding the closest lattice vector to a given vector is NP-complete. Balanced against this is the LLL polynomial-time algorithm of Lenstra, Lenstra, and Lovász [**LLL**] that finds a "moderately small" basis for a lattice L. Before describing the algorithm, we need some definitions.

Let $\mathbf{v}_1, \ldots, \mathbf{v}_n$ be a basis for a lattice L. We apply the Gram-Schmidt orthogonalization process to produce an orthogonal (but not orthonormal) basis of $\mathbb{R}^n$ as follows:

$$\begin{aligned}
\mathbf{v}_1^* &= \mathbf{v}_1 \\
\mathbf{v}_2^* &= \mathbf{v}_2 - \frac{\mathbf{v}_2 \cdot \mathbf{v}_1^*}{\|\mathbf{v}_1^*\|^2}\mathbf{v}_1^* \\
\mathbf{v}_3^* &= \mathbf{v}_3 - \frac{\mathbf{v}_3 \cdot \mathbf{v}_2^*}{\|\mathbf{v}_2^*\|^2}\mathbf{v}_2^* - \frac{\mathbf{v}_3 \cdot \mathbf{v}_1^*}{\|\mathbf{v}_1^*\|^2}\mathbf{v}_1^* \\
&\ \,\vdots \qquad\qquad\qquad\qquad \ddots \\
\mathbf{v}_n^* &= \mathbf{v}_n - \frac{\mathbf{v}_n \cdot \mathbf{v}_{n-1}^*}{\|\mathbf{v}_{n-1}^*\|^2}\mathbf{v}_{n-1}^* - \frac{\mathbf{v}_n \cdot \mathbf{v}_{n-2}^*}{\|\mathbf{v}_{n-2}^*\|^2}\mathbf{v}_{n-2}^* - \cdots - \frac{\mathbf{v}_n \cdot \mathbf{v}_1^*}{\|\mathbf{v}_1^*\|^2}\mathbf{v}_1^*
\end{aligned}$$

The intuition here is:

$$\mathbf{v}_i^* = \text{Projection of } \mathbf{v}_i \text{ onto } \operatorname{Span}(\mathbf{v}_1, \ldots, \mathbf{v}_{i-1})^\perp.$$

If any of the coefficients

$$\frac{\mathbf{v}_i \cdot \mathbf{v}_j^*}{\|\mathbf{v}_j\|^2}$$

has magnitude greater than $1/2$, it's not hard to see that replacing $\mathbf{v}_i$ by a linear combination $\mathbf{v}_i - a\mathbf{v}_j$ for some $a \in \mathbb{Z}$ will make the coefficient smaller. We will say that the basis satisfies the SIZE CONDITION if

$$\text{SIZE CONDITION:} \qquad \frac{|\mathbf{v}_i \cdot \mathbf{v}_j^*|}{\|\mathbf{v}_j\|^2} \le \frac{1}{2} \quad \text{for all } j < i.$$

To balance the size condition, we would also like to impose a condition that forces the basis vectors to be somewhat orthogonal, such as

$$\text{QUASI-ORTHOGONALITY CONDITION:} \qquad \|\mathbf{v}_{i+1}^*\|^2 \ge \frac{3}{4}\|\mathbf{v}_i^*\|^2.$$

THEOREM 2.1 (Hermite). *Every lattice has a basis satisfying the* SIZE CONDITION *and the Quasi-Orthogonality Condition.*

Unfortunately, the only algorithms known for finding such a basis are exponential. So we weaken the orthogonality property as follows:

$$\text{LOVÁSZ CONDITION:} \qquad \|\mathbf{v}_{i+1}^*\|^2 \ge \left(\frac{3}{4} - \left(\frac{|\mathbf{v}_{i+1} \cdot \mathbf{v}_i^*|}{\|\mathbf{v}_i\|^2}\right)^2\right)\|\mathbf{v}_i^*\|^2.$$

This looks somewhat messy, but it can be rewritten as the condition:

$$\text{Projection of } \mathbf{v}_{i+1} \text{ onto } \operatorname{Span}(\mathbf{v}_1, \dots, \mathbf{v}_{i-1})^{\perp} \geq \frac{3}{4} \cdot \text{Projection of } \mathbf{v}_i \text{ onto } \operatorname{Span}(\mathbf{v}_1, \dots, \mathbf{v}_{i-1})^{\perp}.$$

THEOREM 2.2 (Lenstra, Lenstra, Lovász [**LLL**]). *There is a polynomial time algorithm that for any given lattice finds a basis satisfying both the* SIZE CONDITION *and the* LOVÁSZ CONDITION. *Such a basis is a called an* LLL reduced basis.

A brief description of the LLL algorithm is given in Figure 1. For a complete description with implementation suggestions, see for example [**Coh, LP, PZ**].

```
[1]  k = 2
[2]  LOOP WHILE k < n
[3]      Replace v_1,...,v_k by linear combinations so
         that the SIZE CONDITION is true.
[4]      If the LOVÁSZ CONDITION is false
[5]          Swap v_k ↔ v_{k-1} and set k = k - 1
[6]      Else
[7]          Set k = k + 1
[8]      If k = n, then done (basis is LLL reduced)
[9]  END LOOP
```

FIGURE 1. The LLL Algorithm

It is clear that if the algorithm terminates, then the basis is LLL reduced. It is not hard to prove that the algorithm executes Step (5) only finitely often, and indeed the maximum number of executions is at most polynomial in $\dim(L)$.

The LLL algorithm is guaranteed to find a basis whose smallest element, say $\mathbf{v}_1$, satisfies

$$\|\mathbf{v}_1\| \leq 2^{(n-2)/2}\lambda_1(L),$$

where $\lambda_1(L)$ is the length of smallest non-zero vector in L. In practice, LLL generally does better than this; but also in practice, if n is large, then LLL will not find a vector whose length is just a few times larger than the smallest vector.

There have been several combinatorial improvements to LLL due principally to Euchner, Hörner, and Schnorr, see [**Sch1, Sch2, SE**]. These methods include:

Deep Insertion Method: Instead of swapping $\mathbf{v}_k$ and $\mathbf{v}_{k-1}$ in Step (5), it may be better to insert $\mathbf{v}_k$ between two $\mathbf{v}_{i-1}$ and $\mathbf{v}_i$. The resulting algorithm is no longer polynomial time in the worst case, but in practice may give better results.

Block Reduction Method: Rather than swapping $\mathbf{v}_k$ and $\mathbf{v}_{k-1}$ in Step (5) if this decreases the length of $\mathbf{v}^*_{k-1}$, one instead takes a block of vectors $\mathbf{v}_i, \mathbf{v}_{i+1}, \dots, \mathbf{v}_{i+b-1}$ and replaces it with vectors so as to minimize $\mathbf{v}^*_i$. The resulting algorithm is exponential in the blocksize b, but will yield better bases as b is increased. More precisely, for moderately large b, one is guaranteed that

$$\|\mathbf{v}_1\| \leq \left(\frac{b}{\pi e}\right)^{\frac{n-1}{b-1}} \lambda_1(L),$$

but the running time is multiplied by b^{cb}.

Pruning Method: The time consuming part of the Block Reduction Method is completely enumerating all short lattice vectors in the sublattice spanned by a block. The Pruning Method cuts off the enumeration process if the Gaussian volume heuristic predicts that the current list of vectors has probability smaller than 2^{-p} of being completed to a basis of smallest vectors.

The study of lattice reduction and the problem of finding short and/or closest vectors remains an active area of current research. For a sample of recent results, see for example [**BS, CN, DKS, GG, GMSS, Mio**].

3. Lattice Reduction in Cryptanalysis

Lattice reduction has been used as a tool for breaking cryptosystems. One of the earliest public key cryptosystems proposed was the Merkle-Hellman Knapsack Cryptosystem [**MH**]. This system is based on the *Knapsack* (or more precisely the *Subset Sum*) *Problem*:

The Subset Sum Problem

Let $S = \{s_1, \ldots, s_n\}$ be a set of positive integers. For a given integer t, find a subset of S satisfying

$$s_{i_1} + \cdots + s_{i_k} = t.$$

The general Subset Sum Problem is NP-complete. To create a cryptosystem, one takes a special case of the problem (e.g., each s_j is larger than $s_{j-1} + \cdots + s_1$) for which the problem is easy to solve, and then transforms the set S in some way (e.g., as $as_i \bmod p$) to conceal the special form of S.

In the special cases of the subset sum problem used to create cryptosystems, it is possible to transform the problem into finding short vectors in a lattice. If the density

$$\text{Density}(S) = \frac{n}{\max \log(s_i)}$$

of the subset sum problem is not too large, which is the case for Merkle-Hellman, then Brickell and Lagarias-Odlyzko showed how to solve the problem in polynomial time using LLL. (See also [**Sh**] for an analysis of the Merkle-Hellman system via integer programming.)

Chor and Rivest proposed a more elaborate use of the subset sum problem whose underlying lattice has higher density, but LLL with block reduction and the Schnorr-Hörner pruning method was able to break the Chor-Rivest system with $n = 103$, and even in about 10% of the cases with $n = 151$. (See [**Ste**] for an overview of lattices in cryptography.)

REMARK 3.1. Lattice reduction has also been used for cryptanalysis of RSA in certain situations. Coppersmith [**Cop**] has explained how to use lattice reduction techniques to find small roots to polynomial congruences $f(X) \equiv 0 \pmod{N}$. He uses this construction to break certain instances of RSA when the encryption exponent is small (e.g., 3) and when similar or partially padded messages are broadcast.

4. Lattice-Based Cryptosystems

Despite the failure of the knapsack cryptosystems, there has been continued interest in using hard lattice problems for cryptography because of the potential for greatly improved encryption and decryption speed. Two recent systems have been proposed by Atjai-Dwork [**AD**] and Goldreich-Goldwasser-Halevi [**GGH**]. The AD system is notable in that its security is provably equivalent to the *worst case complexity* of a certain class of lattice problems. The GGH system is similar and more practical, so we sketch how it works.

Outline of the GHH Public Key Cryptosystem

(1) Choose a quasi-orthogonal set of vectors $\mathcal{B} = \{\mathbf{v}_1, \ldots, \mathbf{v}_n\}$ and let L be the lattice they generate. The basis $\mathcal{B}$ is the private key.

(2) Apply an invertible linear transformation to the basis elements of $\mathcal{B}$ to form a new basis

$$\mathcal{B}' = \{\mathbf{w}_1, \ldots, \mathbf{w}_n\}.$$

Note that $\mathcal{B}'$ will be quite non-orthogonal.

(3) To encrypt a (binary) message $\mathbf{m} = (m_1, \ldots, m_n)$, choose a small random vector $\mathbf{r}$ and form the vector

$$\mathbf{e} = m_1\mathbf{w}_1 + \cdots + m_n\mathbf{w}_n + \mathbf{r}.$$

Note that $\mathbf{e}$ will not be in the lattice L.

(4) To decrypt $\mathbf{e}$, one solves the CVP. In other words, one finds the vector $\mathbf{u} \in L$ that is closest to $\mathbf{e}$. Then almost certainly

$$\mathbf{u} = \mathbf{e} - \mathbf{r},$$

so expressing $\mathbf{u}$ in terms of the basis $\mathcal{B}' = \{\mathbf{w}_1, \ldots, \mathbf{w}_n\}$ allows recovery of $m_1, \ldots, m_n$.

(5) Security depends on the fact that it is difficult to solve CVP using the basis $\mathcal{B}'$. Decryption depends on the fact that it is relatively easy to solve CVP using the quasi-orthogonal basis $\mathcal{B}$. The basic idea is to use elementary linear algebra to write

$$\mathbf{e} = \mu_1\mathbf{v}_1 + \cdots + \mu_n\mathbf{v}_n \qquad \text{with } \mu_1, \ldots, \mu_n \in \mathbb{R}.$$

If $\mathbf{v}_1, \ldots, \mathbf{v}_n$ are sufficiently orthogonal and if the random noise $\mathbf{r}$ is small enough, then one can show that

$$\lfloor \mu_1 \rceil \mathbf{v}_1 + \cdots + \lfloor \mu_n \rceil \mathbf{v}_n \in L$$

will be the closest vector to $\mathbf{e}$.

The advantage of the GGH system is that its operating speed is $O(n^2)$, as compared to an operating speed of $O(n^3)$ for RSA and ECC systems. However, the drawback of GGH is its key size. A key for the GGH cryptosystem consists of an entire basis for L, so requires n vectors of dimension n. It thus consists of n^2 (small) numbers. This means that GGH is not practical once n gets larger than (say) 100 to 200, but Nguyen, Stern, and others (see [**Ng, Ste**]) have shown that for this range of n, the LLL algorithm (with its modern refinements) is quite capable of finding a basis for L that is sufficiently orthogonal to break GGH. Thus a secure GGH system probably requires n in the range of 500 to 1000, leading to impractical megabit size keys!

	NTRU	RSA	ECC
Encrypt/Decrypt Speed	$O(n \log n)$	$O(n^3)$	$O(n^3)$
Key Size (bits)	$\approx \frac{1}{2} n \log_2 n$	n	n
Key Creation Speed	$O(n^2)$	—	$O(n^3)$
Typical n	526	1024	168

TABLE 1. Asymptotic Speed and Key Size Comparison

5. The NTRU Public Key Cryptosystem

The GGH cryptosystem corrects small perturbations in a lattice vector using a quasi-orthogonal basis for the lattice. A cryptosystem called NTRU (proposed by Hoffstein, Pipher, and Silverman [**HPS**]) corrects large perturbations in a lattice vector via congruences. This allows the underlying basis to be cyclically generated from a single vector, which means that the keys have length n while retaining the speed advantages over RSA and ECC. A brief asymptotic speed and key size comparison is contained in Table 1.

The NTRU cryptosystem is most easily explained using the ring

$$R = \frac{\mathbb{Z}[X]}{(X^N - 1)}$$

rather than the underlying lattice. Notice that multiplication in R is simply a finite convolution on the polynomial coefficients:

$$\left(\sum_{i=0}^{N-1} a_i X^i\right)\left(\sum_{j=0}^{N-1} b_j X^i\right) = \left(\sum_{k=0}^{N-1} c_k X^k\right) \quad \text{with} \quad c_k = \sum_{i+j \equiv k \pmod{N}} a_i b_j.$$

The naive computation of a product $a(X)b(X)$ take $O(N^2)$ steps, but fancier methods (e.g., Karatsuba multiplication or FFTs) reduce this to $O(N \log N)$ steps.

We will say that a polynomial is *trinary* if all of its coefficients are chosen from the set $\{-1, 0, 1\}$. We will say that the trinary polynomial is *small* if there are some further restrictions on the number of non-zero coefficients.

The NTRU cryptosystem depends on three integer parameters

$$N, \qquad p, \qquad q, \qquad \text{with } \gcd(p, q) = 1.$$

Typical choices would be

$$(N, p, q) = (263, 3, 128) \qquad \text{and} \qquad (N, p, q) = (503, 3, 256).$$

Outline of the NTRU Public Key Cryptosystem

Key Creation: Choose random small trinary polynomials $f, g \in R$. Compute

$$F_p \equiv f^{-1} \pmod{p} \qquad \text{and} \qquad F_q \equiv f^{-1} \pmod{q}.$$

Let

$$h \equiv gF_q \pmod{q}.$$

The public key is the polynomial h, the private key is the pair (f, F_p).

Encryption: The message m is a polynomial with mod p coefficients. Choose a random small trinary polynomial r. The encrypted message is

$$e \equiv prh + m \pmod{q}.$$

Decryption: Compute

$$a \equiv ef \pmod{q},$$

choosing the coefficients of a to lie between $-q/2$ and $q/2$. Next reduce the coefficients of a modulo p and compute

$$b \equiv aF_p \pmod{p}.$$

Then $b \equiv m \pmod{p}$.

Why It Works: The polynomial a is equal to

$$\begin{aligned} a &\equiv ef \equiv (rh + m)f \pmod{q} \\ &\equiv \big(r(pgF_q) + m\big)f \pmod{q} \\ &\equiv rpgF_q f + mf \pmod{q} \\ &\equiv prg + mf \pmod{q}. \end{aligned}$$

The coefficients of the polynomials r, m, f and g and the number p are all quite small compared to q, so the polynomial a is in fact exactly equal to the product $prg + mf$. In other words,

$$a = prg + mf \quad \text{in } \frac{\mathbb{Z}[X]}{(X^N - 1)},$$

without having to reduce the coefficients modulo q. Hence when we reduce a modulo p, we get $a \equiv mf \pmod{p}$, and multiplying by $F_p = f^{-1} \bmod p$ recovers m.

The above description of NTRU does not seem to have much to do with lattices, but we can rephrase the problem in the following manner. The NTRU public key is

$$h \equiv f^{-1}g \pmod{q},$$

so NTRU can be broken by finding the small polynomial f with the property that $fh \bmod q$ is also small. This leads to the NTRU lattice L^{NT} generated by the rows of the following $2N$-by-$2N$ matrix:

$$L^{\mathrm{NT}} = \left(\begin{array}{cccc|cccc} \lambda & 0 & \cdots & 0 & h_0 & h_1 & \cdots & h_{N-1} \\ 0 & \lambda & \cdots & 0 & h_1 & h_2 & \cdots & h_0 \\ \vdots & \vdots & \ddots & \vdots & \vdots & \vdots & \ddots & \vdots \\ 0 & 0 & \cdots & \lambda & h_{N-1} & h_0 & \cdots & h_{N-2} \\ \hline 0 & 0 & \cdots & 0 & q & 0 & \cdots & 0 \\ 0 & 0 & \cdots & 0 & 0 & q & \cdots & 0 \\ \vdots & \vdots & \ddots & \vdots & \vdots & \vdots & \ddots & \vdots \\ 0 & 0 & \cdots & 0 & 0 & 0 & \cdots & q \end{array}\right)$$

REMARK 5.1.

- The number λ is a balancing constant that is chosen to maximize LLL efficiency.

- The shortest vector in L^{NT} is (probably) the vector
$$(\lambda f_{N-1}, \ldots, \lambda f_0, g_0, \ldots, g_{N-1}).$$
Thus solving the SVP for L will reveal the NTRU private key f.
- The NTRU lattice L^{NT} may also be used to search for the plaintext m. More precisely, the vector in L^{NT} closest to the known vector
$$(0, \ldots, 0, e_0, \ldots, e_{N-1})$$
is (probably) the vector
$$(\lambda r_{N-1}, \ldots, \lambda r_0, e_0 - m_0, \ldots, e_{N-1} - m_{N-1}).$$
It is thus clear that recovery of the plaintext m is *equivalent* to solving the CVP for $(0, \ldots, 0, e_0, \ldots, e_{N-1})$ in L^{NT}.
- Extensive testing using LLL indicates that the NTRU lattices L^{NT} are no easier to solve than random lattices.
- In practice, LLL seems to return a "q-vector" (i.e., one of the last N rows) until the block size is sufficiently large, at which point it finds the target vector. The necessary block size appears to increase linearly with N, and thus the running time increases exponentially with N. For details and the results of extensive experiments, see [**HPS**].
- It is interesting to observe that the given basis for L^{NT} is already quite orthogonal. Indeed, the q-vectors already give an orthogonal basis for half the space, and the fact that the h_i's are randomly distributed between $-q/2$ and $q/2$ gives a moderately orthogonal basis for the other half of the lattice.
- A detailed analysis of the NTRU algorithm is given in [**HPS**]. Further work on the NTRU system is given in the paper of Coppersmith and Shamir [**CS**] and the thesis of Alex May [**Ma**].

6. The PASS Authentication and Signature Scheme

The PASS authentication and signature scheme is a construction based on similar lattices to those underlying NTRU, but built out of a slightly different problem. The NTRU problem is based on the difficulty of factoring a polynomial $h \bmod q$ into a product $f^{-1}g \bmod q$ using small polynomials f and g. The PASS problem is based on the difficulty of finding a small polynomial $f \bmod q$ that takes on prescribed values at a collection of numbers $X = \alpha_1, \ldots, X = \alpha_{N/2}$. One version of PASS was presented at CrypTEC '99 and published in [**HLS**]. We will describe a modified version that appears to be equally secure and has improved operating characteristics.

A PASS scheme depends on two integer parameters q and N, where q is a prime and N divides $q-1$. Typical values for q would be $q = 769$, $q = 929$, or $q = 1053$ with $N = q - 1$ in each case. (We will assume, in any case, that N is even.) Let
$$R_q = \frac{\mathbb{F}_q[X]}{(X^N - 1)}$$
be the quotient ring of polynomials with coefficients in the finite field $\mathbb{F}_q$. We observe that $\mathbb{F}_q$ contains a primitive N^{th}-root of unity, so the polynomial $X^N - 1$ factors completely into linear factors over $\mathbb{F}_q$

The other public parameter for PASS is a set

$$S = \{\alpha_1, \alpha_2, \dots, \alpha_t\}$$

of distinct non-zero elements of $\mathbb{F}_q$ with the property that if $\alpha \in S$, then also $\alpha^{-1} \in S$. For example, one could take

$$S = \{w^i : (N-t)/2 \le i \le (N+t)/2\}$$

for a fixed N^{th}-root of unity $w \in \mathbb{F}_q$, so there is no need to actually store the set S.

For simplicity, we will assume henceforth that $N = q - 1$. Suppose that $f(X) \in R_q$ is a polynomial, and suppose that you are given the full set of values $\{f(\alpha) : \alpha \in \mathbb{F}_q\}$. It is then a simple matter to reconstruct the polynomial $f(X)$. Briefly, if w is a primitive N^{th}-root of unity in $\mathbb{F}_q$, and if $f(X) = \sum a_i X^i$, then the ith coefficient of $f(X)$ may be recovered from the values of $f(X)$ via the well-known formula

$$a_i = \frac{1}{N} \sum_{j=0}^{N-1} f(w^j) w^{-ij}.$$

In standard terminology, the set of values of $f(X)$ is its *Discrete Fourier Transform* (DFT), and the above formula says that taking the DFT twice gives back the original function (up to a factor of the dimension).

Now suppose, instead, that you are only given some of the values of f, for example the set of values

$$f(S) = \{f(\alpha) : \alpha \in S\}.$$

Simple linear algebra (treating the coefficients of $f(X)$ as unknowns) shows that there are a large number of polynomials taking this set of values on S. Precisely, one expects approximately q^{N-t} solutions, since the set of all possible f's with no constraints is an $\mathbb{F}_q$-vector space of dimension N, and each specified value $f(\alpha)$ imposes one linear constraint on the coefficients.

Finally, suppose that $f(X)$ is not a random polynomial, but instead that it is a small polynomial, that is, a polynomial in R_q with small coefficients (e.g., all coefficients equal to 0, 1, or -1). It then becomes a difficult task to find, amongst the q^{N-t} polynomials that take the correct values on all $\alpha \in S$, the specific polynomial with small coefficients. This, then, is the problem underlying PASS. The Prover performs the following steps:

- The Prover chooses a small polynomial $f(X) \in R_q$.
- She publishes the partial set of values $f(S)$.
- She proves her identity by demonstrating that she knows a small polynomial that takes the correct values on S.

Of course, the Prover's demonstration of knowledge must be indirect, since she does not want to give away information that might let an adversary reconstruct the polynomial $f(X)$. We describe below how this is done.

REMARK 6.1. The security of PASS is based on the fact that it is difficult to simultaneously control both the values and the coefficients of a polynomial over a finite field. As indicated above, the values and the coefficients of a polynomial are (discrete) Fourier transforms of one another. Thus underlying PASS is the mathematical principle that it is difficult to simultaneously control the values of a function and the values of its Fourier transform. This principle is the discrete analogue (for finite fields) of the Heisenberg Uncertainty Principle.

Outline of the PASS Authentication and Signature Scheme

Key Creation: The Prover selects a small polynomial $f(X) \in R_q$. She publishes the set of values $f(S) = \{f(\alpha) : \alpha \in S\}$.

Commitment: In the commitment step, the Prover selects another small polynomial $g_1(X) \in R_q$. She computes and sends to the Verifier the set of values $g_1(S)$.

Challenge: In the challenge step, the Verifier selects two extremely small polynomials $c_1(X)$ and $c_2(X)$ (say with between four and eight nonzero coefficients) and sends them to the Prover. For security reasons, it is also important that $c_1(X)$ have no nonzero roots $\alpha \in \mathbb{F}_q$ with $\alpha \notin S$. (There is at least a 50% chance that this will be true for a randomly chosen c_1.)

Response: In the response step, the Prover selects a third small polynomial $g_2(X) \in R_q$. She computes in R_q and sends to the Verifier the polynomial

$$h(X) = \big(f(X) + c_1(X)g_1(X) + c_2(X)g_2(X)\big)g_2(X). \tag{6.1}$$

Verification: The Verifier performs the following two checks to verify the Prover's identity:

(A) The Verifier checks that the polynomial $h(X)$ is moderately small, which is measured by computing the sum of the squares of its coefficients and checking if that sum is smaller than a certain prespecified bound.

(B) For each $\alpha \in S$, the Verifier computes the quantity

$$\big(f(\alpha) + c_1(\alpha)g_1(\alpha)\big)^2 + 4c_2(\alpha)h(\alpha) \in \mathbb{F}_q \tag{6.2}$$

and checks that it is a square in $\mathbb{F}_q$.

If the polynomial h passes tests (A) and (B), then the Verifier accepts the Prover's identity.

Why It Works: First we note that the definition (6.1) says that the polynomial $h(X)$ is as a simple combination of the polynomials f, g_1, g_2, c_1, c_2, all of which are small, so it is clear that the coefficients of $h(X)$ will also be moderately small. It is a simple matter to experimentally find a norm bound that the coefficients of h will almost always satisfy.

Next we observe that the verifier is able to compute the quantity (6.2), since he knows the polynomials $h(X), c_1(X), c_2(X)$ and he knows the values of $f(X)$ and $g_1(X)$ for all $\alpha \in \mathbb{F}_q$. To see why (6.2) is a square, we use the definition (6.1) of $h(X)$ to compute

$$\begin{aligned}(f + c_1g_1)^2 + 4c_2h &= (f + c_1g_1)^2 + 4c_2\big((f + c_1g_1 + c_2g_2)g_2\big)\\ &= f^2 + 2c_1g_1 + c_1^2g_1^2 + 4c_2g_2f + 4c_1c_2g_1g_2 + 4c_2^2g_2^2\\ &= (f + c_1g_1 + 2c_2g_2)^2\end{aligned} \tag{6.3}$$

Thus $\big(f(X) + c_1(X)g_1(X)\big)^2 + 4c_2(X)h(X)$ is actually the square of a polynomial in R_q, so it certainly gives a square in $\mathbb{F}_q$ when evaluated at any $\alpha \in \mathbb{F}_q$.

REMARK 6.2.

- The polynomial identity (6.3) may seem mysterious, but it has a very simple explanation. Directly from the definition (6.1) of $h(X)$, we see that the equation

$$c_2T^2 + (f + c_1g_1)T - h = 0$$

has the solution $T = g_2$ in R_q. The quadratic formula says that the solution is given by

$$g_2 = \frac{-(f + c_1g_1) \pm \sqrt{(f + c_1g_1)^2 + 4c_2h}}{2c_2}.$$

Solving for the quantity under the square root sign shows that it is equal to $(f + c_1g_1 + 2c_2g_2)^2$.

- The best way to create the challenge polynomials c_1 and c_2 is for the Verifier to choose a random string (of 80 to 160 bits) and send it to the Prover. They then both apply a common hash function to the random string in order to create c_1 and c_2. This has the advantage of cutting down the number of bits transmitted, as well as making it more difficult for the Verifier to mount any sort of attack based on choosing c_1 and c_2 to have a particular form.
- We have presented PASS as an authentication scheme, but any authentication scheme that includes a challenge step can be combined with a hash function to create a signature scheme. Thus if a digital document D is to be signed, it is sent through a standard hash function to obtain a small bit string (say 80 to 160 bits), and this bit string is hashed to create the challenge polynomials c_1, c_2. The Signer then publishes the values $g_1(S)$, $h(X)$, and D. (We assume, of course, that the Signer's public key $f(S)$ is already in the public domain.) Anyone wishing to verify the signature can use D and the hash functions to recreate c_1 and c_2, and then he has enough information to perform the verification step described above.
- The computation of the sets of values $f(S)$, $g_1(S)$, and $h(S)$ required during the PASS process can be performed extremely rapidly. Even a direct computation takes only $O(N^2)$ steps, but it is even more efficient to use FFT, especially if N is highly divisible by 2 (as are the suggested values $N = 768$, $N = 928$, and $N = 1052$), in which case the computation is reduced to $O(N \log N)$ steps. Note that the field $\mathbb{F}_q$ contains a primitive N^{th}-root of unity, so in this setting one can compute Fast Fourier Transforms using only integer arithmetic; there is no need to use complex numbers or floating point numbers.
- In verification step (B), the Verifier needs to check if certain quantities are squares. This can be done rapidly using quadratic reciprocity, although if a little extra storage is available, it's even quicker to simply precompute a table of squares.
- If the computational capabilities of the Prover and the Verifier are different, for example, if one of them is a low power smart card, then it is quite easy to transfer much of the computational requirements to the more powerful processor.

Just as with the NTRU public key cryptosystem, it is easy to formulate a lattice problem whose solution will recover the private key (or another polynomial that can be used to forge a signature). To construct this lattice, we proceed as follows.

For each $1 \le i \le t$, let β_i be the known value of $f(\alpha_i)$. Then if we treat the coefficients $a_0, \ldots, a_{N-1}$ of $f(X)$ as unknown variables, we have a system of t linear congruences

$$a_0 + a_1\alpha_i + a_2\alpha_i^2 + \cdots + a_{N-1}\alpha_i^{N-1} \equiv \beta_i \pmod{q} \qquad \text{for } 1 \le i \le t$$

for the unknown a_j's. Solving these congruence for the final t values $a_{N-t}, \ldots, a_{N-1}$ in terms of the other a_j's yields

$$a_{N-1-t+i} \equiv \sum_{j=0}^{N-1-t} \gamma_{ij} a_j + \delta_i \pmod{q} \qquad \text{for } 1 \le i \le t. \tag{6.4}$$

In other words, we know the values of the γ_{ij}'s and the δ_i's, and we are searching for small numbers $a_0, \ldots, a_{N-1-t}$ so that when we plug those numbers into (6.4), then the numbers $a_{N-t}, \ldots, a_{N-1}$ are also small.

This problem leads to the PASS lattice L^{PA} generated by the rows of the following N-by-N matrix:

$$L^{\mathrm{PA}} = \left(\begin{array}{cccc|cccc} 1 & 0 & \cdots & 0 & \gamma_{10} & \gamma_{20} & \cdots & \gamma_{t0} \\ 0 & 1 & \cdots & 0 & \gamma_{11} & \gamma_{21} & \cdots & \gamma_{t1} \\ \vdots & \vdots & \ddots & \vdots & \vdots & \vdots & \ddots & \vdots \\ 0 & 0 & \cdots & 1 & \gamma_{1,N-1-t} & \gamma_{2,N-1-t} & \cdots & \gamma_{t,N-1-t} \\ \hline 0 & 0 & \cdots & 0 & q & 0 & \cdots & 0 \\ 0 & 0 & \cdots & 0 & 0 & q & \cdots & 0 \\ \vdots & \vdots & \ddots & \vdots & \vdots & \vdots & \ddots & \vdots \\ 0 & 0 & \cdots & 0 & 0 & 0 & \cdots & q \end{array}\right)$$

The PASS lattice contains the vector

$$\mathbf{a}^{\mathrm{PA}} = (a_0, a_1, a_2, \ldots, a_{N-t-1}, a_{N-t} - \delta_1, a_{N-t+1} - \delta_2, \ldots, a_{N-1} - \delta_t).$$

This may be seen by taking a_0 times the first row, plus a_1 times the second row, and so on up to a_{N-t-1} times the $(N-t-1)^{\mathrm{st}}$ row, and then adding or subtracting appropriate multiples of the final t rows as determined by the congruences (6.4).

The vector $\mathbf{a}^{\mathrm{PA}}$ is not itself a small vector, but if $a_0, \ldots, a_{N-1}$ are small, then $\mathbf{a}^{\mathrm{PA}}$ will be very close to the known vector

$$(0, 0, \ldots, 0, -\delta_1, -\delta_2, \ldots, -\delta_{N-1}). \tag{6.5}$$

Thus the problem of finding the PASS key has been reduced to the problem of finding the vector in the lattice L^{PA} that is closest to the known nonlattice vector (6.5). In other words, solution of the closest vector problem in L^{PA} will find the PASS key.

REMARK 6.3.

- The NTRU lattice L^{NT} was balanced by using a multiple of the identity matrix in the upper lefthand square. One can check that the PASS lattice L^{PA} is already balanced.
- One way to use a lattice reduction algorithm such as LLL to solve the PASS closest vector problem is to increase the dimension of the lattice

by 1. We simply add an extra row and column,

$$\hat{L}^{\mathrm{PA}} = \left(\begin{array}{cccc|cccc|c} 1 & 0 & \cdots & 0 & \gamma_{10} & \gamma_{20} & \cdots & \gamma_{t0} & 0 \\ 0 & 1 & \cdots & 0 & \gamma_{11} & \gamma_{21} & \cdots & \gamma_{t1} & 0 \\ \vdots & \vdots & \ddots & \vdots & \vdots & \vdots & \ddots & \vdots & \vdots \\ 0 & 0 & \cdots & 1 & \gamma_{1,N-1-t} & \gamma_{2,N-1-t} & \cdots & \gamma_{t,N-1-t} & 0 \\ \hline 0 & 0 & \cdots & 0 & q & 0 & \cdots & 0 & 0 \\ 0 & 0 & \cdots & 0 & 0 & q & \cdots & 0 & 0 \\ \vdots & \vdots & \ddots & \vdots & \vdots & \vdots & \ddots & \vdots & \vdots \\ 0 & 0 & \cdots & 0 & 0 & 0 & \cdots & q & 0 \\ \hline 0 & 0 & \cdots & 0 & \delta_1 & \delta_2 & \cdots & \delta_t & 1 \end{array}\right)$$

This augmented PASS lattice $\hat{L}^{\mathrm{PA}}$ contains the small vector

$$\hat{\mathbf{a}}^{\mathrm{PA}} = (a_0, a_1, a_2, \ldots, a_{N-t-1}, a_{N-t}, \ldots, a_{N-1}, 1),$$

so solving the shortest vector problem for $\hat{L}^{\mathrm{PA}}$ will probably solve the relevant closest problem for L^{PA}.

- As with all authentication/digital signature schemes, one must study the possibility of using the information in a long transcript to create an attack. The requirement that for each element α in S we have also α^{-1} in S helps to eliminate certain "second moment" information from a transcript. Coppersmith (private communication) has indicated how, in principle, one can use fourth moment information to formulate an attack. However, experiments show that a transcript containing as many as 100 million digital signatures created by a single PASS key are insufficient to gain useful information. For further details, see [**HLS**].

References

[AD] M. Ajtai, C. Dwork, *A public-key cryptosystem with worst case/average case equivalence*, in Proc. 29th ACM Symposium on Theory of Computing, 1997, 284–293.

[BS] J. Blömer, J.-P. Seifert, *On the Complexity of Computing Short Linearly Independent Vectors and Short Bases in a Lattice*, STOC '99

[CN] J.Y. Cai, A.P. Nerukar, *An improved worst-case to average-case reduction for lattice problems*, Proc. 38th Symposium on Foundations of Computer Science, 1997, 468–477

[Ca] J.W.S. Cassels, *An introduction to the geometry of numbers*, Grundlehren der mathematischen Wissenschaften volume 99, Springer-Verlag, Berlin, 1959.

[Coh] H. Cohen, *A course in computational algebraic number theory.* Graduate Texts in Math., vol. 138, Springer Verlag, Berlin, 1993.

[Cop] D. Coppersmith, *Small solutions to polynomial equations and low exponent RSA vulnerabilities*, J. of Cryptology **10** (1997), 233–260.

[CS] D. Coppersmith and A. Shamir, *Lattice attacks on NTRU*, in Proceedings of Eurocrypt 97, Lect. Notes in Comp. Sci. 1233 (W. Fumy, ed.), 1997, 52–61.

[DKS] I. Dinur, G. Kindler, S. Safra, *Approximating CVP to within almost-polynomial factors is NP-hard*, Proc. 39th Symposium on Foundations of Computer Science, 1998, 99–109

[GG] O. Goldreich, S. Goldwasser, *On the limits of non-approximability of lattice problems*, Proc. 39th Symposium on Foundations of Computer Science, 1998, 1–9

[GGH] O. Goldreich, S. Goldwasser, S. Halevi, *Public-key cryptography from lattice reduction problems*, in Proceedings of CRYPTO 97, Lect. Notes in Comp. Sci. 1294, 1997, 112–131.

[GMSS] O. Goldreich, D. Micciancio, S. Safra, J.-P. Seifert, *Approximating shortest lattice vectors is not harder than approximating closest lattice vectors*, Electronic Colloquium on Computational Complexity, TR99-002, 1999

[HPS] J. Hoffstein, J. Pipher, J.H. Silverman, *NTRU: A new high speed public key cryptosystem*, in Algorithmic Number Theory (ANTS III), Portland, OR, June 1998, Lecture Notes in Computer Science 1423 (J.P. Buhler, ed.), Springer-Verlag, Berlin, 1998, 267–288.

[HLS] J. Hoffstein, D. Lieman, J.H. Silverman, *Polynomial Rings and Efficient Public Key Authentication*, in Proceeding of the International Workshop on Cryptographic Techniques and E-Commerce (CrypTEC '99), Hong Kong, (M. Blum and C.H. Lee, eds.), City University of Hong Kong Press.

[LG] C.G. Lekkerkerker, P.M. Gruber, *Geometry of numbers* (2nd edition), North-Holland mathematical library volume 37, Elsevier Science Pub. Co., Amsterdam, 1987

[LLL] A.K. Lenstra, H.W. Lenstra Jr., L. Lovśz, *Factoring polynomials with rational coefficients*, Mathematische Ann. **261** (1982), 513–534.

[LP] H.W. Lenstra Jr., M. Pohst, *Computational Number Theory*, Birkhäuser, Boston, 1992.

[Ma] A. May, *Auf Polynomgleichungen basierende Public-Key-Kryptosysteme*, Johann Wolfgange Goethe-Universitat, Frankfurt am Main, Fachbereich Informatik. (Master Thesis in Computer Science, 4 June, 1999; Thesis advisor, Dr. C.P. Schnorr) Available at <www.mi.informatik.uni-frankfurt.de/research/mastertheses.html>

[MH] R. Merkle, M. Hellman, *Hiding information and signatures in trapdoor knapsacks*, IEEE Trans. Inform. Theory, **IT-24**, September 1978 ,525–530.

[Min] H. Minkowski, *Geometrie der Zahlen*, (David Hilbert and Andreas Speiser, eds.), Teubner, Leipzig, 1910.

[Mio] D. Miociancio, *The shortest vector in a lattice is hard to approximate to within some constant*, Proc. 39th Symposium on Foundations of Computer Science, 1998, 92–98.

[Ng] P. Nguyen, *Cryptanalysis of the Goldreich-Goldwasser-Halevi Cryptosystem from Crypto '97*, Advances in Cryptology—Proceedings of CRYPTO '99, (August 15–19, 1999, Santa Barbara, California), M. Wiener (ed.), Lecture Notes in Computer Science, Springer-Verlag.

[PZ] M. Pohst, H. Zassenhaus, *Algorithmic Algebraic Number Theory*, Cambridge University Press, 1989.

[Sch1] C.-P. Schnorr, *A hierarchy of polynomial time lattice basis reduction algorithms*, Theoretical Computer Science **53** (1987), 201–224.

[Sch2] ——— *A more efficient algorithm for lattice basis reduction*, J. Algorithms **9** (1988), 47–62.

[SE] C.-P. Schnorr and M. Euchner, *Lattice basis reduction: improved practical algorithms and solving subset sum problems*, Math. Programming **66** (1994), 181–199.

[Sh] A. Shamir, *A polynomial-time algorithm for breaking the basic Merkle-Hellman cryptosystem*, in Proceedings of the 23rd IEEE Symposium on Foundations of Computer Science, IEEE, 1982, 145-152.

[Si] C.L. Siegel, *Lectures on the geometry of numbers*, Springer-Verlag, Berlin, 1989.

[Ste] J. Stern, *Lattices and Cryptography: An Overview* in Public Key Cryptography (PKC '98), Lecture Notes in Computer Science 1431, Springer-Verlag, 1998, 50–54.

[Sti] D. Stinson, *Cryptography: Theory and Practice.* CRC Press, 1997.

[dW] B.M.M. de Weger, *Algorithms for Diophantine Equations*, CWI Tract 65, Stichting Math. Centrum, 1989.

NTRU Cryptosystems, Inc. and Brown University, Mathematics Department, Box 1917, Providence, RI 02912 USA

E-mail address: jhs@ntru.com (www.ntru.com), jhs@math.brown.edu

DIMACS Series in Discrete Mathematics
and Theoretical Computer Science
Volume **64**, 2004

Zeta Functions of Graph Coverings

H. M. Stark and A. A. Terras

ABSTRACT. We consider zeta and L-functions of graphs and their similarities to zeta and L-functions of algebraic number fields.

1. Zeta Functions.

Our subject is the graph theoretic analogue of the Riemann and Dedekind zeta functions as well as the Artin L-function. Some references are Bass [**1**], Ihara [**3**], Hashimoto [**2**], Stark and Terras [**5**], [**6**], Sunada [**7**].

Consider a connected finite (not necessarily regular) graph X with vertex set V and undirected edge set E and with fundamental group of rank r. For an example (with loops and multiedges) look at Figure 1. There $|V| = 4$, $|E| = 6$, and $r = 3$. Note that r is the number of edges which must be cut in the graph to produce a spanning tree T (a connected subgraph with the same number of vertices). It follows that $r = |E| - (|V| - 1) = |E| - |V| + 1$.

2000 *Mathematics Subject Classification.* Primary 11M41.

Key words and phrases. Ihara zeta function, Artin L-function of a normal graph covering, induced representation, Galois group of normal graph covering.

Research of the first author at the Institute for Advanced Study was supported by The Weyl Fund.

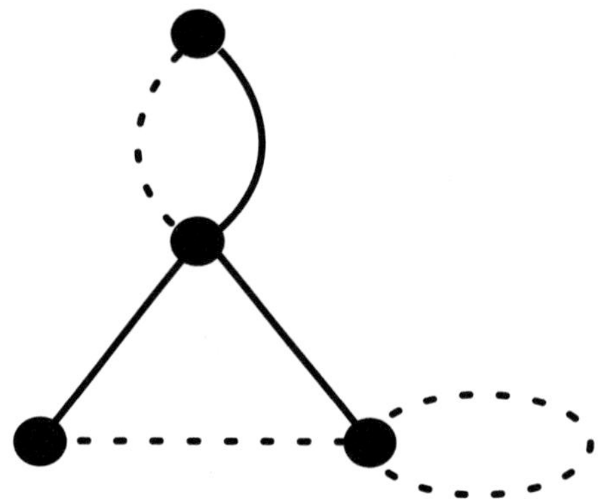

FIGURE 1. A graph X and its spanning tree T. Here 3 edges of X are cut to form T. The edges of the tree are indicated with solid lines and the cut edges with dotted lines.

A **cycle** in the graph is a closed (directed) path. The two types of **elementary cycle equivalences** are:

i) $abb^{-1}cd \sim acd$

ii) $acd \sim cda$, where we have written the cycles in terms of the sequence of edges they traverse.

Equivalence i) eliminates **backtracking**. Use i) and ii) together to eliminate **tails**. Then define $[C]$ to be the **equivalence class** of the cycle C. We always choose C in $[C]$ to have a minimum number of edges. Thus C has **no backtracking or tails**.

The class $[C]$ is **prime** (we say "C is prime") when $C \neq D^m$ $(m > 1)$.

The **zeta functions** of X all have the form:

$$\zeta(\cdot, X) = \prod_{\substack{[C]\ prime \\ in\ X}} (1 - \mathbb{N}(C))^{-1},$$

where $\mathbb{N}(C)$ is a graph theoretic analogue of the norm in number theory, which will be defined in various ways below. The product is infinite unless X is a cycle.

The simplest case is the **Ihara zeta function** (or **vertex** zeta function) for which $\mathbb{N}_V(C) = u^{\nu(C)}$, where u is a complex variable and $\nu(C) =$ length of C. We have

DEFINITION 1.

$$\zeta_V(u, X) = \prod_{\substack{[C]\ prime \\ in\ X}} (1 - u^{\nu(C)})^{-1}.$$

This infinite product (assuming X is not a cycle) is the reciprocal of a polynomial. This is confusing since it also appears to have a pole of infinite order at $u = 1$, but the product has radius of convergence smaller than 1.

The next simplest case is the multiedge zeta function. Here we need to direct the edges of X. We have $2|E|$ directed edges. Then we introduce a complex variable $w(i, j)$ for each pair of directed edges i and j such that i "feeds into" j provided that $i \neq j^{-1}$. Set all remaining $w(i, j) = 0$. The **multiedge matrix** then is

$$W = (w(i,j))_{1 \leq i,j \leq 2|E|}.$$

Suppose the cycle $C = a_1 a_2 \cdots a_n$. Then the **multiedge norm** of C is

$$\mathbb{N}_E(C) = w(a_1, a_2)w(a_2, a_3) \cdots w(a_{n-1}, a_n)w(a_n, a_1).$$

DEFINITION 2. The **multiedge zeta function** is

$$\zeta_E(W, X) = \prod_{\substack{[C] \ prime \\ in \ X}} (1 - \mathbb{N}_E(C))^{-1}.$$

All these zeta functions are reciprocals of polynomials. In particular we have:

(1) $$\zeta_E(W, X) = \det\left(I_{2|E|} - W\right)^{-1},$$

There are also multipath zeta functions attached to the fundamental group of X but we do not discuss them here. See Stark and Terras [**6**].

Next let us consider an **unramified finite covering** graph Y of X. If our graphs had no multiple edges or loops this would simply mean that there is a covering map $\pi : Y \longrightarrow X$ such that π is an onto graph map and for each $x \in X$ and each $y \in \pi^{-1}(x)$, the set of points adjacent to y in Y is mapped by π 1-1, onto the set of points adjacent to x in X. However, to make Galois theory work in general, we need a more precise definition.

Each edge of X is arbitrarily assigned a direction. A covering map π not only takes vertices of Y to vertices of X but also takes edges of Y to edges of X. Directions are then assigned to edges of Y so that π gives a direction preserving map of edges. Further it is required that π gives an isomorphism between neighborhoods of Y and neighborhoods of X. This is critical to handle loops in X. For instance in Figure 2, "circular neighborhoods" of a selected vertex of X and the three vertices of Y projecting down to it are shown. Within the neighborhood of the vertex on X, a loop yields both an "outbound" edge and "inbound" edge in the neighborhood. Thus in each of the three neighborhoods of the lifted vertex in Y, there should be both outbound and inbound edges projecting to the loop.

In analogy to Riemann surface theory, we will think of the inverse images of a fixed spanning tree T of X as the **sheets** of Y and say that Y is an n-sheeted covering of X when there are n sheets in Y.

DEFINITION 3. We say that a graph automorphism $g : Y \longrightarrow Y$ is a **covering automorphism** of Y/X if $\pi g = \pi$, that is the actions of πg and π coincide on vertices as well as on directed edges of Y.

If Y has n sheets, then there are at most n distinct covering automorphisms.

DEFINITION 4. A **normal** n-sheeted **(unramified) graph covering** Y of X is an n-sheeted graph covering with n covering automorphisms $g : Y \longrightarrow Y$. Then $G = Gal(Y/X)$, the **Galois group** of Y/X, is the set of all these $g's$.

Examples can be found in Figures 5 and 6. Actually, the graph covering Y/X in Figure 6 is a redrawn version of the graph Y/X in Figure 5 which better illustrates some of the concepts of Galois theory.

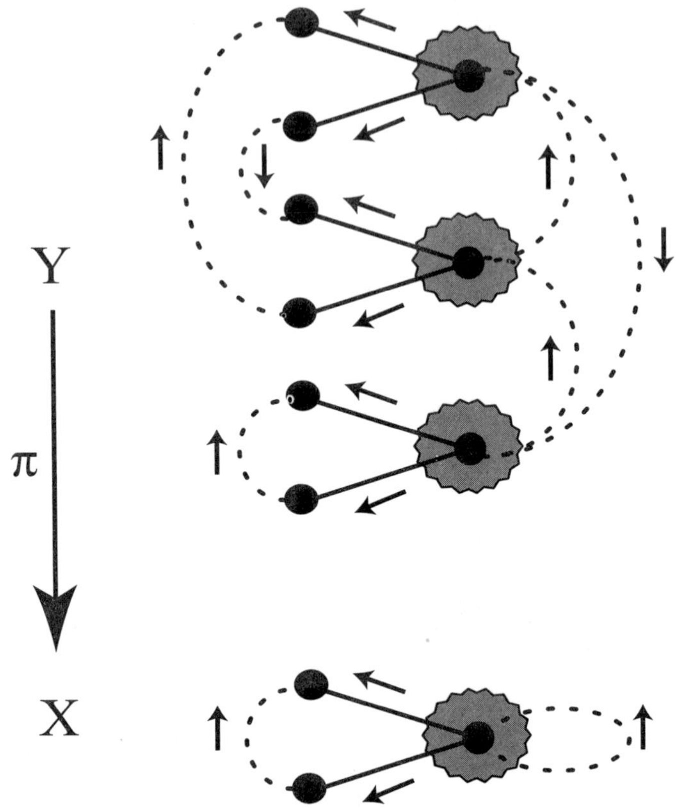

FIGURE 2. A 3-sheeted covering. Shown is a neighborhood of a selected vertex in X together with all of its inverse images under π in Y. Neighborhoods are indicated by zig-zagged circles (shaded grey).

Note that without directed edges, $Gal(Y/X)$ could be much bigger. Also, every group is a Galois group $Gal(Y/X)$ for some X (as long as graphs are connected).

Notation.

If $g, h \in G$, then gh applied to Y means apply h first and then g.

DEFINITION 5. Pick a sheet and call it sheet 1, where 1 denotes the identity of G. If $g \in G$, sheet g is $g \circ (sheet\ 1)$

Basic Calculation:

$$g \circ (sheet\ h) = sheet(gh),$$

since

$$g \circ (h \circ (sheet\ 1)) = gh \circ (sheet\ 1)\,.$$

As a result of our basic calculation, if p is a path on Y starting on sheet a and ending on sheet b and g is in $G = Gal(Y/X)$, then $g \circ p$ starts on sheet ga and ends on sheet gb (see Figure 3).

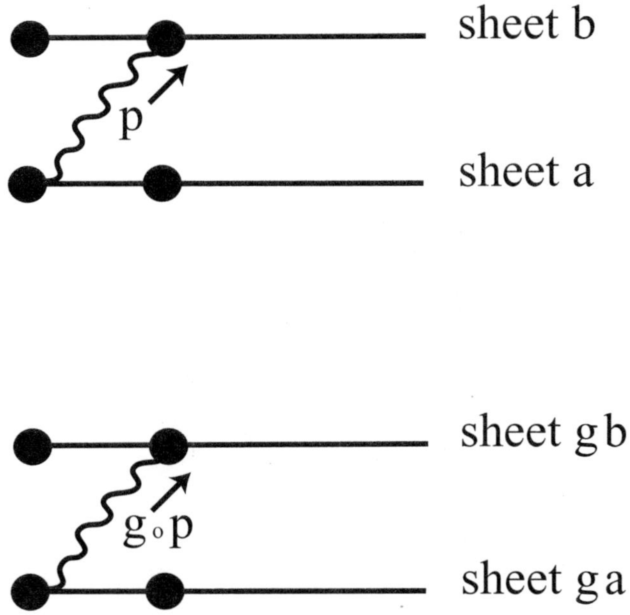

FIGURE 3. The action of the Galois group on paths.

Suppose Y/X is a graph covering with projection map π.We say a graph $\widetilde{X}$ is an **intermediate covering** if there are projection maps $\pi_1 : \widetilde{X} \longrightarrow X$ and $\pi_2 : Y \longrightarrow \widetilde{X}$ such that $\pi = \pi_1\pi_2$. Technically, it is the triple $(\widetilde{X}, \pi_1, \pi_2)$ that gives the intermediate covering. This brings us to the question as to how to distinguish between abstract intermediate coverings.

DEFINITION 6. Suppose i is an isomorphism between $\widetilde{X}$ and $\widetilde{X'}$. If the notation is as in Figure 4, and we have $\pi_1 = \pi_1' i$, then we say i is a **covering isomorphism**. If, in addition, we have $i\pi_2 = \pi_2'$, we say that $\widetilde{X}$ and $\widetilde{X'}$ are **the same** or **equal**.

In [**6**] we show that the Fundamental Theorem of Galois Theory holds. If an intermediate graph $\widetilde{X}$ corresponds to the subgroup H of G, the sheets of $\widetilde{X}$ can be labeled by cosets of H. This is illustrated in Figure 6. Suppose we have the following correspondences between intermediate graphs and subgroups of G:

$$\widetilde{X} \longleftrightarrow H \subset G$$
$$\widetilde{X'} \longleftrightarrow gHg^{-1} \subset G, \text{ for some } g \in G.$$

We say $\widetilde{X}$ and $\widetilde{X'}$ are **conjugate**. We then have the following result.

THEOREM 7. *Intermediate graphs $\widetilde{X}$ and $\widetilde{X'}$ are conjugate if and only if they are covering isomorphic.*

Suppose Y/X is a normal covering with group G. Suppose p is a path of X which lifts to a path $\widetilde{p_i}$ of Y starting on sheet i of Y and ending on sheet j of Y as in Figure 7.

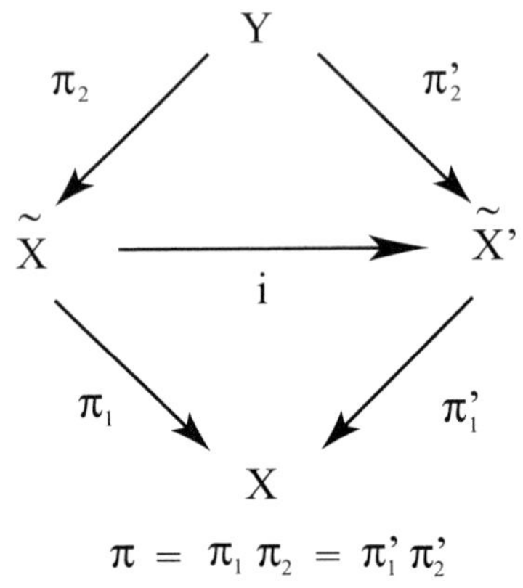

FIGURE 4. A covering isomorphism i of intermediate graphs.

DEFINITION 8. Define the **Frobenius automorphism**

$$Frob(\widetilde{p_i}) = Frob\left(\frac{Y/X}{\widetilde{p_i}}\right) = ji^{-1}.$$

We single out the special case of $i = 1$.

DEFINITION 9. The **normalized Frobenius automorphism** in $G = Gal(Y/X)$ of path p in X is defined to be

$$\sigma(p) = \sigma\left(\frac{Y/X}{p}\right) = Frob(\widetilde{p_1}),$$

which is as pictured in Figure 8.

It is too bad that there does not appear to be an analogue of the normalized Frobenius automorphism for extensions of algebraic number fields. It is what allows us to give a simplified proof of the induction property of Artin L-functions of graph coverings (see the proof of Theorem 14 below), not to mention the evaluation of these Artin L-functions in Theorem 13.

Apply i to $\widetilde{p_1}$ as in Figure 9 and find that if $\sigma = \sigma(p)$,

$$j = i\sigma \quad \text{and} \quad Frob(\widetilde{p_i}) = ji^{-1} = i\sigma i^{-1}.$$

If p and p' are paths on X with p "feeding into" p', then the path pp' means follow p and then follow p'. Suppose $\sigma = \sigma(p)$ and $\sigma' = \sigma(p')$ are the normalized Frobenius automorphisms. Then we have

PROPOSITION 10. $\sigma(pp') = \sigma(p)\sigma(p')$.

PROOF. See Figure 10. □

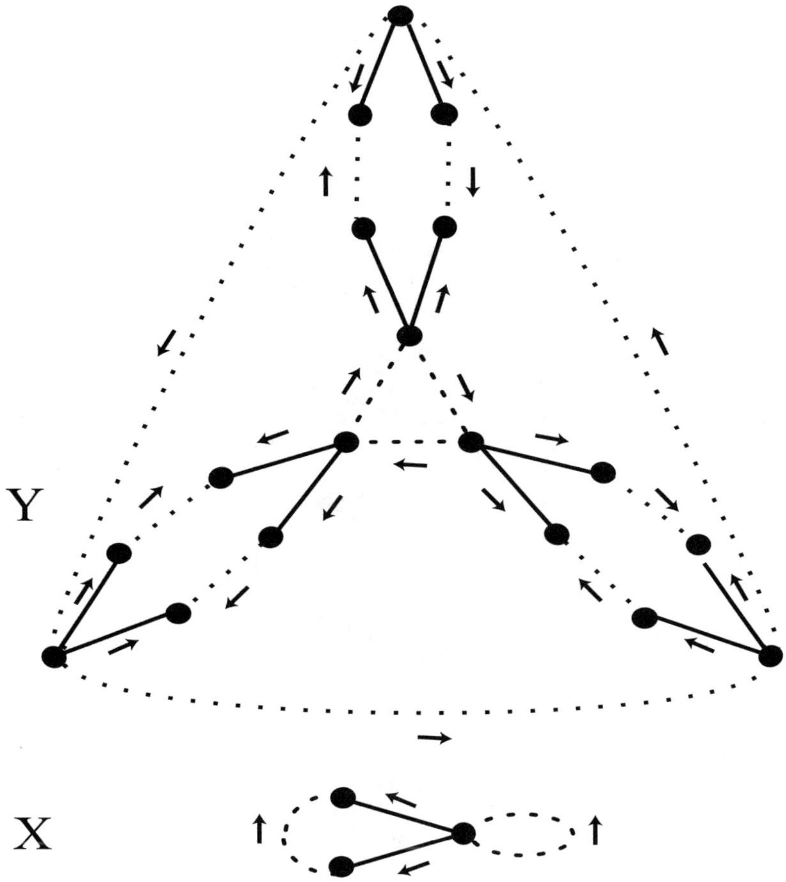

FIGURE 5. A normal covering.

2. Artin L-Functions.

Suppose Y/X is normal with group G and that ρ is a representation of G.

DEFINITION 11. The **vertex Artin L-function** is:

$$L_V(u, \rho, Y/X) = \prod_{\substack{[C] \; prime \\ in \; X}} \det \left(1 - \rho\left(Frob(\widetilde{C})\right) u^{\nu(C)}\right)^{-1}.$$

The **edge Artin L-function** is:

$$L_E(W, \rho, Y/X) = \prod_{\substack{[C] \; prime \\ in \; X}} \det \left(1 - \rho\left(Frob(\widetilde{C})\right) \mathbb{N}_E(C)\right)^{-1}.$$

Here, for each prime C in X, $\widetilde{C}$ denotes any of the lifts of C to Y.

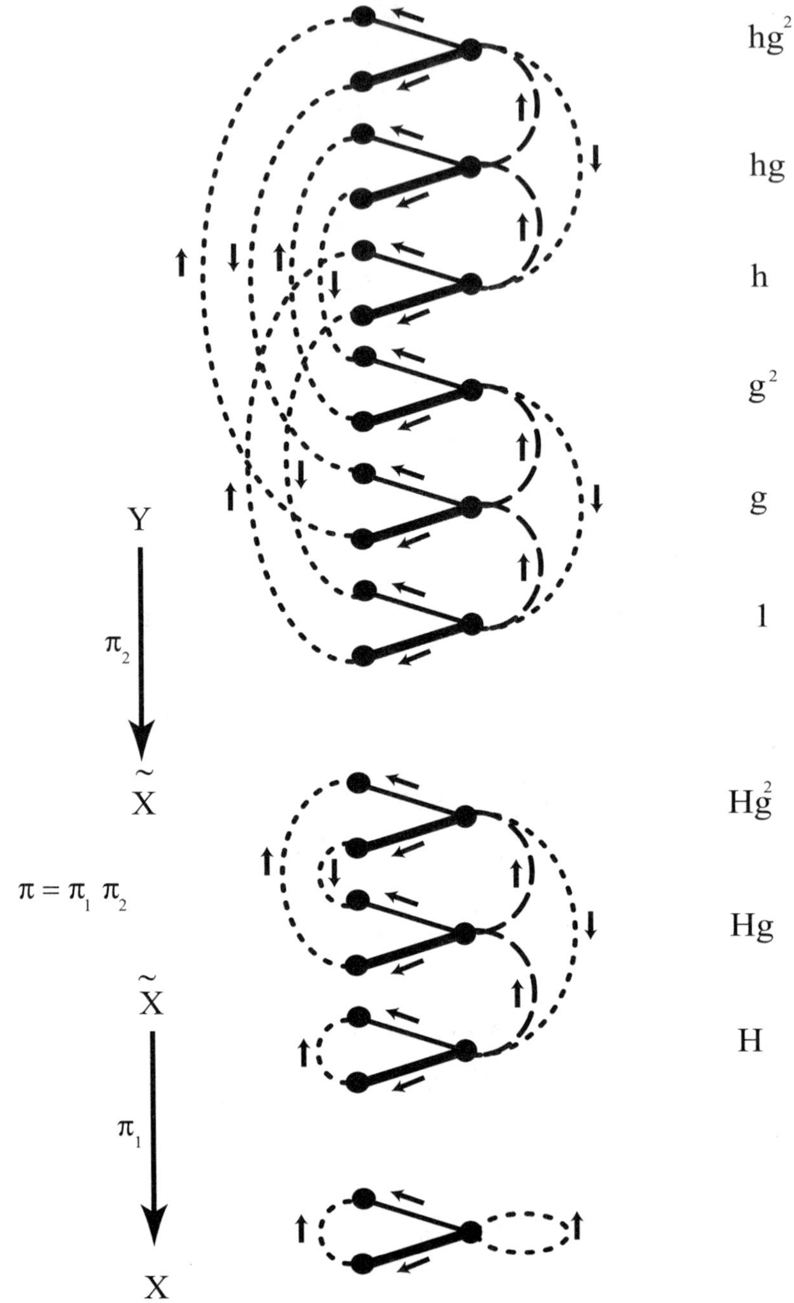

FIGURE 6. A tower of coverings. Y/X is normal with Galois group $G = < g, h > = S_3$ with $g^3 = h^2 = 1$, $gh = hg^2$. The intermediate graph $\widetilde{X}$ corresponds to the subgroup $H = \{1, h\}$ via Galois theory. The tree in X is drawn with solid lines. The tree of $\widetilde{X}$ consists of three copies of the tree of X together with two extra edges shown as long dashed lines. In this way, we can also see the two copies of the tree of $\widetilde{X}$ inside Y. With respect to $\widetilde{X}$, Y has two sheets labeled by the group elements 1 and h of H.

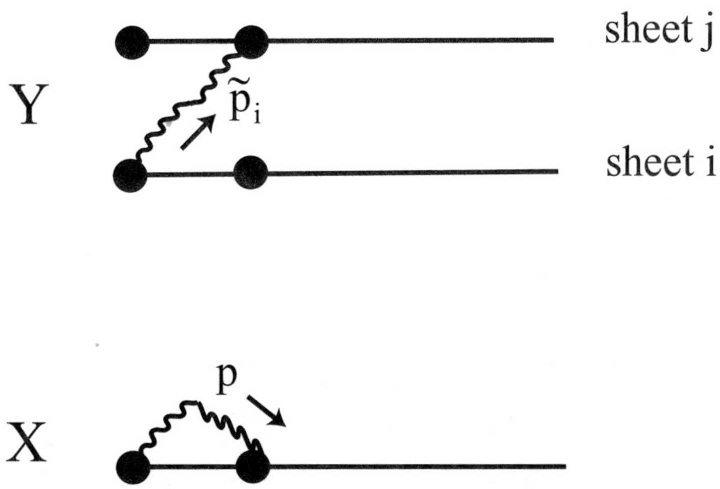

FIGURE 7. Defining the Frobenius automorphism $Frob(\widetilde{p_i})$.

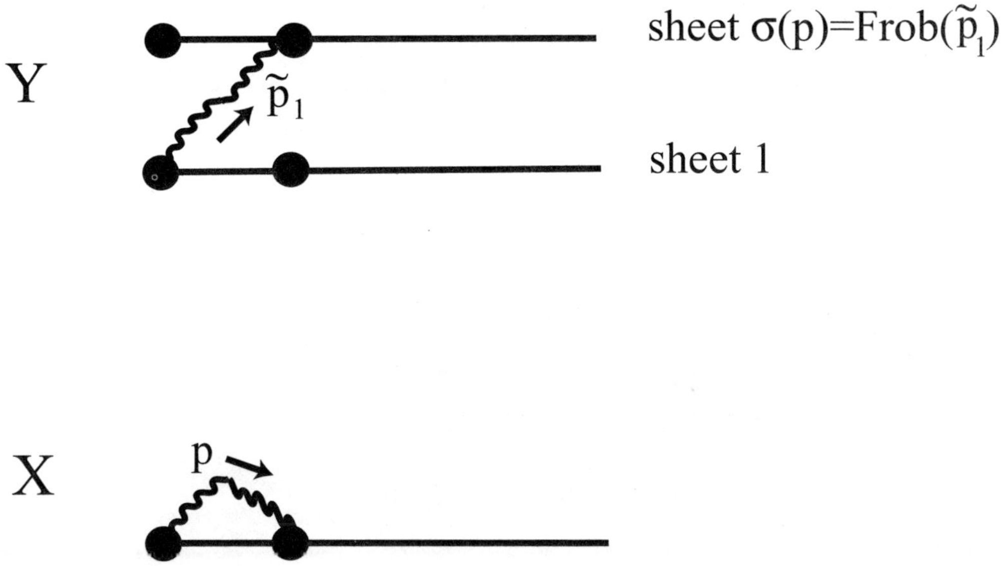

FIGURE 8. The normalized Frobenius automorphism.

Some Properties of the Artin L-Function.

- 1) $L(\cdot, \rho_1, Y/X) = \zeta(\cdot, X)$.
- 2) $L(\cdot, \rho \oplus \rho', Y/X) = L(\cdot, \rho, Y/X)L(\cdot, \rho', Y/X)$

DEFINITION 12. We define $W_\rho = (\rho(\sigma(i))w(i,j))$, a $2|E|d \times 2|E|d$ matrix, where d=degree of ρ.

We have the analogue of formula (1).

FIGURE 9. Dependence of the Frobenius on the beginning sheet of lift.

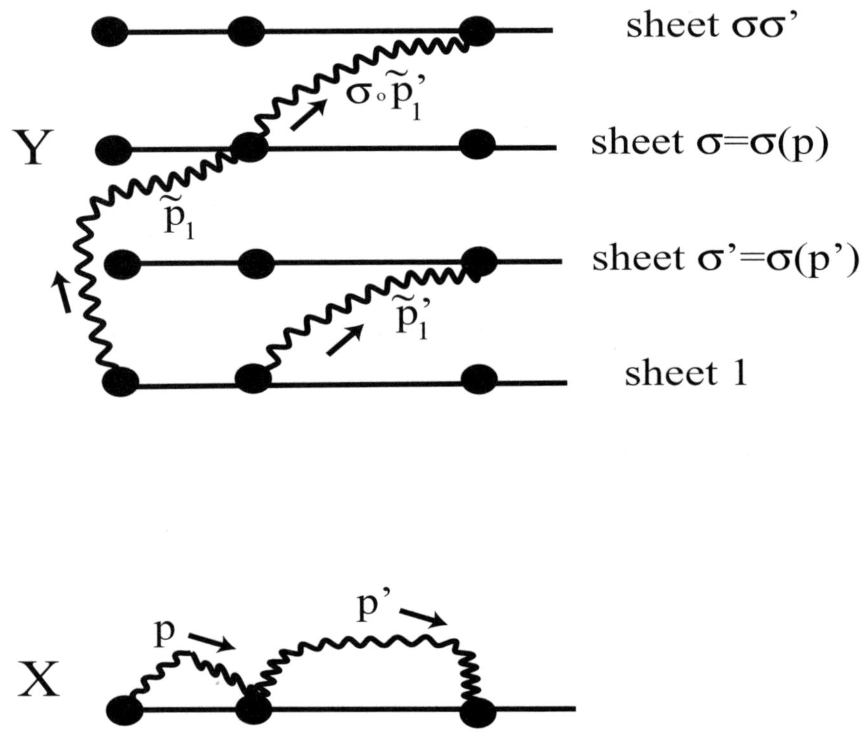

FIGURE 10. Proof of Proposition 10.

THEOREM 13. $L_E(W,\rho,Y/X) = \det(I_{2|E|d} - W_\rho)^{-1}$.

Specialization of the W-Matrix.

We need to consider specialization of the w variables in order to develop the induction property of the multiedge Artin L-functions. Consider Figure 11 where $\widetilde{X}$ covers X and edge $\widetilde{i}$ in $\widetilde{X}$ covers edge i in X. Set $\widetilde{w}\left(\widetilde{i},\widetilde{j}\right)_{spec} = w(i,j)$ if $\widetilde{i}$ feeds into $\widetilde{j}$ as in Figure 11. Then $\widetilde{W}_{spec}$ is the matrix with entries $w(i,j)$ and $\widetilde{W}_{\rho,spec}$ is defined similarly.

THEOREM 14. ***(The Induction Property)***. *Suppose Y/X is a normal graph covering with Galois group G and $\widetilde{X}$ is an intermediate graph corresponding to the*

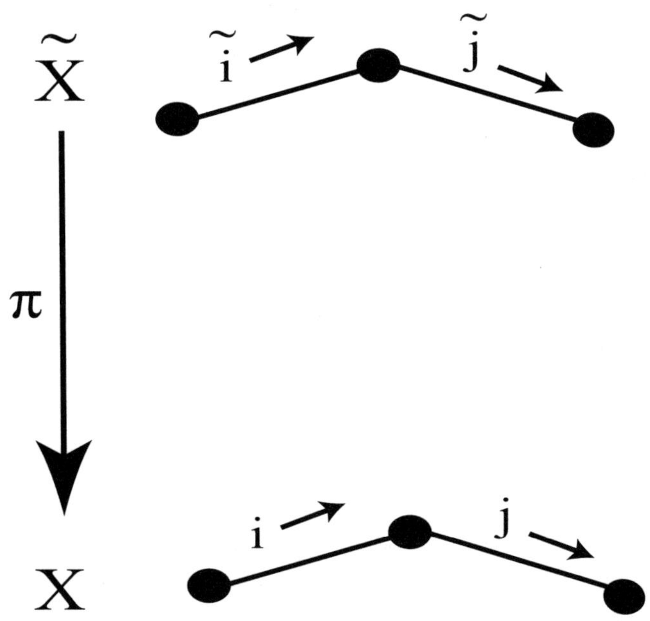

FIGURE 11. Specializing W-variables.

subgroup H of G. Suppose ρ is a representation of H and $\rho^{\#}$ is the corresponding induced representation of G. Then defining $\widetilde{W}_{spec}$ as above:

$$L_E(\widetilde{W}_{spec}, \rho, Y/\widetilde{X}) = L_E(W, \rho^{\#}, Y/X).$$

PROOF. The proof in Stark and Terras [**6**] follows standard number theoretic methods for Artin L-functions. Here we give a direct combinatorial proof by showing that, with the right choice of basis, we have a matrix equality

$$\widetilde{W}_{\rho,spec} = W_{\rho^{\#}}. \tag{2}$$

Theorem 14 will follow immediately from this and the preceding Theorem.

To prove (2), we begin by splitting $\widetilde{W}_\rho$ into $2|E| \times 2|E|$ blocks indexed by the directed edges of X. The ith block row will come from all directed edges of $\widetilde{X}$ projecting to the directed edge e_i of X. Since an edge e_i of X lifts to each sheet of $\widetilde{X}$ and since the sheets of $\widetilde{X}$ are labeled by the cosets Hg_k of H, we can write every edge $\widetilde{e}$ of $\widetilde{X}$ in the form $\widetilde{e_{i,k}}$, where $\widetilde{e}$ projects to e_i and $\widetilde{e}$ has initial vertex on sheet Hg_k.

We must now show that the i,j block $\left(\widetilde{W}_{\rho,spec}\right)_{i,j}$ of $\widetilde{W}_{\rho,spec}$ is nothing more than $\rho^{\#}(\sigma(e_i))\, w_{ij}$. The trivial situation is when $w_{ij} = 0$ which happens when e_i does not feed into e_j. In this case no lift of the e_i feeds into a lift of e_j and therefore the corresponding i,j block of $\widetilde{W}_{\rho,spec}$ is the zero matrix.

The interesting situation is the case that e_i does feed into e_j. In this case, for each lift $\widetilde{e_{i,k}}$ of e_i in $\widetilde{X}$, there is precisely one lift $\widetilde{e_{j,l}}$ of e_j which begins at the terminal vertex of $\widetilde{e_{i,k}}$ and the corresponding variable $\widetilde{w}\,(i,k;j,l)_{spec} = w_{ij}$. Thus $\left(\widetilde{W}_{\rho,spec}\right)_{i,j}$ can in turn be written in block form with rows and columns indexed by the cosets of H where the k,l sub-block is $\rho\left(\widetilde{\sigma}\left(\widetilde{e_{i,k}}\right)\right) w_{ij}$, when $\widetilde{e_{i,k}}$ feeds into $\widetilde{e_{j,l}}$ and is the zero matrix otherwise.

By definition the normalized Frobenius $\widetilde{\sigma}\left(\widetilde{e_{i,k}}\right) = h_{kl} \in Gal(Y/\widetilde{X})$, where e_i lifts to $\widetilde{e_{i,k}}$ on $\widetilde{X}$ and then on to an edge on Y starting on sheet g_k and ending on sheet $h_{kl}g_l$ as in Figure 12. On the other hand, by definition, e_i has a lift to Y starting on sheet 1 of Y and ending on sheet $\sigma\left(e_i\right)$ of Y. Applying g_k to this lifted edge gives the desired relation:

$$g_k\sigma\left(e_i\right) = h_{kl}g_l$$

and hence,

$$\widetilde{\sigma}\left(\widetilde{e_{i,k}}\right) = h_{kl} = g_k\sigma\left(e_i\right) g_l^{-1}.$$

In defining the representation $\rho^{\#}$, it is convenient to extend ρ to a function on G by setting $\rho(g) = 0$, the zero matrix, when g is not in H. We then have

$$\left(\widetilde{W}_{\rho,spec}\right)_{i,j} = \left(\rho\left(g_k\sigma\left(e_i\right) g_l^{-1}\right)\right) w_{ij},$$

where the k,l sub-block on the right is $\rho\left(h_{kl}\right)$ when $\widetilde{e_{i,k}}$ feeds into $\widetilde{e_{j,l}}$ and is the zero matrix otherwise.

It follows from the definition of induced representation $\rho^{\#}$ (see Terras [**8**], for example) that

$$\left(\widetilde{W}_{\rho,spec}\right)_{i,j} = \rho^{\#}\left(\sigma(e_i)\right) w_{ij}$$

in all cases and the proof is complete.

□

COROLLARY 15. ***(Factorization of the Zeta Function).*** *If* $\rho_1^{\#} = \bigoplus_\rho a_\rho\,\rho$, *then*

$$\zeta_E(\widetilde{W}_{spec}, \widetilde{X}) = \prod_\rho L_E(W, \rho, Y/X)^{a_\rho}.$$

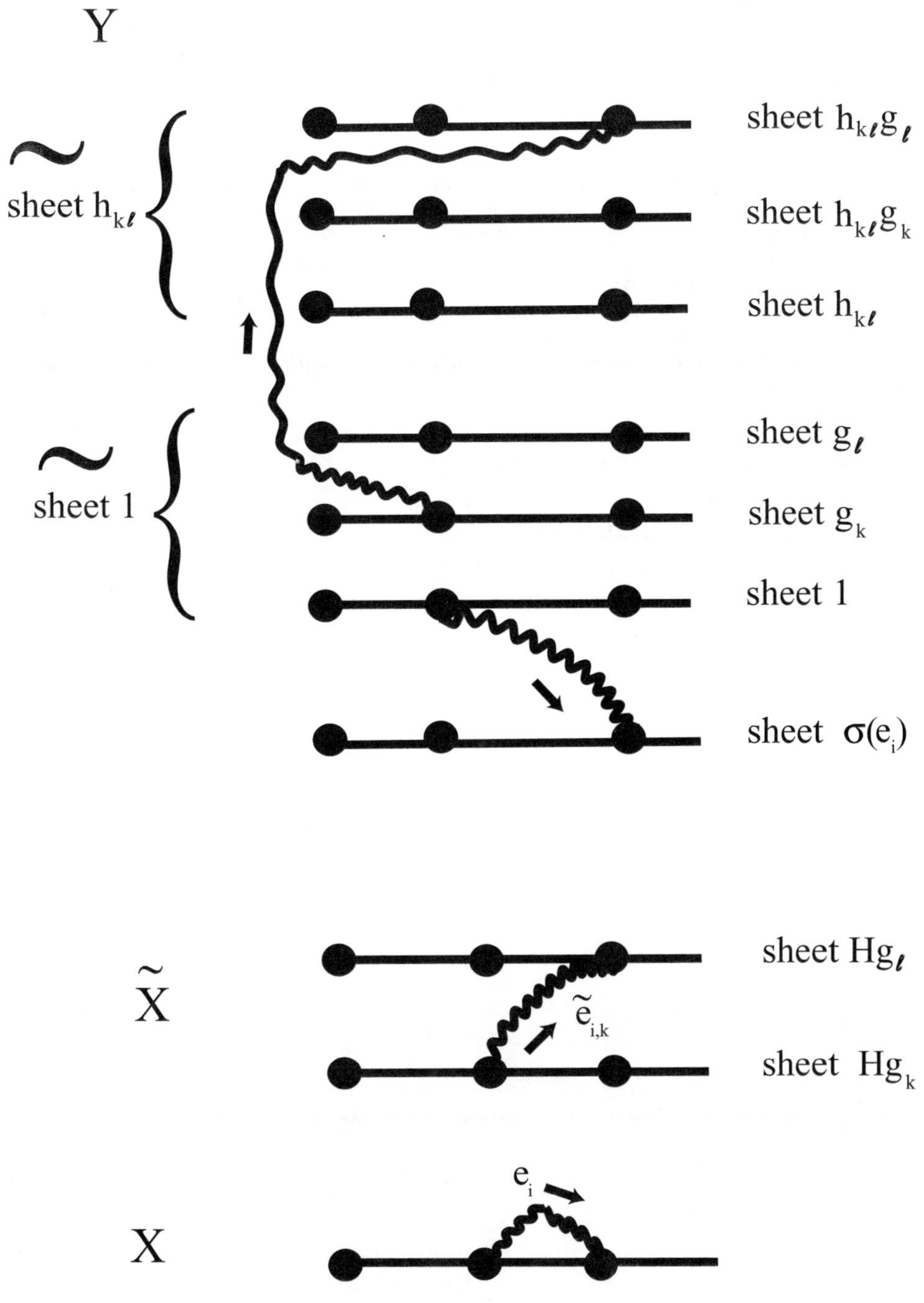

FIGURE 12. Part of the proof of the induction property of multi-edge L-functions.

References

[1] H. Bass, The Ihara-Selberg zeta function of a tree lattice, *Int. J. Math., 3* (1992), 717-797.
[2] K. Hashimoto, Artin type L-functions and the density theorem for prime cycles on finite graphs, *Intl. J. Math., 3* (1992), 809-826.
[3] Y. Ihara, On discrete subgroups of the two by two projective linear group over p-adic fields, *J. Math. Soc. Japan, 19* (1966), 219-235.
[4] H. M. Stark, Galois theory, algebraic number theory and zeta functions, in *From Number Theory to Physics*, M. Waldschmidt et al (Eds.), Springer-Verlag, Berlin, 1992.
[5] H. M. Stark and A. A. Terras, Zeta functions of finite graphs and coverings, *Adv. in Math., 121* (1996), 124-165.
[6] H. M. Stark and A. A. Terras, Zeta functions of finite graphs and coverings, Part II, *Adv. in Math.,154* (2000), 132-195.
[7] T. Sunada, L-functions in geometry and applications, *Lecture Notes in Math., 1201*, Springer-Verlag, N.Y., 1986, 266-284.
[8] A. Terras, *Fourier Analysis on Finite Groups and Applications*, Cambridge U. Press, Cambridge, 1999.

MATH. DEPT., U.C.S.D., SAN DIEGO, CA 92093-0112
E-mail address: stark@math.ucsd,edu

MATH. DEPT., U.C.S.D., SAN DIEGO, CA 92093-0112
E-mail address: aterras@ucsd.edu

DIMACS Series in Discrete Mathematics
and Theoretical Computer Science
Volume **64**, 2004

COMPARISON OF SELBERG'S TRACE FORMULA WITH ITS DISCRETE ANALOGUES

Audrey Terras

ABSTRACT. We compare the original trace formula of Selberg for discrete groups acting on the Poincaré upper half plane with that for trees and finite upper half planes.

1. Introduction.

The spectral theory of operators on Hilbert spaces became important in the 1930's and 40's with Schrödinger's equation in quantum mechanics $\mathcal{L}\phi_n = \lambda_n \phi_n$. Given some favorite operator $\mathcal{L}$; e.g. the Laplacian Δ on $L^2(M)$, for some manifold M, one has favorite questions such as those asked in quantum chaos theory (see Sarnak [**13**] or Hejhal et al [**9**] or the talks given in the spring, 1999 random matrix theory workshop at the web site http://www.msri.org).

Examples of Questions of Interest:

- 1) Does the spectrum of $\mathcal{L}$ determine M? Can you hear the shape of M?
- 2) Give bounds on the spectrum of $\mathcal{L}$.
- 3) Does the histogram of the spectrum approach the Wigner semi-circle distribution [**21**]?
- 4) Arrange the spectrum so that $\lambda_n \leq \lambda_{n+1}$. Consider $\lambda_{n+1} - \lambda_n$ normalized to have mean 1. What is the limiting histogram for the level spacings?
- 5) Behavior of the nodal lines ($\phi_n = 0$) of eigenfunctions.

It would be nice to know the answer to such questions when

$$M = SL(n,\mathbb{Z})\backslash SL(n,\mathbb{R})/SO(n,\mathbb{R}).$$

Only for $n = 2, 3$ is there much information. The trace formula is the main tool. So here we consider replacing this difficult continuous problem with two sorts of discrete ones. Our goal here is to present a dictionary for the 2 types of trace formula. There is, of course, a third type - the adelic trace formula, but we will not go there.

2000 *Mathematics Subject Classification.* 11F72, 11T60, 05C05.

Key words and phrases. Selberg's trace formula, symmetric space, Poincare upper half plane, k-regular tree, finite upper half plane, Selberg zeta function, Ihara zeta function.

References for the continuous trace formula are: Buser [**3**], Elstrodt [**5**], Elstrodt, Grunewald and Mennicke [**6**], Hejhal [**8**], Selberg [**14**] and Terras [**18**]. References for the trace formula on trees are: Ahumada [**1**], Nagoshi [**12**], Terras [**17**] and Venkov and Nikitin [**20**]. A reference for the trace formula on finite upper half planes is Terras [**17**].

2. Comparison of Symmetric Spaces.

We arrange our comparisons of symmetric spaces G/K in a series of figures and tables. Figure 1 shows the 3 spaces with the Poincaré upper half plane H on the upper left, the 3-regular tree T on the bottom, and the finite upper half plane H_3 over the field with 3 elements on the upper right. In contrast to the case of the finite upper half plane, we cannot draw all of the infinite spaces H and T.

In Table 1, the first column belongs to the Poincaré upper half plane H which is the set of complex numbers with positive imaginary part. In the second column is the $(q+1)$-regular tree T, which is a $(q+1)$-regular connected graph without circuits or cycles. In the third column we find the finite upper half plane H_q over $\mathbb{F}_q$. Here $\mathbb{F}_q$ denotes a finite field with an odd number q of elements. Pick a non-square $\delta \in \mathbb{F}_q$. Then H_q consists of points $z = x + y\sqrt{\delta}$ with $x, y \in \mathbb{F}_q$ and y non-zero. Table 1 can be viewed as a dictionary for the 3 languages.

Symmetric spaces G/K can be described in terms of Gelfand pairs (G, K) of group G and subgroup K. This means that the convolution algebra of functions on G which are K bi-invariant is a commutative algebra. Here convolution of functions $f, g : G \longrightarrow \mathbb{C}$ means

$$f * g(x) = \sum_{y \in G} f(y) g(y^{-1}x).$$

For the Poincaré upper half plane the group G is the special linear group $SL(2,\mathbb{R})$ which consists of 2×2 real matrices of determinant 1 acting on H by fractional linear transformation. In the case of the finite upper half plane H_q the group is the general linear group $G = GL(2, \mathbb{F}_q)$ which consists of 2×2 matrices with entries in the finite field and non-0 determinant.

For the tree T, we will emphasize the graph theory rather than the group theory as that would involve p-adic groups which would require more background of the reader. See Nagoshi [**12**], for example, for this point of view. Thus for the $(q+1)$-regular tree we view the group G as a group of graph automorphisms, meaning 1-1, onto maps from T to T preserving adjacency. There are 3 types of such automorphisms of T. See Figá-Talamanca and Nebbia [**7**] for a proof.

AUTOMORPHISMS OF THE TREE T

1) rotations fix a vertex;

2) inversions fix an edge and exchange endpoints;

3) hyperbolic elements ρ fix a geodesic $\{x_n | n \in \mathbb{Z}\}$ and $\rho(x_n) = x_{n+s}$. That is, ρ shifts along the geodesic by $s = \nu(\rho)$. See the definition of geodesic below.

The subgroup K of $g \in G = SL(2,\mathbb{R})$ such that $gi = i$ is easily seen to be the special orthogonal group $SO(2,\mathbb{R})$ of 2×2 rotation matrices. The analogue for the tree is the subgroup of rotations of T fixing the origin $\mathcal{O}$ (which is any vertex of T chosen once and for all). The analogue for $G = GL(2, \mathbb{F}_q)$ consists of elements fixing the origin $\sqrt{\delta}$ in the finite upper half plane H_q and it is the subgroup K of matrices of the form $\begin{pmatrix} a & b\delta \\ b & a \end{pmatrix}$. It is easily seen that the map sending this matrix to $a + b\sqrt{\delta}$ provides a group isomorphism from K to the multiplicative group of $\mathbb{F}_q(\sqrt{\delta})$.

Table 1. Basic geometry of the 3 symmetric spaces. The list starts with the group G which acts on the space, the origin, the subgroup K fixing the origin. The Poincaré upper half plane is in the first column. It is continuous. The $(q+1)$-regular tree is in the second column. It is infinite and discrete. The finite upper half plane is in the third column. We give the formulas for the arc length and Laplacian on the 3 spaces. In the case of the finite upper half plane the arc length does not give a metric as it has values in the finite field. And there are $q-2$ graphs of degree $q+1$ associated to the finite upper half plane using this pseudo-distance. Thus for finite upper half planes there are many Laplacians formed using the adjacency operators of these graphs. We also give formulas for spherical functions, the spherical transform and its inverse, the horocycle transform.

Poincare Upper Half Plane	**$(q+1)$-regular Tree**
space $= H =$	**space** $= T =$
$\{z = x + iy \mid x, y \in \mathbb{R}, y > 0\}$	$(q+1)$-regular connected graph
	with no circuits
group $G = SL(2, \mathbb{R})$	**group** $G =$
$\left\{\begin{pmatrix} a & b \\ c & d \end{pmatrix} \middle\vert\, ad - bc = 1\right\}$	graph automorphisms of T
	rotations, inversions, shifts
group action $gz = \frac{az+b}{cz+d}$, $z \in H$	**group action** gx
origin $=$ i	**origin** $=$ any point $\mathcal{O}$
subgroup K fixing origin	**subgroup** K fixing origin
$SO(2, \mathbb{R}) = \{g \in G \vert {}^t gg = I\}$	rotations about $\mathcal{O}$
$\mathbf{H} \cong \mathbf{G/K}$	$\mathbf{T} \cong \mathbf{G/K}$
arc length $ds^2 = \frac{dx^2+dy^2}{y^2}$	**graph distance**=# edges in path
	joining 2 points $x, y \in T$
Laplacian $\Delta = y^2 \left(\frac{\partial}{\partial x^2} + \frac{\partial}{\partial y^2}\right)$	$\Delta = A - (q+1)I$,
	A=**adjacency operator** for T
spherical function	**spherical function**
$h_s(z) = \int_K (\mathrm{Im}(ka))^s dk$,	$p_s(x) = q^{-sd(x,\mathcal{O})}$
$\Delta h_s = s(s-1)h_s$	$h_s(x) = c(s)p_s(x) + c(1-s)p_{1-s}(x)$
spherical transform	**spherical transform**
of $f : K\backslash G/K \longrightarrow \mathbb{C}$	of $f : K\backslash G/K \longrightarrow \mathbb{C}$
$\widehat{f}(s) = \int_H f(z)\overline{h_s(z)} \frac{dxdy}{y^2}$	$\widehat{f}(s) = \sum_{x\in T} f(x)\overline{h_s(x)}$
inversion $f(z) =$	**inversion** $f(x) =$
$\frac{1}{4\pi} \int_{\mathbb{R}} \widehat{f}(\frac{1}{2} + it) h_{\frac{1}{2}+it}(z) t \tanh \pi t\, dt$	$\int\limits_{-2\sqrt{q}}^{2\sqrt{q}} \widehat{f}(\frac{1}{2} + it) h_{\frac{1}{2}+it}(x) \frac{\sqrt{4q-t^2}}{(q+1)^2-t^2} dt$
horocycle transform	**horocycle transform**
	$\mathfrak{h} =$ horocycle on T
$F(y) = y^{-1/2} \int_{\mathbb{R}} f(x + iy)dx$	$F(\mathfrak{h}) = \sum_{x\in\mathfrak{h}} f(x)$
invertible	invertible

Finite Upper Half Plane
space $= H_q =$
$\{z = x + \sqrt{\delta} y \mid x, y \in \mathbb{F}_q, y \neq 0\}$
where $\delta \neq a^2$, for any $a \in \mathbb{F}_q$
group $G = GL(2, \mathbb{F}_q)$
$\left\{ \begin{pmatrix} a & b \\ c & d \end{pmatrix} \middle\vert ad - bc \neq 0 \right\}$
group action $gz = \frac{az+b}{cz+d}$, $z \in H_q$
origin $= \sqrt{\delta}$
subgroup K fixing origin $=$
$K = \left\{ \begin{pmatrix} a & b\delta \\ b & a \end{pmatrix} \in G \right\} \cong \mathbb{F}_q(\sqrt{\delta})^*$
$\mathbf{H \cong G/K}$
pseudo-distance $d(z, w) = \frac{N(z-w)}{\operatorname{Im} z \operatorname{Im} w}$,
$Im(x + y\sqrt{\delta}) = y, \ \ Nz = x^2 - y^2\delta$
$X_a = X_q(\delta, a)$, **graph** - vertices $z \in H_q$
edge between z and w if $d(z, w) = a$
$\Delta = A_a - (q+1)I$, if $a \neq 0$ or 4δ
A_a $=$ **adjacency operator** for X_a
spherical function
$h_\pi(z) = \frac{1}{\lvert K\rvert} \sum_K \chi_\pi(kz)$, $\pi \in \widehat{G}^K$,
i.e., π occurs in $Ind_K^G 1, d_\pi = \deg \pi$
spherical transform
of $f : K\backslash G/K \longrightarrow \mathbb{C}$
$\widehat{f}(\pi) = \sum_{z \in H_q} f(z)\overline{h_\pi(z)}$
inversion
$f(z) = \frac{1}{\lvert G\rvert} \sum_{\pi \in \widehat{G}^K} d_\pi \widehat{f}(\pi) h_\pi(z)$
horocycle transform
$F(y) = \sum_{x \in \mathbf{F}_q} f(x + y\sqrt{\delta})$
not invertible

In Figure 2 we show the finite upper half plane graph for $q = 3$ and one of the 3 different finite upper half plane graphs for $q = 5$. These graphs are defined in the third column of Table 1 using a G-invariant analogue of the distance in H.

In Figure 3, we illustrate some of the geometry of the symmetric spaces. Geodesics in the Poincaré upper half plane are curves which minimize the Poincaré distance ds from Table 1. It is not hard to see that the points in H on the y-axis form a geodesic and thus so do the images of this curve under fractional linear transformations from $G = SL(2, \mathbb{R})$ since ds is G-invariant. These are semi-circles and half-lines orthogonal to the real axis. These are the straight lines of a non-Euclidean geometry.

Geodesics in the tree are paths $\{x_n \mid n \in \mathbb{Z}\}$ which are infinite in both directions (i.e., x_n is connected to x_{n+1} by an edge).

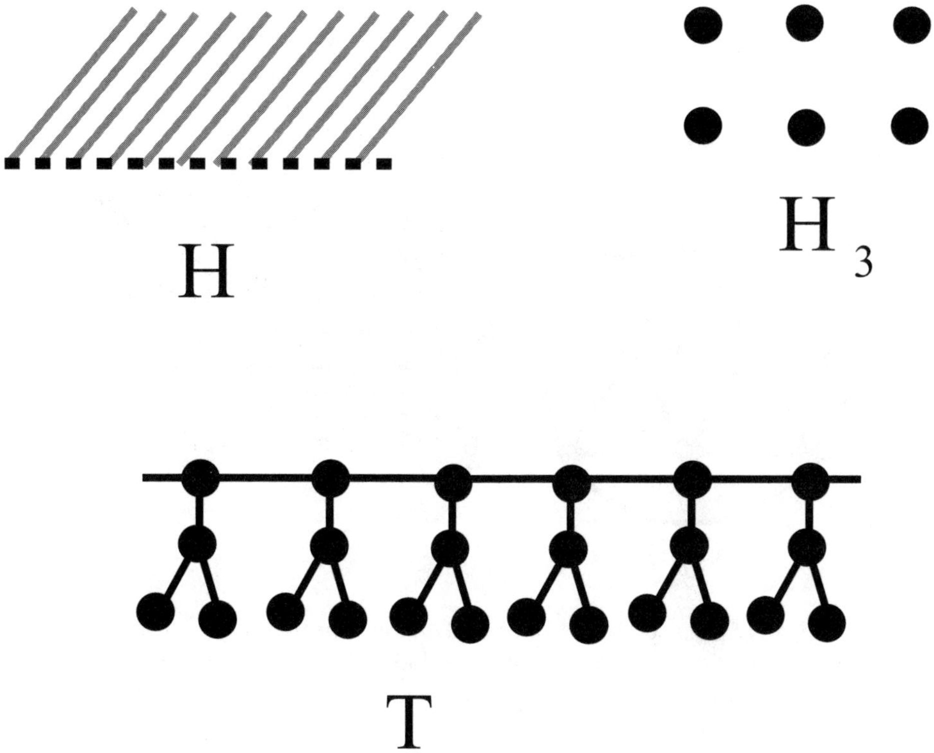

FIGURE 1. The 3 Symmetric Spaces. The Poincaré upper half plane H is on the upper left. The 3-regular tree T is on the bottom and the finite upper half plane H_3 over the field with 3 elements is on the upper right.

We define geodesics in the finite upper half plane to be images of the analogue of the y-axis under fractional linear transformation by elements of $GL(2, \mathbb{F}_q)$. For example, one geodesic in H_q is the set of points $y\sqrt{\delta}$, for $y \in \mathbb{F}_q^* = \mathbb{F}_q - \{0\}$.

Horocycles are curves orthogonal to the geodesics as in Figure 4. So in the Poincaré upper half plane they are lines parallel to the x-axis and their images by fractional linear transformation from $SL(2, \mathbb{R})$. For the finite upper half plane we make the analogous definition of horocycles. One example of a horocycle in H_q is the set of points of the form $x + y_0\sqrt{\delta}$, for $x \in \mathbb{F}_q$. Here y_0 is fixed in $\mathbb{F}_q^*$.

In the tree, horocycles are obtained by fixing a half geodesic or chain $C = [\mathcal{O}, \infty] = \{x_n \mid n \in \mathbb{Z},\ n \geq 0\}$. Then the chain connecting a point x in the tree to infinity along C is called $[x, \infty]$. If x and y are points of T, then $[x, \infty] \cap [y, \infty] = [z, \infty]$. We say x and y are equivalent if $d(x, z) = d(y, z)$. Horocycles with respect to C are these equivalence classes.

Next we consider spherical functions on the symmetric spaces. These are K-invariant eigenfunctions of the Laplacian(s) which are normalized to have the value 1 at the origin. They are indeed analogues of the classical spherical harmonics of Laplace. Since the power function $y^s = \mathrm{Im}(z)^s$ is an eigenfunction of the Laplacian

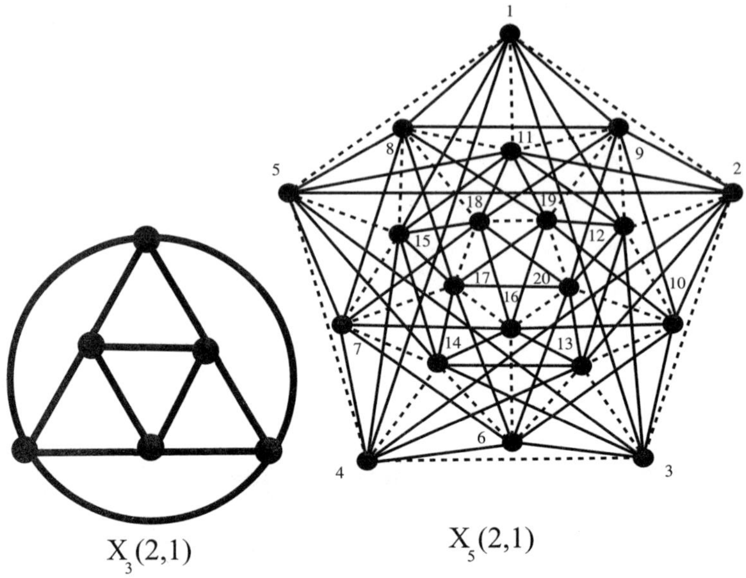

FIGURE 2. Some finite upper half plane graphs for $q = 3$ and 5. The graph for $q = 5$ is shown on a dodecahedron whose edges are indicated by dotted lines while the edges of the graph $X_5(2,1)$ are given by solid lines.

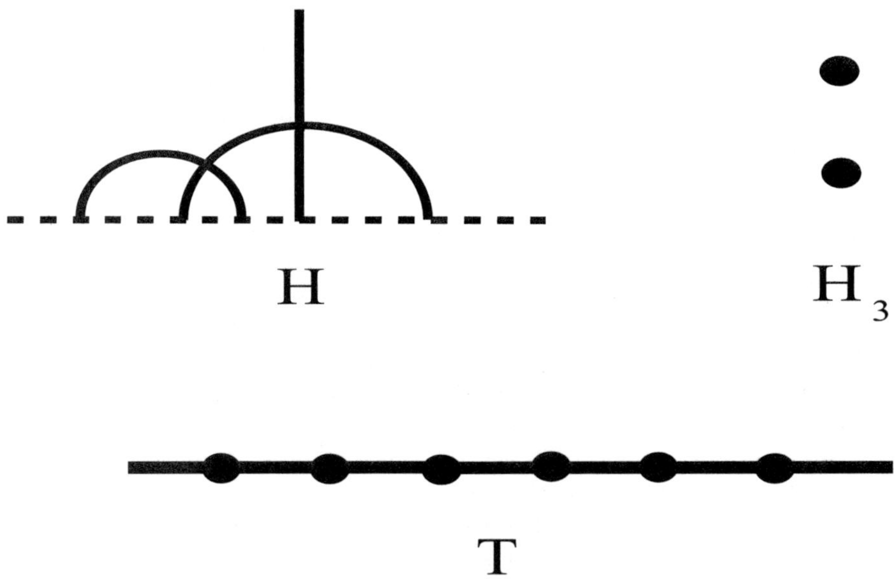

FIGURE 3. Geodesics in the 3 Symmetric Spaces.

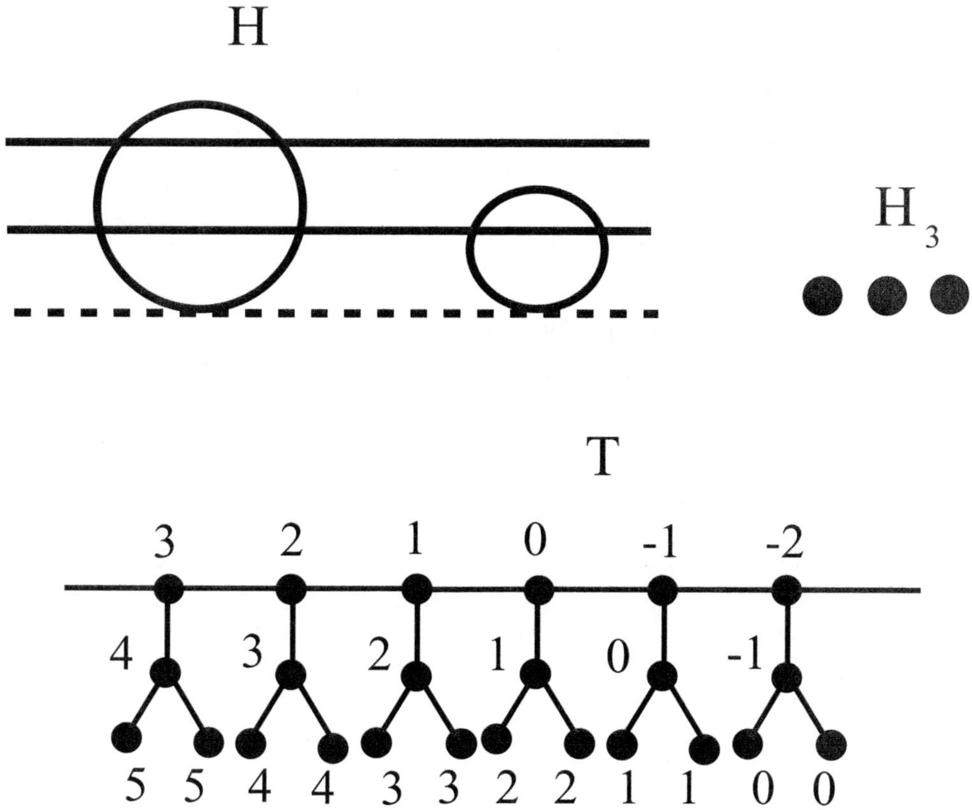

FIGURE 4. Horocycles in the 3 Symmetric Spaces. For the tree, points in the nth horocycle are labeled n

on H corresponding to the eigenvalue $s(s-1)$, we can build up spherical functions from the power function by integration over K. Here dk denotes the Haar measure given by $d\theta$, if $k = \begin{pmatrix} \cos\theta & \sin\theta \\ -\sin\theta & \cos\theta \end{pmatrix}$. See Terras [**18**] for more information on spherical functions for continuous symmetric spaces.

A similar method works for the tree except that the analogue of the power function $p_s(x) = q^{-sd(x,\mathcal{O})}$ is not quite an eigenfunction for the adjacency operator on T with eigenvalue $\lambda = q^s + q^{1-s}$. This is fixed by writing

$$h_s(x) = c(s)p_s(x)+c(1-s)p_{1-s}(x), \quad \text{with} \quad c(s) = \frac{q^{s-1} - q^{1-s}}{(q+1)(q^{s-1} - q^{-s})}, \quad \text{if} \quad q^{2s} \neq 1.$$

Take limits when $q^{2s-1} = 1$. More information on spherical functions on trees can be found in references [**2**], [**4**], [**7**],[**17**], [**19**].

For the finite upper half plane one can build up spherical functions from characters of inequivalent irreducible unitary representations π of G occurring in the left regular representation of G on functions in $L^2(G/K)$. This is the same thing as saying that π occurs in the induced representation $Ind_K^G 1$ defined in Chapter 16 of Terras [**17**]. We use $\widehat{G}^K$ to denote the set of such representations. Then the spherical function h_π corresponding to π is obtained by averaging the character $\chi_\pi = Trace(\pi)$ over K. See Terras [**17**] for more information on these finite spherical functions and the representations of $GL(2, \mathbb{F}_q)$.

The spherical transform of a function f in $L^2(K\backslash G/K)$; i.e., a K-bi-invariant function on G, is obtained by integrating (summing) f times the spherical function over the symmetric space G/K. These transforms are invertible.

The horocycle transform of f is obtained by integrating or summing f over a horocycle $\mathfrak{h}$ in the symmetric space. These horocycle transforms are invertible in 2 out of 3 cases.

3. Comparison of Trace Formulas.

The Selberg trace formula involves another subgroup Γ of G. This subgroup should be discrete and, for ease of discussion, have compact quotient $\Gamma\backslash G/K$. This rules out the subgroups like the modular group $\Gamma = SL(2, \mathbb{Z})$ for which the fundamental domain illustrated in Figure 5 on the left is not compact. For examples of Γ such that the fundamental domain is compact, see Katok [**10**].

Figure 5 also shows an example of a finite graph (the tetrahedron) which is a quotient of the 3-regular tree by its fundamental group Γ. And finally on the right in Figure 5 is a quotient of the finite upper half plane H_9 by $GL(2, \mathbb{F}_3)$. Here $\mathbb{F}_9$ is identified with $\mathbb{F}_3(\sqrt{-1})$ and we write $\sqrt{-1} = i$. The non-square element δ of $\mathbb{F}_3(i)$ is chosen to be $\delta = \sqrt{1+i}$. And H_9 consists of elements $z = x + y\sqrt{1+i}$ with $x, y \in \mathbb{F}_3(i)$ and $y \neq 0$.

In Table 2 we give a dictionary of trace formulas for the 3 types of symmetric spaces considered here. The first column corresponds to the Poincaré upper half plane H, the second to the tree T, and the third to the finite upper half plane H_q.

In column 1 of Table 2, note that there are only two types of conjugacy classes in Γ. The hyperbolic conjugacy classes $\{\gamma\}$ correspond to γ with diagonal Jordan form with distinct diagonal entries t and $1/t$ with $t > 1$. The norm of such an element is $N\gamma = t^2$. The centralizer Γ_γ of γ is a cyclic group with generator the primitive hyperbolic element γ_0. The numbers $logN\gamma_0$ represent lengths of closed prime geodesics in the fundamental domain $\Gamma\backslash H$. The set of these numbers is the length spectrum. The length spectrum can be viewed as analogous to the set of primes in $\mathbb{Z}$.

The spectrum of the Laplacian on the fundamental domain $\Gamma\backslash H$ consists of a sequence of negative numbers λ_n with $|\lambda_n| \longrightarrow \infty$. And $\lambda_0 = 0$ corresponds to the constant eigenfunction. We write $\lambda_n = s_n(1 - s_n)$. The eigenfunction φ_n corresponding to λ_n is called a Maass wave form. They are much like classical holomorphic automorphic forms. The Selberg trace formula given in Table 2 column 1 involves the spherical transform of a function f in $L^2(K\backslash G/K)$ as well as the horocycle transform of f. Since there are only two types of conjugacy classes in Γ when the fundamental domain is compact and Γ acts without fixed points, there are only two sorts of terms on the right-hand side of the formula. For a Γ like $SL(2, \mathbb{Z})$ with non-compact fundamental domain having fixed points of some elements of Γ (such as i), there will also be parabolic and elliptic conjugacy classes.

The Selberg trace formula on H provides a correspondence between the length spectrum and the Laplacian spectrum. It thus gives analogues of many of the explicit formulas in analytic number theory. There are many applications; e.g. to prove the Weyl law for the number of $|\lambda_n| \leq x$ as $x \longrightarrow \infty$. And one can deduce the prime geodesic theorem. Here we only mention the application to the Selberg zeta function defined in Table 2. The trace formula can be used to show that the Selberg zeta function has many of the properties of the Riemann zeta function and moreover one has the Riemann hypothesis for the Selberg zeta function, saying that the non-trivial zeros are on the line $Res = \frac{1}{2}$.

The middle column of Table 2 gives the graph theoretic analogue of all of this. Here Γ is the fundamental group of the finite connected $(q+1)$-regular graph $X = \Gamma\backslash T$. And the tree T is the universal covering graph of X. The non-central conjugacy classes in Γ are hyperbolic. Again the hyperbolic γ fix a geodesic in T and operate by shifting along the geodesic by $\nu(\gamma)$

The spectrum of the adjacency operator on X consists of eigenvalues $\lambda_n = q^{s_n} + q^{1-s_n}$. If $|\lambda_n| \neq q+1$, then $0 < Res_n < 1$. The graph X is defined by Lubotzky, Phillips and Sarnak [**11**] to be Ramanujan if $|\lambda_n| \neq q+1 \Longrightarrow |\lambda_n| \leq 2\sqrt{q}$. This is equivalent to saying $Res_n = \frac{1}{2}$.

Again the Selberg trace formula in the middle column of Table 2 involves the spherical transform of a function f in $L^2(\Gamma\backslash T)$ as well as the horocycle transform. And it has only two sorts of terms because again there are only two sorts of conjugacy classes in Γ - central and hyperbolic. We give in the table the relationship between the horocycle transform defined in the last rows of Table 1 and that appearing in the trace formula.

Once more there is an application to zeta functions. In this case it is the Ihara zeta function of the finite graph X. That function is defined in column 2 of Table 2 as a product over primitive hyperbolic elements of Γ. This product can also be viewed as one over closed primitive backtrackless, tailless cycles in X. The trace formula gives an explicit formula for the function as the reciprocal of a polynomial. There are also direct combinatorial proofs of this fact that work more generally for irregular graphs. This has a long history in papers of Bass, Hashimoto, Sunada and others. See Stark and Terras [**15**], [**16**] and the references given there. The direct combinatorial proofs of the explicit formula for the Ihara zeta function suggest that there exists a more general version of the Selberg trace formula for irregular graphs. Note that the analogue of the Riemann hypothesis holds for the Ihara zeta function if and only if X is a Ramanujan graph. Of course, some graphs are Ramanujan and others are not. See Terras [**17**] for some examples and references. The finite upper half plane graphs mentioned above are all Ramanujan, for example.

Finally the third column of Table 2 concerns the trace formula for finite upper half planes. Here it is not clear what sort of subgroups Γ one should consider. We look mainly at the case $G = GL(2, \mathbb{F}_q)$ and $\Gamma = GL(2, \mathbb{F}_p)$, where $q = p^n$. There are 4 types of conjugacy classes $\{\gamma\}$ in Γ, according to the Jordan form of γ. These are listed in the 3rd column of Table 2. Then we give the trace formula only in the case that $q = p^2$. The general case $q = p^n$ can be found in Terras [**17**].

Again we see that the left hand side of the trace formula is a sum of spherical transforms of f, this time over irreducible representations of G appearing in the induced representation $Ind_\Gamma^G 1$. The right hand side involves horocycle transforms of f and now there are 4 types of terms because there are 4 types of conjugacy classes. The parabolic terms are not so hard to deal with as in the continuous case since there can be no divergent integrals. The elliptic terms are a bit annoying because they behave differently depending on whether q is an even or odd power of p.

We put ? at the end of the 3rd column of table 2 because we haven't found an analogue of the Selberg zeta function for finite upper half plane graphs. We have worked out a number of examples of the formula in [**17**].

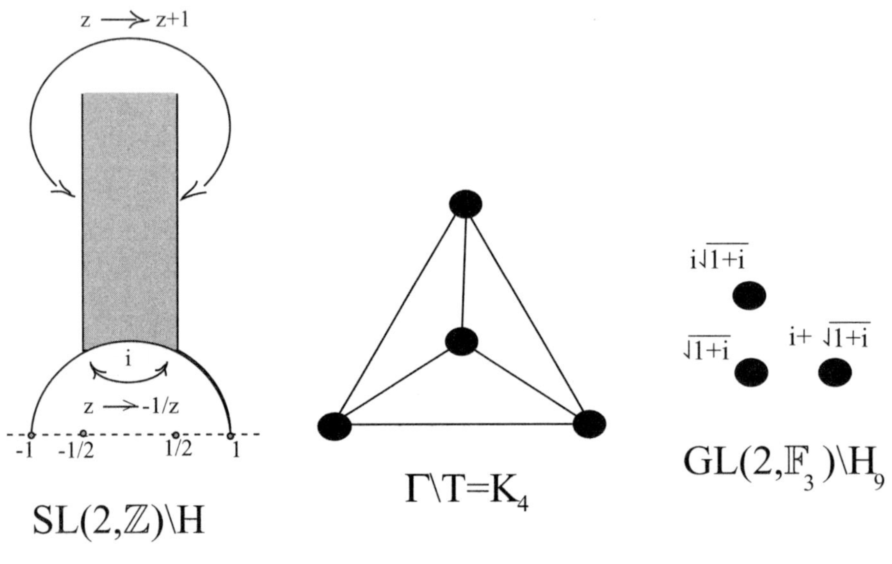

FIGURE 5. Fundamental domains for some discrete subgroups Γ.

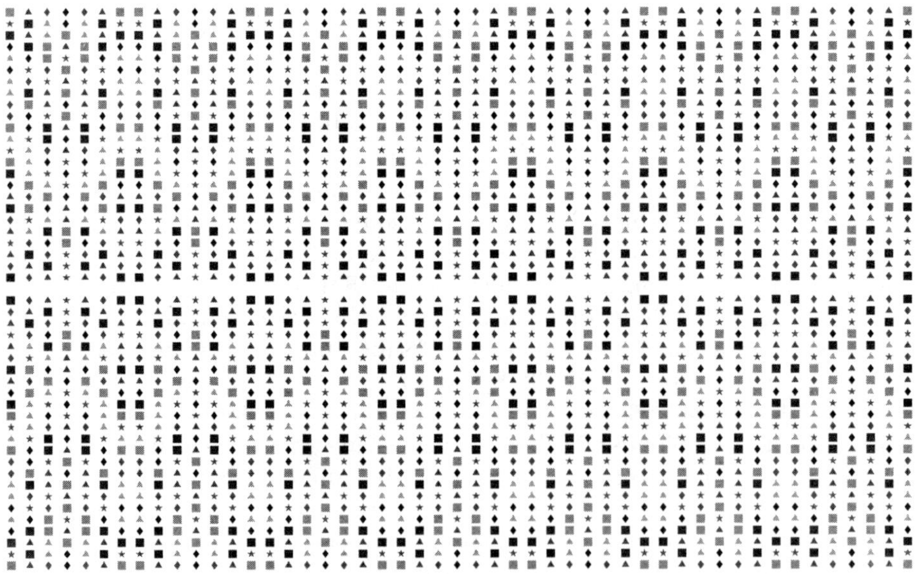

FIGURE 6. The tessellation of H_{49} by $GL(2, \mathbb{F}_7)$. Any collection of 7 different points in color (shape) is a fundamental domain for $GL(2, \mathbb{F}_7)\backslash H_{49}$.

Table 2. Trace formulas for the 3 symmetric spaces. There are three columns as in table 1. In the third column we consider $q = p^2$.

Γ **discrete subgroup** of G	$\Gamma \subset G$, fundamental group of X
with $\Gamma\backslash H$ compact,	$X = \Gamma\backslash T$ finite
no fixed points	$(q+1)$-regular connected graph
conjugacy classes in Γ	**conjugacy classes** in Γ
$\{\gamma\} = \{x\gamma x^{-1} \mid x \in \Gamma\}$	
center $\{\pm I\}$	**identity**
hyperbolic $\gamma \backsim \pm \begin{pmatrix} t & 0 \\ 0 & t^{-1} \end{pmatrix}$,	**hyperbolic** γ fixes geodesic C
$t > 1, t \neq t^{-1}$	shifts along C by $\nu(\gamma)$
centralizer $\Gamma_\gamma = \{x \in \Gamma \mid x\gamma = \gamma x\}$	$\Gamma_\gamma = \{x \in \Gamma \mid x\gamma = \gamma x\}$
=cyclic $=< \gamma_0 >$	$=< \gamma_0 >$
γ_0 **primitive hyperbolic**	γ_0 **primitive hyperbolic**
Spectrum Δ on $L^2(\Gamma\backslash G/K)$	$A\varphi_n = \lambda_n\varphi_n,\ \lambda_n = q^{s_n} + q^{1-s_n}$
$\Delta\varphi_n = \lambda_n\varphi_n,\ n = 0, 1, 2, \ldots,$	$\lvert\lambda_n\rvert \neq q+1 \Longrightarrow 0 < \operatorname{Re} s_n < 1$
$\lambda_n = s_n(1-s_n) \leq 0$	X Ramanujan
	$\Longleftrightarrow$ for $\lvert\lambda_n\rvert \neq q+1, \operatorname{Re} s_n = \frac{1}{2}$
Selberg trace formula	**Selberg trace formula**
	$f(x) = f(d(x, \mathcal{O}))$
$\sum\limits_{\lambda_n = s_n(1-s_n)} \widehat{f}(s_n) = area(\Gamma\backslash H) f(i)$	$\sum\limits_{i=1}^{\lvert X\rvert} \widehat{f}(s_i) = f(\mathcal{O})\lvert X\rvert$
$+ \sum\limits_{\{\gamma\}\ hyperbolic} \frac{\log N\gamma_0}{N\gamma^{\frac{1}{2}} - N\gamma^{-\frac{1}{2}}} F(N\gamma)$	$+ \sum\limits_{\substack{\{\rho\}\ primitive \\ hyperbolic}} \nu(\rho) \sum\limits_{e \geq 1} Hf(e\nu(\rho))$
$\Gamma_\gamma =< \gamma_0 >$	$F(\mathfrak{h}_n) = c_n Hf(n)$,
	$c_n = \begin{cases} 2^n, n > 0 \\ 1, n \leq 0 \end{cases}$
	$Hf(n) = f(\lvert n\rvert)$
	$+(q-1) \sum\limits_{j \geq 1} q^{j-1} f(\lvert n\rvert + 2j)$
Selberg zeta function	**Ihara zeta function**
$Z(s) = \prod\limits_{\{\gamma_0\}} \prod\limits_{j \geq 0} (1 - N\gamma_0^{-s-j})$	$Z(s) = \prod\limits_{\substack{\{\rho_0\} \\ prim.\ hyp.}} (1 - u^{\nu(\rho)})^{-1}, \quad u = q^{-s}$
non-trivial zeros correspond	$Z(s)^{-1} =$
to spectrum Δ on $L^2(\Gamma\backslash G/K)$	$(1-u^2)^{r-1} \det(I - Au + qu^2 I)$
except for finite # of zeros	$r = rank\ \Gamma,\ r - 1 = \frac{\lvert X\rvert(q-1)}{2}$
Riemann Hypothesis	$Z(s)$ satisfies Riemann Hypothesis
$s \in (0,1), \operatorname{Re} s = \frac{1}{2}$	$\Longleftrightarrow X$ is Ramanujan

$\Gamma \subset G$; $\Gamma = GL(2, \mathbb{F}_p)$
$\lvert\Gamma\backslash H_{p^2}\rvert = p$
conjugacy classes in Γ
central $\begin{pmatrix} a & 0 \\ 0 & a \end{pmatrix}, a \in \mathbb{F}_p^*$
hyperbolic $\begin{pmatrix} a & 0 \\ 0 & b \end{pmatrix}$,
$a \neq b \in \mathbb{F}_p^*$
parabolic $\begin{pmatrix} a & 1 \\ 0 & a \end{pmatrix}, a \in \mathbb{F}_p^*$
elliptic $\begin{pmatrix} a & b\xi \\ b & a \end{pmatrix}$
$a, b \in \mathbb{F}_p, b \neq 0$,
$\xi \neq u^2$, $u \in \mathbb{F}_p$
Selberg trace formula
$(\mathbf{q} = \mathbf{p}^2)$
$\sum_{\pi \in \widehat{G}^K} \widehat{f}(\pi) mult(\pi, Ind_\Gamma^G 1)$
$= \lvert\Gamma\backslash G\rvert(p-1) f(\sqrt{\delta})$
$+ \frac{(q+1)(q-1)^2}{2(p-1)} \sum_{\substack{c \in F_p^* \\ c \neq 1}} F(c)$
$+ \frac{q(q^2-1)}{p} \{F(1) - f(\sqrt{\delta})\}$
$+ \frac{q^2-1}{2} \sum_{\substack{a,b \in \mathbb{F}_p \\ b \neq 0}} F(\frac{a+b\eta}{a-b\eta})$,
where $\eta^2 = \xi$, $\eta \in \mathbb{F}_q$
?

As a last note, we consider finite analogues of tessellations. In Figure 6 we show the tessellation of the finite upper half plane over the field with 49 elements corresponding to the action of $GL(2, \mathbb{F}_7)$ on its fundamental domain. Those points in a fundamental domain are given 7 different characteristic colors (shapes). Then we let $\gamma \in GL(2, \mathbb{F}_7)$ act on these points z and color the resulting points γz the same as z. The result is Figure 7. Any collection of 7 points with different colors (shapes) is a fundamental domain for $GL(2, \mathbb{F}_7)\backslash H_{49}$.

References

[1] G. Ahumada, Fonctions periodiques et formule des traces de Selberg sur les arbres, *C. R. Acad. Sci. Paris, 305* (1987), 709-712.

[2] R. Brooks, The spectral geometry of k-regular graphs, *J. d'Analyse, 57* (1991), 120-151.

[3] P. Buser, *Geometry and Spectra of Compact Riemann Surfaces*, Birkhäuser, Boston, 1992.

[4] P. Cartier, Harmonic analysis on trees, *Proc. Symp. Pure Math., 26*, Amer. Math. Soc., Providence, 1973, pp. 419-423.

[5] J. Elstrodt, Die Selbergsche Spurformel für kompakte Riemannsche Flächen, *Jber. d. Dt. Math.-Verein., 83* (1981), 45-77.

[6] J. Elstrodt, F. Grunewald, and J. Mennicke, *Groups acting on Hyperbolic Space*, Springer-Verlag, Berlin, 1998.

[7] A. Figá-Talamanca and C. Nebbia, *Harmonic Analysis and Representation Theory for Groups acting on Homogeneous Trees*, Cambridge U. Press, Cambridge, 1991.

[8] D. A. Hejhal, The Selberg trace formula and the Riemann zeta function, *Duke Math. J., 43* (1976), 441-482.

[9] D.A. Hejhal, J. Friedman, M.C. Gutzwiller and A. M. Odlyzko (Eds.), *Emerging Applications of Number Theory, IMA Volumes in Math. and its Applications, Vol. 109*, Springer-Verlag, N.Y., 1999.

[10] S. Katok, *Fuchsian Groups*, Univ. of Chicago Press, Chicago, 1992.

[11] A. Lubotzky, R. Phillips, and P. Sarnak, Ramanujan graphs, *Combinatorica, 8* (1988), 261-277.

[12] H. Nagoshi, The Selberg zeta functions over function fields, preprint.

[13] P. Sarnak, Arithmetic quantum chaos, *Israel Math. Conf. Proc., 8* (1995), 183-236 (published by the A.M.S.).

[14] A. Selberg, Harmonic analysis and discontinuous groups in weakly symmetric Riemannian spaces with applications to Dirichlet series, *J. Ind. Math. Soc., 20* (1954), 47-87.

[15] H.M. Stark and A. Terras, Zeta functions of finite graphs and coverings, *Advances in Math., 121* (1996), 124-165.

[16] H.M. Stark and A. Terras, Zeta functions of finite graphs and coverings, Part II, *Advances in Math., Vol. 154* (2000), 132-195.

[17] A. Terras, *Fourier Analysis on Finite Groups and Applications*, Cambridge Univ. Press, Cambridge, U.K., 1999.

[18] A. Terras, *Harmonic Analysis on Symmetric Spaces and Applications, I,II*, Springer-Verlag, N.Y., 1985, 1988.

[19] A. Terras and D. Wallace, Ergodic properties of quotients of flows for trees, preprint.

[20] A. B. Venkov and A. M. Nikitin, The Selberg trace formula, Ramanujan graphs and some problems of mathematical physics, *Petersburg Math. J., 5* (1994), No. 3, 419-484.

[21] E. P. Wigner, Random matrices in physics, *S.I.A.M. Review, Vol. 9, No. 1* (1967), 1-23.

MATH. DEPT., U.C.S.D., SAN DIEGO,CA 92093-0112
E-mail address: aterras@ucsd.edu
URL: http://math.ucsd.edu/~aterras/

DIMACS Series in Discrete Mathematics
and Theoretical Computer Science
Volume **64**, 2004

Unusual Applications of Quadratic Gaussian Sums

Carlos Julio Moreno and Aaron Wan

This paper is dedicated to the memory of Bernie Dwork.

ABSTRACT. This article is an exposition of some unusual aplications of quadratic Gaussian sums to quadratic reciprocity law, representation theory, staionnary phase, p-adic analysis, and error correcting codes. A detailed study is made of the so-called local quadratic Gaussian sums.

Introduction. Gaussian sums, disguised in the form of Lagrange resolvents, were introduced by Gauss in the algebraic study of the uniform division of the circle, a theory which lies at the foundations of the theory of cyclotomic fields (see Gauss's *Disquisitiones Arithmeticae*, Arts. 343., 356.) The sums studied in this paper are those associated with the general operation of squaring and, as we shall see, are of great importance in number theory and its modern applications. The classic result of Gauss, a paradigm whose form and significance still drives much of the modern theory, is the evaluation of the **Quadratic Gaussian Sum**:

$$\sum_{s\in\mathbf{Z}/n\mathbf{Z}} e^{-2\pi i s^2/n} = \frac{1+i^n}{1+i}\sqrt{n}, \tag{1}$$

a sum which figures prominently in one of Gauss's many proofs of the quadratic reciprocity law (Gauss's Diary, Entries Nos. 118, 123). We shall be concerned with the evaluation of analogous sums in the particularly difficult case of local fields with even residue field characteristic. The underlying theme for much of this work will be the determination of the **sign** of the general quadratic Gaussian sum, and a more appropriate title for this article could perhaps be "**On the ubiquity of quadratic Gaussian sums.**"

For the purpose of clarity of exposition we have divided the paper into two parts, the first of which aims at defining the problem and describing its relevance in areas outside of number theory. The second part, a section that can be skipped by those whose main interests lie in the applications, covers in greater depth the advanced results obtained about the evaluation of quadratic Gauss sums.

2000 *Mathematics Subject Classification.* Primary 11L05, 11A15; Secondary 11B65, 11F70.
Key words and phrases. Number theory, representation theory, L-functions.
The first author was supported in part by PSC-CUNY Grant #64480-00 33.

PART I: Unusual Applications

The Quadratic Reciprocity Law. Among the many well known elementary proofs of this fundamental result of number theory, perhaps the most unusual is the one found by Gauss which uses the q-binimial coefficients, also known as **Gaussian polynomials**: for integers m, and n satisfying $0 \leq n \leq m$,

$$\begin{bmatrix} m \\ n \end{bmatrix} := \frac{(1-q^n)(1-q^{n-1})\cdots(1-q^{n-m+1})}{(1-q)(1-q^2)\cdots(1-q^m)}.$$

The theory of these polynomials is a well known chapter in the elementary theory of partitions (see Andrews [1], chapt. 3). These remarkable polynomials, which in a very precise sense can be thought of as "*universal deformations*" of the ordinary binomial coefficients, share many of the properties of the latter with which we have grown accustumed: i.e., (a) boundary values $\begin{bmatrix} m \\ 0 \end{bmatrix} = \begin{bmatrix} m \\ m \end{bmatrix} = 1$, (b) symmetry $\begin{bmatrix} m \\ n \end{bmatrix} = \begin{bmatrix} m \\ m-n \end{bmatrix}$, (c) recursive relation

$$(*) \qquad \begin{bmatrix} m \\ n \end{bmatrix} = \begin{bmatrix} m-1 \\ n-1 \end{bmatrix} + q^n \begin{bmatrix} m-1 \\ n \end{bmatrix},$$

and, above all, a **q-Pascal's Triangle**:

$$\begin{array}{ccccccc}
 & & & \begin{bmatrix} 0 \\ 0 \end{bmatrix} & & & \\
 & & \begin{bmatrix} 1 \\ 0 \end{bmatrix} & & \begin{bmatrix} 1 \\ 1 \end{bmatrix} & & \\
 & \begin{bmatrix} 2 \\ 0 \end{bmatrix} & & \begin{bmatrix} 2 \\ 1 \end{bmatrix} & & \begin{bmatrix} 2 \\ 2 \end{bmatrix} & \\
\begin{bmatrix} 3 \\ 0 \end{bmatrix} & & \begin{bmatrix} 3 \\ 1 \end{bmatrix} & & \begin{bmatrix} 3 \\ 2 \end{bmatrix} & & \begin{bmatrix} 3 \\ 3 \end{bmatrix} \\
\bullet & \bullet & & \bullet & & \bullet & \bullet
\end{array}$$

Where the row formation is to be understood in the sense of the recursive relation (*). As in the usual case, many of the basic properties of these coefficients can be found by a careful combinatorial analysis of the q-Pascal's triangle. One has for intance the **q-binomial expansion**:

$$(1+y)(1+xy)(1+x^2y)\cdots(1+x^{m-1}y) = \sum_{\nu=0}^{m} \begin{bmatrix} m \\ \nu \end{bmatrix} x^{\nu(\nu-1)/2} y^{\nu}.$$

The alternating sum of the entries of an odd numbered row in the q-Pascal's triangle add up to zero as one is lead to expect from $(1-1)^m = 0$. On the other hand, the alternating sum of the terms in an even numbered row, taken from left to right, yield the identity

$$\sum_{\nu=0}^{2n} (-1)^{\nu} \begin{bmatrix} 2n \\ \nu \end{bmatrix} = (1-q)(1-q^3)\cdots(1-q^{2n-1}),$$

the left hand side of which we denote by $f(q, 2n)$. If we now take a positive odd integer $k = 2n+1$ and specialize q to be a primitive kth root of unity, we obtain readily from the definition of the q-binomial coefficients, using that q^{-2} is also a primitive kth root of unity, and the simple identity $\frac{1-q^{k-j}}{1-q^j} = -q^{-j}$, that

$$f(q^{-2}, k-1) = \sum_{\nu=0}^{k-1} q^{\nu(\nu-1)},$$

a result which, when combined with the earlier product formula for the left hand side, yields the identity

$$(2.)\qquad \sum_{j \text{ mod. } k} q^{j^2} = (-1)^{(k-1)/2}(q^2-q^{-2})(q^4-q^{-4})\cdots(q^{k-1}-q^{-(k-1)}),$$

an expression which coinsides with the Gausssian sum in (1) when $q = e^{-2\pi i/k}$. Armed with this expression, Gauss was then able to show that the value of the left hand side in the equation (2) is

$$(3.)\qquad \sqrt{k}(-i)^{(k-1)/2}\cdot sign \prod_{j=1}^{(k-1)/2} \sin\frac{4\pi j}{k}.$$

Since the sin function $\sin\frac{4\pi j}{k}$ is positive for j in the range $0 < j < \frac{k}{4}$ and negative in the range $\frac{k}{4} < j < \frac{k}{2}$, one sees that the sign of the product in (3.) is equal, by a simple application of Gauss Lemma, to the quadratic residue symbol

$$\Big(\frac{k}{2}\Big) = (-1)^{(k^2-1)/8}.$$

Thus the value of (3.) is $i^{[(k-1)/2]}\sqrt{k}$, and this agrees with with that given in (1) for n odd. The even case can be treated by elementary means. For example if $n \equiv 2$ mod. 4, then $\exp\frac{2\pi i}{n}\{s+\frac{n}{2}\}^2 = -\exp\frac{2\pi i}{n}s^2$, and hence all the terms in the sum can be grouped in pairs consisting of equal elements of opposite signs thus giving the value zero to the sum. The case $n \equiv 0$ mod. 4 can be treated in a similar way.

If h and k are positive integers which are relatively prime, and if q denote the primitive k-th root of unity $e^{2\pi ih/k}$, then, denoting by $g(h,k)$ the expression obtained from the sum in (2) with this value of q, one obtains without difficulty that, with $ha + kb = 1$,

$$(4.)\qquad g(h,k) = \big(\frac{h}{k}\big)g(1,k), \qquad g(1,hk) = \big(\frac{a}{k}\big)g(1,k)\cdot\big(\frac{b}{h}\big)g(1,h).$$

(Here $\big(\frac{a}{k}\big) = \big(\frac{h}{k}\big)$, and $\big(\frac{b}{h}\big) = \big(\frac{k}{h}\big)$, denote the Legendre symbols.) From (4) one obtains inmediately the usual statements of the quadratic resiprocity law:

$$\Big(\frac{h}{k}\Big)\Big(\frac{k}{h}\Big) = (-1)^{\frac{k-1}{2}\cdot\frac{h-1}{2}}.$$

(The details of this proof are given with insightfull clarity in the elegant book by Rademacher [35], pp. 83-96.)

Among the many proofs of the law of quadratic reciprocity, it may be worth while to compare the above one with the analytic approach as developped by Hecke, Siegel and Weil. The inspiration for Hecke's proof comes from the following Gaussian sum reciprocity relation due to Krazer:

$$(5.)\qquad G(a,b) = \zeta_8^{w(ab)-ab}G(b,-a),$$

where $\zeta_8 = e^{i\pi/4}$, $w(r) = sgn\, r$, a and b are odd integers, and

$$G(a,b) = |a|^{-1/2}\sum_{k(mod.a)} E\{\frac{b}{a}k^2 + bk\}.$$

($E\{v\} = e^{\pi iv}$.) Let k be a number field of degree n over $\mathbf{Q}$ with m real embeddings $k^{(1)},\ldots,k^{(m)}$. Let $\mathcal{O}_k$ be the ring of integers in k, $\mathcal{D}_k$ the absolute different of k

and $Tr = Tr_{k/\mathbf{Q}}$ the trace from k to $\mathbf{Q}$. Denote by $\psi_k = E \circ Tr$ the composite of the trace function with the character of second degree $E\{v\}$ defined above. Let A be an ideal in $\mathcal{O}_k$ and ω a non-zero element of k for which the fractional ideal $(\omega)\mathcal{D}_k = \frac{A}{B}$ with A and B relatively prime integral ideals not divisible by 2. Suppose furthermore that $\delta \in \mathcal{D}_k$ is such that the integral ideal $(\delta)\mathcal{D}_k = C$ is relatively prime to $(2)B$. Let τ be a number in k which is such that $Tr(\omega\xi^2) + Tr(\tau\omega\xi)$ is even for all $\xi \in A$. Let $w = w(\omega)$ denote the signature of the quadratic form $Tr(\omega\xi^2)$: $w(\omega) = \sum_{p=1}^{m} sgn\, \omega^p$, where ω^p is the p-th real conjugate of ω. With this notations we can now formulate the *quadratic sum reciprocity relation* for the number field k:

$$(5'.) \qquad G(\omega) = \zeta_8^{w(\omega)-Tr(\omega\tau^2)} G(\omega^*), \quad \omega\omega^* = -\delta^2,$$

where the quadratic Gaussian sum is defined by

$$G(\omega) = (N(A))^{-1/2} \sum_{x \in \mathcal{O}_k/A} \psi_k(\omega x^2 + \tau\omega x).$$

Siegel's derivation of (5′.) from (5.) is a short and elegant induction based on a diagonalization process. His proof of (5.), using Kronecker's analytic method, occupies a mere 14 lines, being possibly the simplest application of Cauchy's residue theorem to prove an important result in number theory. Following Gauss' proof, Siegel is then able to establish the full reciprocity law for quadratic residue symbols in k in the form given to it by Hasse:

$$\left(\frac{\alpha}{\beta}\right)\left(\frac{\beta}{\alpha}\right) = (-1)^{g+Tr(\frac{\alpha-1}{2}\frac{\beta-1}{2})}, \quad (\alpha \equiv \beta \equiv 1(\bmod.\ 2)),$$

where

$$g = \sum_{p=1}^{m} \frac{sgn\, \alpha^{(p)} - 1}{2} \frac{sgn \beta^{(p)} - 1}{2}.$$

As in the 1-variable case, special choices of ω's lead to evaluations of $G(\omega)$. Siegel's version [1] in comparison to Hecke's original approach [2] has the merit that it avoids the use of the theta transformation formulas and the required multi-dimensional Poisson summation formula. It also brings out the true algebraic nature of the reciprocity relation with respect to the underlying quotients of lattices in n-space, duality and the character of second degree $E\{v\}$. It is a pleasant excercise in adelic analysis to transform Siegel's version of the reciprocity law into the familiar reciprocity law for the Hilbert norm residue symbol. The fundamental fact that emerges from these approaches to quadratic reciprocity is that the Gaussian quadratic sums generate 1-dimensional vector spaces which afford a representation of $k_{\mathbf{A}}^{\times}/(k_{\mathbf{A}}^{\times})^2$. Four years after Siegel's publication, these ideas reached a high level of development in the monumental work of Weil "**Sur Certain Groupes des Operateurs Unitaires** [3]" in which the group theoretic nature of quadratic Gaussian sums and their relation to the automorphic nature of theta functions is clarified. In Weils version, which is still of a global nature, the essential fact is that the local analogue of the Gaussian sum is in fact a character of the Witt group of a local field (Proposition 3. in Weil's paper).

[1] Siegel's *Collected Works*, vol.III, paper # 74.

[2] E. Hecke's *Lectures on the Theory of Algebraic Numbers*, Springer-Verlag, 1981

[3] *Acta Mathematica*, vol. 111, (1964)

Weil's version of the quadratic reciprocity law can be stated briefly as the identity

$$1 = \prod_v W_v(\chi_v, \psi_v),$$

where the product runs over all the primes of k, including the archimedean ones,

$$\chi = \prod_v \chi_v$$

is the local product decomposition of a Hecke character of $k_{\mathbf{A}}^{\times}$ of order two, $\psi_k = \prod_v \psi_v$ is the local decomposition of a non-trivial additive character of $k_{\mathbf{A}}^{+}$, and

$$W_v(\chi_v, \psi_v) = \gamma(\psi_v \circ N_{K_v/k_v})$$

is the Weil index of the character of degree two obtained by composing ψ_v with the norm N_{K_v/k_v}. (See Weil [28], §30, Prop. 5; the definition of $W_v(\chi_v, \psi_v)$, as a local quadratic Gaussian sum, is also given in Part II of this article). The translation into the more traditional form

$$1 = \prod_v (a, a')_v, \quad (a, a' \in k^{\times}),$$

involving the Hilbert norm residue symbol, requires the use of the simple identity

$$\frac{W_v(\chi_v, \psi_v) W_v(\chi', \psi)}{W_v(\chi_v \chi_v')} = (a, a')_v,$$

where χ_v, resp. χ_v', are the quadratic characters associated to the local extensions $k_v(\sqrt{a})$, $k_v(\sqrt{a'})$.

In so far as the quadratic reciprocity law is concerned, Weil's adelic approach is no more and no less general than Siegel's [4]. The latter actually brings out more clearly the inductive approach and requires no version of the Stone-von Neuman theorem (See also Weil's remarks in Art. 30 of his paper). If one observes that Krazer's identity (5.) can be proved by evaluating both sides with the appropriate versions of (1.) together with the ordinary quadratic reciprocity law for $\mathbf{Q}$, it then becomes clear that Siegel's proof of the general quadratic reciprocity law for number fields is truly elementary. In fact, Siegel's proof is a prime example of **hard symplectic geometry** in the sense of Gromov[5], while Weil's is certainly an example of **soft symplectic geometry**.

Despite the great advances in our conceptual understanding of the role which Gaussian sums play in the proof of quadratic reciprocity, as is brought out in the work of Hecke, Siegel, and Weil, we do not yet see clearly the role played by the Gaussian polynomials. The latter appear prominently in various aspects of the theory of group representations, as in the theory of deformations of Lie algebras, and in quantum groups. The Gaussian polynomials also control various counting arguments in the study of the distribution of square matrices over finite fields. For

[4] "*Malgré les apparences, la démonstracion de la loi de réciprocité quadratic exposée ci-dessus ne diffè pas substanciellement de la démonstration classique au moyen des fonctions thêta et des sommes de Gauss;...*" (See Weil [28], §30)

[5] See Gromov's lecture "*Soft and hard symplectic geometry*" (Proc. International Congress of Mathematicians, (1986), Berkeley, California, Vol. I, p. 96)

instance, if $\mathbf{F}_q$ is a finite field of odd characteristic, the number of solutions in $GL_m(\mathbf{F}_q)$ of the equation $g^2 = 1$ is given by

$$\sum_{k=0}^{m} \left[\frac{m}{k}\right] q^{k(m-k)}.$$

(See John H. Hodges, Amer. Math. Monthly, vol. 65 (1958), pp. 518-520). These ideas appear in a number of works by Hodges and by Carlitz in which Gaussian and Kloostermann sums defined over matrix rings appear in a role analogous to that in Siegel's work. The group theoretic reconsideration of their work from the optic of Weil's metapletic group, and a conceptual clarification of the appearance of Gaussian polynomials will be a worthwhile goal.

The approach to the evaluation of Gaussian sums using the q-binomial coefficients has one advantage over the analytic approach that should be noted. A local field k whose residue class field $\mathbf{F}_q$ satisfies $q \equiv 1$ mod. m contains primitive m-th roots of 1 and for these, the analogue of identity (2.) holds. These new identities open the way to a p-adic study of quadratic Gaussian sums, a topic which will be pursued elsewhere.

Finally, a very unusual application of Feyman's sum over paths method has been used to prove a version of the Gaussian sum reciprocity formula by Vernon Armitage and Alice Rogers[6]. In this approach, a formula such as (5.) is obtained by realising one side of the equality as the trace of an evolution operator on a quantized torus while the other side results from an adaptation of Feyman's sum over paths method to a discrete system, thus obtaining a finite version of Kelvin's method of images. Armitages approach to quadratic reciprocity appears to be purely combinatorial and rather elementary. Again one would like to have a group theoretic explanation of these ideas whose motivation realy comes from quantum mechanics. We are reminded here of the appearance of Gaussian polynomials in the statement of partition identities in statistical mechanics. Identities such as (2.) seem to suggest that the underlying principles in the evaluation of Gaussian sums is the convergence of bosonic and fermionic ideas, a notion which may not be out of place when discussing the appearance of theta functions in quantum mechanics.

We end this section by formulating a tantalizing question:

Problem 1. *Is there a multidimensional version of the Gaussian polynomials and their accompanying identities which is capable of yielding a direct evaluation of the quadratic Gaussian sums over number fields which appear in Hecke's proof of the "quadratic reciprocity law?"*

REMARK. A very readable and highly motivated presentation of the analytic proof of the quadratic reciprocity law and its modern connections with the representation theory of the metaplectic group can be found in the book by Michael Berg "**The Fourier-Analytic Proof of Quadratic Reciprocity**". The book also discusses the major open problems concerning the extension of the analytic method to prove the higher reciprocity laws for the norm residue symbol; the book has an ample bibliography and can serve as an introduction to Weil's difficult *Acta Paper.*

[6]See *Gauss Sums and Quantum Mechanics*, Los Alamos Preprint, February, 2000

Classical Stationary Phase. Let M be Riemaniann manifold of dimension n with volume element dy. For the purposes of illustration, we can take M to be an sphere or an ellipsoid embedded in 3-space. Let f be a non-linear function with the property that the **critical set** $C_f := \{s \in M : (df)_s = 0\}$, the set of points where the gradient of f vanishes, consists of finitely many points. Suppose that at each point $s \in C_f$ the determinant of the Hessian $H(s) := (d^2 f)_s$, the matrix of coefficients of the quadratic form arising from the second order terms in the Taylor expansion of f about the point s, is non-zero. We assume furthermore that $a(y)$ is a smooth function defined on M and with compact support.

The principle of stationary phase in its simplest formulation, as is usually presented in geometry and in physics, is the following striking asymptotic relation:

$$\int_M a(y)e^{2\pi ikf(y)}dy = \Big\{ \sum_{s\in C_f} a(s)\zeta_8^{sgn\, H(s)} \frac{e^{2\pi ikf(s)}}{\sqrt{|\det H(s)|}} \Big\} k^{-n/2} + \mathcal{O}(k^{-(n+1)/2}),$$

(See Hormander [6], p. 220; Guillemin-Sternberg [5], p. 6; Duistermaat [7], p. 21). Here the quantity $sgn\, H(s)$ denotes the signature of the quadratic form $H(s)$. The presense of the eight-th root of unity $\zeta_8 = e^{\frac{\pi i}{4}}$ suggests an intimate relation of the expression inside the curly brackets with "quadratic Gaussian sums" associated to archimedean primes. To understand the number theoretic significance of this relation one must go back to one of Riemann's proofs of the functional equation of the zeta function $\zeta(s)$, the one which involves countour integrals of quadratic exponential functions (see Deuring [25]), as well as to Hecke's analytic proof of quadratic reciprocity law using the asymptotic behavior of theta functions as the group parameter approaches a rational cusp. As we shall give a more natural number theoretic version of the above formula in the next section, we limit our remarks to the fact that the main tool in its proof is a suitable version of **Morse's Lemma**, which in its essential points states that a smooth function whose behavior near the origen is given by the Taylor expansion with real coefficients a_i

$$f(x) = a_2x^2 + a_3x^3 + \cdots, \qquad a_2 > 0$$

has a representation in the form $f(x) = (g(x))^2$ for x in a neighbourhood of the origen. This allows one, after an integration by parts, to concentrate, for large values of the (frequency-)parameter k, the values of the integral around the points in the critical set C_f.

One of the most interesting aspects of the principle of stationary phase is Harish-Chandra's and Gelfand's realization that for certain classes of semi-simple Lie groups, e.g. $SL_2(\mathbf{C})$, the comparison of the leading terms in certain oscillatory integrals leads to a derivation of the Plancherel formula (See Guillemin-Sternberg [5], Appendix to Chap. VI). Similarly, Shalika's p-adic germ expansions have been of fundamental importance in the harmonic analysis of p-adic Lie groups (See Shalika [23]).

The deeper aspects of the meaning of the term $\zeta_8^{sign\, H(s)}$ in the formula for stationary phase as a type of **Maslov index** lies in the degenerate cases, i.e., those where the determinant of the Hessain $H(s)$ does not have a full rank, in which case the formula has to be modified in a way which reflects some structural information about the nature of the singular critical point. A particularly interesting example is that in which the rank of the Hessian $H(s)$ is one less than the dimension of the

ambient space. This particular case, which involves the singularity theory of " fold points", has been analysed extensively by Ludwig and Guillemin-Schaeffer (see [5], p. 384). At the heart of this analysis lies the important Airy function defined by the integral formula

$$Ai(t) = \frac{1}{2\pi}\int_{-\infty}^{\infty} e^{i(t\theta+\theta^3/3)}d\theta.$$

The same type of functions enter in Kubota's analytic approach to the higher reciprocity laws (in this case n=3) through the definition of higher order metapletic groups (see Kubota [31]). The p-adic analogues of these results have not yet been made explicit in the literature, but their elucidation, particularly from the optic of Igusa's asymptotic theory of "complex powers", promisses to be essential to an understanding of the still conjectural analytic proof of the higher reciprocity laws. Although we shall describe later on Igusa's p-adic formula of stationary phase, it may be worth mentioning the need to interpret the interesting work of Livne concerning the Sato-Tate distribution results for sums of the type

$$S(a,b) = \sum_{s=0}^{p-1} e^{2\pi i\{as^3+bs\}/p}$$

in terms of Igusa's stationary phase formula (see Livne [46]), and the appropriate p-adic analogues of Airy's function.

P-adic Stationary Phase. The strong number theoretic significance of the classical formula of stationary phase would not have much merit were it not for the companion p-adic analogue as originally discovered by L.K. Hua in the context of exponential sums and by the deep and rather extensive developpment given to the idea by Igusa in his study of generalized Poisson-summation formulas.

We begin our description of this aspect of stationary phase by recalling first the one variable case, better known to number theorists as the **Fundamental Lemma** in the additive theory of numbers. We give in this section a brief introduction to the method of stationary phase as it applies to exponential sums of Carlitz-Uchiyama type with functions of polynomial type. These exponential sums have many applications to error correcting codes and to signal processing (m-sequences, etc.). Our presentation is a modification of a well known method in the additive theory of numbers first introduced by L.K.Hua and extensively considered by Vinogradov. The main result is in all its essentials a primitive version of the one-variable version of the method of stationary phase for functions of several variables which we discuss in the next section.

Let p be a prime and ℓ a positive integer; let $R_\ell = \mathbf{Z}/p^\ell\mathbf{Z}$, and define an additive character on R_ℓ by $\psi_\ell(s) = e^{2\pi is/p^\ell}$. We shall be considering polynomials of degree k with integer coefficients

$$f(x) = b_0 + b_1x + b_2 + \cdots + b_kx^k.$$

Without loss of generality we assume that the leading coefficient $b_0 = 0$. The exponential sums of Carlitz-Uchiyama type are defined by

$$G(p^\ell, f) := \sum_{s\in R_\ell} e^{2\pi if(s)/p^\ell}.$$

The special case of $\ell = 1$, which to some extent is the more difficult case, is dealt with adequately by Weil's estimates of the roots of the L-functions of exponential sums over finite fields. We therefore make the assumption $\ell \neq 1$. We now make a definition which makes precise the notion of multiplicity for the roots of arithmetic polynomials.

DEFINITION. *A polynomial congruence*

$$f(x) \equiv 0 \ mod.\ p$$

has a root of multiplicity m *at* μ *when* $f(x) = (x-\mu)^m f_1(x) + pf_2(x)$ *with* f_1, f_2 *polynomials with the property that* $p \not| f_1(\mu)$.

To define the critical set for the polynomial $f(x)$ we assume that $f(x)$ is not trivial modulo p, that is to say, $p \not| (b_1, b_2, \dots, b_k)$. Let a be the exact power of p which divides the derivative of f, namely $p^a \parallel (b_1, 2b_2, \dots, kb_k)$. We define the critical set for the polynomial f by

$$C_f(p) := \{s : p^{-a} f'(s) \equiv 0 \text{ mod. } p\}.$$

If m_μ denotes the multiplicity of an element $\mu \in C_f(p)$, then by Lagrange's theorem, the sum of the m_μ's is bounded by $k-1$. Given an element μ in the critical set $C_f(p)$ we consider the polynomial $f(px+\mu) - f(\mu) = c_1 x + \cdots + c_k x^k$ in which the coefficients are obtained from those of f by the recursive relation

$$c_i = p^i \sum_{h=i}^{k} b_h \binom{h}{i} \mu^{h-1}, \qquad 1 \leq i \leq k.$$

We denote by $\rho(\mu)$ the largerst power of p which divides all the coefficients c_i: $p^{\rho(\mu)} \parallel (c_1, \dots, c_k)$. Clearly $\rho(\mu) \geq 1$. With the point μ and the integer $\rho(\mu)$, we define a new polynomial of degree k:

$$f_\mu(x) := p^{-\rho(\mu)}(f(px+\mu) - f(\mu)).$$

There is another important exponent associated to the points in the critical set which is defined by the requirement that $p^{\sigma(\mu)}$ is the highest power of p that devides the derivative of $f'_\mu(x)$; it satisfies $1 \leq \sigma(\mu) \leq k$. The basic principle behind the method of stationary phase for sums of the Carlitz-Uchiyama type is contained in the following result which is an elaboration of Hua's proof.

FUNDAMENTAL LEMMA. *With notations and assumptions as above, we have*

$$\text{(6.)} \qquad \sum_{s \in R_\ell} \psi_\ell(f(s)) = \sum_{\mu \in C_f(p)} \psi_\ell(f(\mu)) p^{\rho(\mu)-1} \sum_{r \in R_{\ell-\rho(\mu)}} \psi_{\ell-\rho(\mu)}(f_\mu(r)).$$

Furthermore, if a critical point $\mu \in C_f(p)$ *has multiplicity* m_μ, *then the number of solutions of the congruence* $p^{-\sigma(\mu)} f'_\mu(x) \equiv 0(\ mod.\ p)$, *counting multiple solutions with their multiplicities, is at most* m_μ; *that is, the critical set for the polynomial* $f_\mu(x)$ *satisfies* $Card\, C_{f_\mu}(p) \leq m_\mu$.

The proof of this result is an easy modification of the original argument given by Hua. A very clear and lucid presentation of Hua's result can be found in Vaughan's tract on Additive Number Theory ([4], §7.2). One point that merits mention in the proof of the above formula which is not stressed in Hua's original argument is the appearence of an "**Arithmetic Morse Lemma** " in which the iteration

$$f \mapsto f_\mu$$

corresponds to a change of coordinates near the critical point μ so that the end result is a quadratic function in the new variables.

In many cases of interest for applications, if the integer power ℓ is odd, a number of iterations of the recursive formula in the Fundamental Lemma leads to an inner sum which is basically a quadratic Gaussian sum, whose evaluation can be obtained from the original Gaussian sum in (1). As a particular application of the Fundamental Lemma we point out here that for an arbitrary polynomial in one variable one derives from it a representation for the exponential sum in the form

$$\sum_{s \in R_\ell} e^{2\pi i f(s)/p^\ell} = \sum_j a_j (\sqrt{p})^{w(j)},$$

where the complex numbers a_j have absolute value 1 and the "weight exponents" $w(j)$ are integers which in the "Morse Case" lie in the set $\{0, 1, \ldots, \ell\}$. The p-adic expansions of Igusa for polynomials in several variable are a generalization of the above representation. We shall discuss these ideas of Igusa. In this section we review the p-adic methods introduced by Igusa in his development of p-adic asymptotic expansions for local Mellin transforms and which are capable of yielding fairly precise results on the asymptotics of exponential sums for polynomials in several variable.

We work over the local ring $\mathbf{Z}_p$ of p-adic integers and take for local uniformizing parameter $\pi = p$ so that the residual rings are given by $\mathbf{Z}_p/p^\ell\mathbf{Z}_p = \mathbf{Z}/p^\ell\mathbf{Z} = R_\ell$. We denote the canonical reduction map by

$$\mathbf{Z}_p \mapsto \mathbf{F}_p, \quad x \mapsto \bar{x}.$$

Let f be a polynomial in $\mathbf{Z}_p[x]$ and make the assumption that not all its coefficients lie in $p\mathbf{Z}_p$ so that its reduction $\bar{f}(x)$ mod.p is not the zero polynomial. For any $\mathbf{F}_p$-algebra A, we denote by $V_f(A)$, the set of A-valued points, and by $C_f(A)$ the A-valued critical set, that is to say, the A-valued singular points. We let $T \subset \mathbf{Z}_p$ be a fixed set of representatives of $\mathbf{F}_p$ in $\mathbf{Z}_p$ (a Teichmuller system).

We now consider a polynomial f in $\mathbf{Z}_p[x_1, x_2, \ldots, x_n]$ and a point $\mu_1 = (y_1, \ldots, y_n)$ in affine n-space $\mathbf{A}^n(\mathbf{Z}_p) = (\mathbf{Z}_p)^n$. In analogy with the definition of the polynomial $f_\mu(x)$, we introduce the polynomial

$$f_{\mu_1}(x) = p^{-e_{\mu_1}} f(px + \mu_1),$$

called the dilatation of f at μ_1, and where e_{μ_1} is the minimum order of p in expansion of the polynomial $f(px+\mu_1)$. We call the $\mathbf{Q}_p$-hypersurface $V_{f_{\mu_1}}$ the dilatation of V_f at μ_1, and the set $C(f_{\mu_1})$, the liftings of $C_{f_{\mu_1}}(\mathbf{F}_p)$, the first generation of descendants of μ_1. Given a sequence $(\mu_i)_{i\geq 1}$, we define inductively $e_{\mu_1\cdots\mu_k}$ and $f_{\mu_1\cdots\mu_k}(x)$, by the following procedure:

$$f_{\mu_1\cdots\mu_k}(x) = \begin{cases} f(x) & \text{if } k = 0 \\ p^{-e_{\mu_1\cdots\mu_k}} f_{\mu_1\cdots\mu_{k-1}}(px + \mu_k) & \text{if } k \geq 1. \end{cases}$$

The union of the sets $C(f_{\mu_1\cdots\mu_k})$, as the μ_k runs through the descendants of the (k-1)-th generation of μ_1, is called the k-th generation of descendants of μ_1.

Let $\bar{X}$ be a subset of affine n-space $\mathbf{A}^n(\mathbf{F}_p)$ and X its pre-image under the canonical map $\mathbf{Z}_p \to \mathbf{F}_p$. Denote by $C(f,X)$ the subset of T^n (T the Teichmuller representatives) which maps bijectively to the set $C_f(\mathbf{F}_p) \cap \bar{X}$. We define two integers associated to the reduced polynomial $\bar{f} \equiv f$ mod. p:

$$\nu(\bar{f}) := \text{Card}\,\{\bar{\mu} \in \bar{X} : \bar{\mu} \notin V_f(\mathbf{F}_p)\}$$

and

$$\sigma(\bar{f}) := \text{Card}\,\{\bar{\mu} \in \bar{X} : \bar{\mu} \in C_{\bar{f}}(\mathbf{F}_p\}.$$

The following is Igusa's version of stationary phase for p-adic integrals.

STATIONARY PHASE FORMULA. (Igusa [8], P. 177) *With notations as above we have*

$$\int_X |f(x)|_p^s dx = \nu(\bar{f}) + \sigma(\bar{f})\frac{(1-p^{-1})p^{-s}}{(1-p^{-1}p^{-s})} + \sum_{\mu \in C(f,X)} q^{-n-e_\mu s} \int_{\mathbf{Z}_p^n} |f_\mu(x)|_p^s dx,$$

where s is a complex number with $Re(s) > 0$.

The essential idea of the proof is not difficult to describe. Suppose the set $\bar{X} = \{\bar{\mu}_1, \ldots, \bar{\mu}_N\}$ and let μ_i be the lifting of $\bar{\mu}_i$. Then the set X is the disjoint union $\cup_\mu X_\mu$, where $\bar{\mu} = (\bar{y}_1, \ldots, \bar{y}_n) \in \bar{X}$ and X_μ is defined as

$$X_\mu = \{(x_1, \ldots, x_n) \in X : x_i = y_i + pz_i,\ z_i \in \mathbf{Z}_p,\ i = 1, \ldots, n\}.$$

With this notation we can now write

$$\int_X |f(x)|_p^s dx = \sum_{\mu \in X_\mu} \int_{X_\mu} |f(x)|_p^s dx = \sum_{\mu \in X_\mu} \int_{\mathbf{Z}_p^n} |f(px+\mu)|_p^s dx$$

$$= \sum_{\mu \in X_\mu} p^{-e_\mu s} \int_{\mathbf{Z}_p^n} |f_\mu(x)|_p^s dx.$$

The determination of the integrals arising from the points μ for which $\bar{\mu} \notin V_f(\mathbf{F}_p)$ are easily computable. The integrals corresponding to the $\mu's$ for which $\bar{\mu}$ is a non-singular point of $V_f(\mathbf{F}_p)$ are calculated by using an appropriate version of Hensel's Lemma together with the implicit function theorem (see [8], p. 177).

Remark. A rather nice proof of the formula for stationary phase and applications can be found in the article by Wilson Zuniga (see [45], §3.)

The p-adic integrals which arise in the applications to exponential sums over finite p-adic rings are of a slightly more general shape. The integrand contains an additional term of the form $\psi(f(x))$; since these are locally constant functions, suitable partitions of the domain of integration will lead to integrals similar to those in Igusa's formula. Let V be a smooth variety defined by polynomial equations over $\mathbf{Z}_p$ of relative dimension n and f a function on V. We suppose that f is a Morse function, i.e., the critical set

$$C_f := \{s \in V : \text{grad}_s f = 0\}$$

is finite and etale[7] set over $\mathbf{Z}_p$. If ψ denotes the canonical additive character on $\mathbf{Q}_p^+$ and Φ denotes a Schwartz function on the $\mathbf{Q}_p$-points of V, then we define a

[7]This is a condition on the relative smoothness of the Galois action of $Gal(\bar{\mathbf{Q}}_p/\mathbf{Q}_p)$.

function F^*_Φ on $\mathbf{Q}_p$ by

$$F^*_\Phi(i^*) = \int_{V(\mathbf{Q}_p)} \Phi(x)\psi(i^* f(x))dx, \quad (i^* \in \mathbf{Q}_p).$$

It is easy to see that if $V = \mathbf{A}^n$ is affine n-space and Φ is the characteristic function of the set $\mathbf{A}^n(\mathbf{Z}_p)$, and f is a polynomial in $\mathbf{Q}_p[x_1, \ldots, x_n]$, then

$$F^*_\Phi(i^*) = p^{-nm} \sum_{i \in \mathbf{A}^n(\mathbf{Z}_p/p^m\mathbf{Z}_p)} \psi(i^* f(i)), \qquad i^* = u \cdot p^{-m},$$

where $u \in \mathbf{Z}_p^\times$.

Several important results of Igusa revolve around the asymptotic expansions of the integrals $F^*_\Phi(i^*)$. Two particularly interesting cases arise as follows. When the critical set $C_f(\mathbf{Z}_p)$ is not empty and $i^* = up^{-m}$ with $m \geq 2$, the integral $F^*_\Phi(i^*)$ can be expressed as a sum of local terms indexed by the finite number of elements in $C_f(\mathbf{Z}_p)$ analogous to those that appear in the classical stationary phase. For example if m is even, we actually get an exact formula

$$\text{(7.)} \qquad \int_{V(\mathbf{Z}_p)} \psi(i^* f(x))dx = \{ \sum_{s \in C_f(\mathbf{Z}_p)} \psi(i^* f(s))\} |i^*|_p^{-n/2}.$$

For the case $i^* = up^m$, $m \geq 3$ odd, we have

$$\text{(8.)} \qquad \int_{V(\mathbf{Z}_p)} \psi(i^* f(x))dx = \{ \sum_{s \in C_f(\mathbf{Z}_p)} \zeta_8^{sgn\, H(s)} \psi(i^* f(s))\} |i^*|_p^{-n/2},$$

where $H(s)$ is the Hessian matrix $(d^2 f)_s$ of f at the critical point $s \in C_f(\mathbf{Z}_p)$, and $\zeta_8^{sgn\, H(s)}$ is just a symbolic way of writing the fourth root of unity whose value is given by the multidimensional Gauss sum

$$\zeta_8^{sgn\, H(s)} = p^{-n/2} \sum_{z \in (\mathbf{Z}_p/p\mathbf{Z}_p)^n} \psi(p^{m-1} i^* H(z)).$$

The extensive work done on these sums in recent years is of particular interest with respect to the determination of the rationality of Igusa's zeta function and to the monodromy properties of the set $C_f(\mathbf{Z}_p)$. See Denef's Bourbaki Report [10].

Error Correcting Codes. There are many applications of the theory of exponential sums to error correcting codes, particularly to their weight distribution. Among these, the recent use of quadratic Gaussian sums in the study of codes over the quaternary ring $R = \mathbf{Z}/4\mathbf{Z}$ is the most unusual and promisses to be a very fertile ground for further investigation of exponential sums. We shall describe below how these sums appear in the study of one particular, but very remarkable, class of error correcting codes over R -the Preparata and Kerdock codes. First we make some observations which are special to codes over the quaternary ring R.

The Gray Map and the Lee Weight. [8] The expression that appears in Gauss's evaluation of the sum in (1.)

$$\varepsilon(n) := \frac{1}{\sqrt{n}} \sum_{s \in \mathbf{Z}/n\mathbf{Z}} e^{-2\pi i s^2/n} = \frac{1+i^n}{1+i},$$

defines a complex valued function on the quaternary ring

$$\varepsilon(n) : \mathbf{Z}/n\mathbf{Z} \longrightarrow \mathbf{C}$$

with some very nice properties. We list below its table of values

$$\begin{array}{ll} \varepsilon(0) = 1 - i & |\varepsilon(0)|^2 = 2 \\ \varepsilon(1) = 1 & |\varepsilon(1)|^2 = 1 \\ \varepsilon(2) = 0 & |\varepsilon(2)|^2 = 0 \\ \varepsilon(0) = \frac{1}{i} & |\varepsilon(3)|^2 = 1. \end{array}$$

The **Gray map** is defined by the function

$$\varphi : \mathbf{Z}/4\mathbf{Z} \longrightarrow (\mathbf{Z}/2\mathbf{Z}) \times (\mathbf{Z}/2\mathbf{Z})$$

whose values are given by

$$\begin{array}{l} \varphi(0) = (0,0) \\ \varphi(1) = (0,1) \\ \varphi(2) = (1,1) \\ \varphi(3) = (1,0). \end{array}$$

This is the most natural map that allows one to construct binary codes from codes over the quaternary ring R. Finally we give the definition of the **Lee weight**, a distance measuring function for spaces over R which behaves almost as " *the square of the Euclidcan distancc*":

$$\ell : \mathbf{Z}/4\mathbf{Z} \longrightarrow \{0,1,2\}$$

whose values are given by

$$\begin{array}{l} \ell(0) = 0 \\ \ell(1) = 1 \\ \ell(2) = 2) \\ \ell(3) = 1. \end{array}$$

It is not difficult to derive the following relation between the function $\varepsilon(n)$ and the Lee weight

$$\ell(a) = |\varepsilon(a+2)|^2 = \frac{2 - i^a - i^{\ a}}{2}.$$

Note that this is simply the real part of an eighth root of unity. We now proceed to describe the Preparata code.

There are several ways of describing the Preparata code. The one given here is due to Duursma and has the merit that it brings forth its true progeny, the Witt

[8] According to Assmus, "It has not been generally recognized that the Hamming and Lee weights are both aspects of the same general phenomenon. A 'weight' is defined by choosing a subgroup of the multiplicative group of the underlying field; the Hamming weight chooses the trivial subgroup, the Lee weight chooses the subgroup of order 2 (when such exists)." (See Assmus' review of *the Mathematical Theory of Coding* by I. Blake and R. Mullin in Siam Review, 1976.)

vectors. We start with the binary field $\mathbf{F}$ of cardinality $N = 2^m$, and consider vectors of length N indexed by the elements of $\mathbf{F}$ with respect to a fixed ordering:

$$c = (c_x)_{x \in \mathbf{F}}.$$

whose coordinates c_x are elements in the quaternary ring $R = \mathbf{Z}/4\mathbf{Z}$. The collection of all such vectors can be viewed simply as a free R-module of rank N. The **Preparata Code** $\mathcal{C}_N$ is the subset of this module consisting of vectors $c = (c_x)_{x \in \mathbf{F}}$ whose components satisfy the congruence relation

$$\prod_{x \in \mathbf{F}^\times} (1 + xT)^{c_x} \equiv 1(\text{ mod. } T^3\mathbf{F}[T])$$

together with the parity condition

$$\sum_{x \in \mathbf{F}} c_x = 0, \qquad \text{in } R = \mathbf{Z}/4\mathbf{Z}.$$

Here T is an indeterminate and $\mathbf{F}[T]$ is the polynomial ring in T over the field $\mathbf{F}$. It is not difficult to obtain from this description the dimension over R of this code, and hence also its rate of transmission.

If $c = (c_1, c_2, \ldots, c_N) \in \mathcal{C}_N$ is a code word in the Preparata code, then its Lee weight is simply defined as the sum of the Lee weights of its coordinates: $\ell_{wt}(c) = \sum_{j=1}^{N} \ell(c_j)$. From the description given above of the Lee weight we see that $\ell_{wt}(c) = N - Re(Q(c))$, where $Q(c)$ is the quadratic Gaussian sum

$$Q(c) = \sum_{j=1}^{N} i^{c_j}.$$

The possibility of investigating this and related sums by purely elementary combinatorial arguments is not quite obvious. We shall concentrate here in exhibiting $Q(c)$ as a true "quadratic Gaussian sum".

The **Kerdock code** $\mathcal{K}_N$ of length N can be defined simply as the dual of the Preparata code in $(\mathbf{Z}/4\mathbf{Z})^N$ with respect to the quadratic form $x \cdot y = x_1y_1 + \ldots + x_Ny_N$. We shall give a different description which, although of a more complicated nature, turns out to be more suitable with respect to the investigation of quadratic Gaussian sums defined over local fields. In this setting we let k be the unique unramified extension of the field of dyadic numbers $\mathbf{Q}_2$ whose residue field is the finite field $\mathbf{F}$ of cardinality $N = 2^m$. Inside k there is a replica of the finite field $\mathbf{F}$, e.g., those elements of k which satisfy the equation $x^{2^m} - x = 0$ and which preserve the cyclic multiplicative nature of $\mathbf{F}^\times$. We call these elements of k the **Teichmuller set** $\mathcal{T}$. If $\mathcal{O}_k$ is the ring of integers of the local field k, we need to consider the residue ring $R_N := \mathcal{O}_k/2^2$. This is basically the elements obtained after truncating the dyadic expansion of the integers in $\mathcal{O}_k$ after the second power of the local uniformizing parameter $t = 2$. On this ring we can define a trace map by letting the Frobenious map σ act on $\mathcal{T}$ by $\sigma(t) = t^2$ and on the local uniformizing parameter by $\sigma(2) = 2$. We denote this trace map by

$$Tr : \mathcal{O}_k/2^2 \longrightarrow \mathbf{Z}/4\mathbf{Z}.$$

The **Kerdock code** $\mathcal{K}_N$ is defined by the set of vectors of length N:

$$\mathcal{K}_N := \Big\{c(a,b) : \quad a \in R_N, \quad b \in \mathbf{Z}/4\mathbf{Z}\Big\},$$

where each code vector $c(a,b) = (c(a,b)_t)_{t\in\mathcal{T}}$ has coordinates indexed by the Teichmuller elements $t \in \mathcal{T}$ and given by

$$c(a,b)_t = Tr(at) + b.$$

After these preparations, and with the definition of the Lee weight indicated earlier,

$$\ell_{wt}(c(a,b)) := \sum_{t\in\mathcal{T}} \ell(c(a,b)_t),$$

we can then see that the Lee weight of a code word has the simple representation

$$\ell_{wt}(c(a,b)) = \frac{2|\mathcal{T}| - Q(c(a,b)) - \bar{Q}(c(a,b))}{2},$$

where $Q(c(a,b))$ is the **quadratic Gaussian sum**

$$Q(c(a,b)) = \sum_{t\in\mathcal{T}} i^{Tr(at)+b}.$$

For lack of space we do not go into the remarkable results that have been developed in recent years concerning codes over the quaternary ring R. The interested reader would do well to start with the excellent catalitic paper by Hammons, Kumar, Calderbank, Sloane, and Solé (Bulletin of the American Mathematical Society, vol. 29(1993),pp. 218-222) which initiated a great deal of activity in this area.

Group Representations. If there were to be an area of mathematics that can claim to be the natural home of the quadratic Gaussian sums, it would have to be the theory of group represention. It was André Weil's insight that destilled from the monumental work of Carl Ludwig Siegel on quadratic forms, the presense of a remarkable group, **the Metapletic group**, which controls a great deal of the relations behind the automorphic nature of theta functions and Eisenstein series. We shall describe here the simplest possible manifestation of this phenomenon by indicating how the quadratic Gaussian sums enter into the definition of the Weil representation on the group SL_2 of 2×2 matrices with entries in a local field.

Let k be a local field, which for the purposes of illustration can be taken to be a finite extension of the field of p-adic numbers $\mathbf{Q}_p$. Let $\mathcal{O}_k$ be its ring of integers, and $\underline{p}$ its maximal ideal; we denote by $\mathbf{F}_k = \mathcal{O}_k/\underline{p}$ its residue class field, whose cardinality is denoted by $N(\underline{p})$. For $x \in k$, we denote its absolute value by $|x| = N(x)^{ord(x)}$. Let ψ denote a non-trivial additive character of k, and note that any other additive character is obtained by twisting by elements $a \in k$: $\psi_a(x) = \psi(ax)$. Let K/k be a quadratic extension, and denote its ring of integers by $\mathcal{O}_K$ and its maximal ideal by $\underline{P}$. If σ is the non-trivial automorphism of K/k, $N(x) = N_{K/k}(x) = x\cdot x^\sigma$ is the norm map, which when viewed as defined on the two dimensional vector space K over k is a non-degenerate quadratic form whose associate symmetric bilinear form is

$$B(x,y) = N(x+y) - N(x) - N(y) = Tr(x\cdot y^\sigma).$$

Associated with the quadratic extension K/k there is a unique character χ_K defined on $k^\times$, and characterized by the property that its kernel coincides with those elements in $k^\times$ which are norms from K: $\ker\chi_K = N_{K/k}(K^\times)$. If $\mathcal{D} = \mathcal{D}_{K/k}$ is the relative differrent of K/k, its norm $D = N(\mathcal{D})$ is the discriminant of K/k which also happens to be the conductor of χ_K. Viewing K as a topological vector space over k, we define its Schwarz space $\mathcal{S}(K)$ to be the space of locally constant, complex valued functions on K of compact support. We define the twisted Fourier transform on $\mathcal{S}(K)$ by

$$\hat{h}(x) = \int_K \psi(B(x,y))h(y)dy,$$

where dy is the unique Haar measure on K such that $\hat{\hat{h}}(x) = h(-x)$. Note that it depends on the choice of ψ. The Fourier transform is unitary with respect to the L^2-norm defined by dy and is stable on $\mathcal{S}(K)$. Weil has defined the relevant quadratic Gaussian sum as the Fourier transform ot the character of second degree associated to the norm map $N_{K/k} \longrightarrow k$ (see p. 19, Theoreme 2. in A. Weil, " Sur Certains Groupes D'operateurs Unitaires", Acta Math. vol. 111(1964), pp. 143-211). For our purposes, we can describe this more simply by attaching to the quadratic extension K/k and the choice of non-trivial additive character ψ, the expression ("**Quadratic Gaussian Sum**")

$$\gamma_K = \lim_{n\to\infty} \int_{\underline{P}^{-n}} \psi(N(x))dx.$$

The evaluation of this constant is facilitated by the fact that the value of the integral in independent of n for n sufficiently large. Another equivalent definition is the one given by Jacquet and Langlands and which is more suitable to the study we present in Part II. Let U_k be the group of units in k, let $n = n(\psi)$ be the conductor of ψ, i.e., the largest integer such that ψ is trivial on the ideal $\underline{p}^{-n}$. Let $m = m(\chi_K)$ be the conductor of χ_K, in this case the relative differental exponent of K/k, the order of the relative different $\mathcal{D}_{K/k}$. Let c be an element in k of order $n(\psi) + m(\chi_K)$. With this notation we can then write

$$\gamma_K = \frac{\int_{U_k} \chi_k(u/c)\psi(u/c)du}{|\int_{U_k} \chi_k(u/c)\psi(u/c)du|}.$$

It is not difficult to show that $\gamma_K^2 = \chi_K(-1)$. We are now in a position to describe the Weil representation

Proposition. *There is a unique representation $\pi(K,\psi)$ of $SL_2(k)$ on $\mathcal{S}(K)$ such that*

(a) $\pi\begin{pmatrix} a & 0 \\ 0 & a^{-1}\end{pmatrix} h(x) = \chi_K(a)|a|h(ax)$

(b) $\pi\begin{pmatrix} 1 & a \\ 0 & 1\end{pmatrix} h(x) = \psi(aN(x))h(x)$

(c) $\pi\begin{pmatrix} 0 & 1 \\ -1 & 0\end{pmatrix} h(x) = \gamma_K\hat{h}(x)$

The representation $\pi(K,\psi)$ is unitary, and extends to a continuous representation of $SL_2(k)$ on $L^2(K)$.

The most remarkable property of the representation $\pi(K,\psi)$ is that it commutes with the action of the subgroup

$$N^1 := \{x \in K : N_{K/k}(x) = 1\}$$

and conjugation by the non-trivial automorphism σ of K/k, the action being defined for $\eta \in N^1$, and $h \in \mathcal{S}(K)$, by $h_\eta(x) = h(\eta x)$ and $h^\sigma(x) = h(x^\sigma)$. If ρ is a character of N^1, the subspace

$$\mathcal{S}(\rho) := \{f \in \mathcal{S}(K) : f(\eta x) = \rho(\eta) f(x), \text{ for all } \eta \in N^1\},$$

is stable under $SL_2(k)$. The resulting representation $\pi_{\mathcal{S}}(K, \psi, \rho)$ of $SL_2(k)$ on $\mathcal{S}(\rho)$ is irreducible when ρ is not of order 2. In the latter case $\mathcal{S}(\rho)$ breakes into the direct sum of two irreducible subspace, the $+1$ and the -1 subspaces under the action of σ. These will be denoted below by $\pi^+(K, \psi, \rho)$ and $\pi^-(K, \psi, \rho)$.

The richness of this construction lies in that it provides a large collection of representations of $SL_2(k)$. In case k is a finite extension of the p-adic field $\mathbf{Q}_p$ with p odd, the tame case, actually all the rrepresentations arise in this way. In the case where k is a finite extension of $\mathbf{Q}_2$, in which case most of the quadratic extensions are wildly ramified, all the representations, except a finite number of exceptinal representations, are also obtained in this way. The exceptional ones are intimately connected with certain A_4- and S_4-extensions of k.

The study, described in the next paragraph, of the possible equivalences between the irreducible components of Weil representations associated to distinct quadratic extensions and different choices of additive characters rests on properties that can be traced back to the properties of quadratic as well as non-quadratic Gaussian sums. This is made quite evident in the statement and proof of the **Plancherel formula** for $SL_2(k)$, and certainly in the formulation given by Sally and Shalika in which the form of the characters of irreducible representations depends on values of Gaussian sums. The internal structure of this formula, the individual terms each of which is associated with a conjugacy class of Cartan subgroups, depends on the distinct classes of quadratic extension. This is an aspect of the study of harmonic analysis on SL_2 that will be pursued at greater depth in future publications. The aim will be to separate those components of the Plancherel formula which are parametrized by quadratic extensions of Artin-Screier type (those of even conductor) from those of Kummer type (the ones of odd conductor). It is expected that this study will reveal some of the finer aspect of the Langlands' correspondence, at least as it concerns the group $SL_2(k)$.

EQUIVALENCES. To describe the possible equivalences between the Weil representations, we use the isomorphism

$$N^1/(N^1)^2 \equiv NK^\times/(k^\times)^2, \tag{9.}$$

which allows us to identify a character τ of $NK^\times$ trivial on $(k^\times)^2$ with a character $\rho_{\prime}$ of N^1 of order 2.

The representation $\pi(K, \psi, 1)$ associated to the trivial representation of N^1 is irreducible and belongs to the principal series of representations of $SL_2(k)$. If ρ is a non-trivial character of N^1 of order 2, then both components $\pi^+(K, \psi, \rho)$ and $\pi^-(K, \psi, \rho)$ are irreducible. If ρ is not of order 2, then $\pi(K, \psi, \rho)$ is irreducible.

The following four isomorphisms form a complete set of generators for the isomorphisms between the representations $\pi(K, \psi, \rho)$:

($\mathcal{A}$) $\pi(K, \psi, \rho) \equiv \pi(K, \psi, \rho^{-1})$,
($\mathcal{B}$) $\pi(K, \psi, \rho) \equiv \pi(K, \psi_a, \rho)$ if $a \in NK^\times$,
($\mathcal{C}$) $\pi^+(K, \psi, \rho) \equiv \pi^{sgn}(K, \psi_a, \rho)$, if ρ is of order 2, τ is the corresponding character of $NK^\times/(k^\times)^2$ and $sgn = sign(\tau(a))$.

To describe the fourth isomorphism we need to introduce some notations. Consider the diamond of quadratic extensions of local fields:

$$\begin{array}{ccccc} & & L & & \\ & / & & \backslash & \\ K_1 & & & & K_2 \\ & \backslash & & / & \\ & & k & & \end{array}$$

Since $k^\times/N_{L/k}(L^\times) \equiv (\mathbf{Z}/2)^2$, the following two quotient groups

$$NK_i^\times/(NK_1^\times \cap NK_2^\times), \quad i = 1, 2,$$

are of order 2. Denote by τ_1 and τ_2, respectively their non-trivial characters of order 2. Let ρ_1 (resp. ρ_2) be the character of N_1^1 (resp. N_2^1) of order 2 corresponding to τ_1, τ_2 under the isomorphism (1.) The following statement completes the list of equivalences:

($\mathcal{D}$) With notations and assumptions as above we have

$$\pi^+(K_1, \psi, \rho_1) \equiv \pi^+(K_2, \psi, \rho_2).$$

These are well known results in the representation theory of $SL_2(k)$. (See Casselman [24], p.814, Th. 1.9). Our interest here is with the statement ($\mathcal{D}$) which is closely related to one of the most fundamental (and non-trivial) properties of quadratic Gaussian sums. Because of its intrinsic interest we shall recall below its proof following the elegant argument of Casselman (see [24], pp. 823-824).

Let $K_i^+ = N_{L/K_i}(L^\times)$, $i = 1, 2$, and note that we have an isomorphism

(10.) $$\Omega : K_1^+/\ker(\rho_1) \longrightarrow K_2^+/\ker(\rho_2).$$

Explicitly this is given by sending the class of $x \cdot \ker(\rho_1)$ in K_1^+ to the class of $N_{L/K_2}(y) \cdot \ker(\rho_2)$ in K_2^+ obtained by choosing a $y \in L^\times$ such that $N_{L/K_1}(y) = x$. This group isomorphism induces a topological isomorphism between the two ideals

$$\mathcal{S}^+(\rho_i) = \mathcal{S}(\rho_i) \cdot \mathrm{Char}_{K_I^+}, \quad i = 1, 2,$$

where

$$\mathrm{Char}_{K_i^+} = \frac{1}{2}(1 + \chi_{L/K_i}), \quad i = 1, 2$$

is the characteristic function of the set K_i^+. Here χ_{L/K_i} denotes the character of $K_i^\times$ which by class field theory corresponds to the quadratic extension L/K_i, $i = 1, 2$. By duality, the isomorphism (2.) induces an isomorphism

(11.) $$\Omega^t : \mathcal{D}^+(\rho_2) \longrightarrow \mathcal{D}^+(\rho_1)$$

between the spaces of distributions $\mathcal{D}^+(\rho_2)$ (resp. $\mathcal{D}^+(\rho_1)$) on $\mathcal{S}^+(\rho_2)$ (resp. $\mathcal{S}^+(\rho_1)$).

Note that for $x \in NK_1^\times \cap NK_2^\times$, Ω maps the set $N_{K_1/k}^{-1}(x)$ to the set $N_{K_2/k}^{-1}(x)$. This shows that Ω commutes with the action of $\mathcal{N} = \left\{\begin{pmatrix} 1 & x \\ 0 & 1 \end{pmatrix} : x \in k\right\}$. Since Ω commutes with the action of $A = \left\{\begin{pmatrix} a & 0 \\ 0 & a^{-1} \end{pmatrix} : a \in k^\times\right\}$, to prove ($\mathcal{D}$) it clearly suffices to show that it commutes with the action of $\begin{pmatrix} 0 & 1 \\ -1 & 0 \end{pmatrix}$. This amounts to showing that for λ_2 a quasi-character of $K_2^+/\ker(\rho_2)$, with $\lambda_1 = \lambda_2 \cdot \Omega$ a character of $K_1^+/\ker(\rho_1)$, we have

$$\Omega^t\left\{\pi\begin{pmatrix} 0 & 1 \\ -1 & 0 \end{pmatrix} \cdot \lambda_2\right\} = \pi\begin{pmatrix} 0 & 1 \\ -1 & 0 \end{pmatrix}\Omega^t(\lambda_2). \tag{12.}$$

If K is a local field, χ is a quasi-character of $K^\times$, and f is a locally constant function defined on $K^\times$ with ordinary (not twisted!) Fourier transform $\hat{f}$, then we can write Tate's functional equation in the form

$$\frac{\int_{K^\times} f(x)\chi(x)d^\times x}{L(\chi)} = \varepsilon(\chi, \psi)\frac{\int_{K^\times} \hat{f}(x)\hat{\chi}(x)d^\times x}{L(\hat{\chi})}, \tag{Tate's FE}$$

where $\hat{\chi}(x) = |x|\chi(x)^{-1}$ and $L(\chi)$ (resp. $L(\hat{\chi})$) is the local L-factor associated to χ (resp. $\hat{\chi}$). (See Deligne [11], §3, (3.3.1)). If we denote by $\mathcal{L}_\chi$ the distribution defined by χ, then we can write Tate's functional equation in the equivalent form

$$\mathcal{L}_\chi(f) = \Gamma(\chi, \psi)\mathcal{L}_{\hat{\chi}}(\hat{f}), \tag{Tate's FE*}$$

with

$$\Gamma(\chi) := \Gamma(\chi, \psi) = \varepsilon(\chi, \psi)\frac{L(\chi)}{L(\hat{\chi})}$$

denoting the "*standard p-adic Gamma function*". (See Sally-Tableson [22]).

We note that since $\mathcal{S}^+(\rho_i)$, $i = 1, 2$, is stable under the action of $G = SL_2(k)$, and since any f in $\mathcal{S}^+(\rho_i)$ corresponds to a function on $\mathcal{S}(\rho_i)$ with support on $K_i^+ = \ker(\chi_{L/K_i})$, the Fourier transform of an f in $\mathcal{S}^+(\rho_i)$ has its support $supp(\hat{f})$ also contained in K_i^+. This implies that

$$\Gamma(\lambda_i \cdot \chi_{L/K_i}) = \Gamma(\lambda_i), \quad i = 1, 2.$$

To see this apply (Tate's FE^*) to λ_i and a function f in $\mathcal{S}^+(\rho_i)$, noting that $\chi_{L/K_i}^{-1} = \chi_{L/K_i} = 1$ on K_i^+.

We now apply these ideas to the distributions $\mathcal{L}_i$ defined respectively by integration against the functions

$$\lambda_i \cdot \mathrm{Char}_{K_i^+} = \lambda_i \cdot \frac{1}{2}(1 + \chi_{L/K_i}), \quad i = 1, 2.$$

If we denote by $\mathcal{L}_i^*$ the distribution defined by integration against the function

$$\hat{\lambda}_i(x^{\sigma_i})\mathrm{Char}_{K_i^+}(x^{\sigma_i}), \quad i = 1, 2,$$

(here σ_i is the non-trivial automorphism of $Gal(K_i/k)$), then we have

$$\pi\begin{pmatrix} 0 & 1 \\ -1 & 0 \end{pmatrix}\mathcal{L}_i = W(\chi_{K_i/k}, \psi)\Gamma(\lambda_i)\mathcal{L}_i^*, \quad i = 1, 2.$$

Since we also have by definition

$$\Omega^t\mathcal{L}_2 = \mathcal{L}_1, \quad \Omega^t\mathcal{L}_2^* = \mathcal{L}_1^*,$$

it follows that $\pi\begin{pmatrix} 0 & 1 \\ -1 & 0 \end{pmatrix}$ and Ω^t commute if and only if the following identity holds:

$$W(\chi_{K_1/k}, \psi)\Gamma(\lambda_1) = W(\chi_{K_2/k}, \psi)\Gamma(\lambda_2).$$

One checks without much effort that this is a slight modification of **Dwork's Second Identity**:

$$W(\chi_{K_1/k}, \psi_k)W(\lambda_1, \psi_{K_1/k}) = W(\chi_{K_2/k}, \psi_k)W(\lambda_2, \psi_{K_2/k}).$$

Here ψ_k is the standard additive character of k and $\psi_{K_i/k} = \psi_k \circ Tr_{K_i/k}$. (See Dwork [13], p. 462,Th. 4(b); see also Langlands' [16], Second Fundamental Lemma) [9] This completes the proof of the equivalence ($\mathcal{D}$).

The main interest in the above derivation lies in the following question.

Problem 2. *Can one find a character-theoretic proof of* ($\mathcal{D}$) *which does not involve the use of Dwork's Second Identity, or for that matter, the Dwork-Langlands theory of local constants?*

An answer to this question which does not involve analytic techniques like those which appear in Deligne's theory of local constants would help simplify the purely local study of root numbers, and most significantly the part that depends on Dwork's difficult second identity.

There are many other applications of the quadratic Gaussian sum $W(\chi, \psi)$ to the representation theory of $SL_2(k)$ and $GL_2(k)$. The origen of many of these go back to the use of $W(\chi, \psi)$ as a **Weil Index** when describing the action of the Weyl element $w = \begin{pmatrix} 0 & 1 \\ -1 & 0 \end{pmatrix}$ in terms of the Fourier transformation. For example in Langlands' theory of characters and local lifting which provides the crucial results about orbital integrals needed in the study of base change for $GL_2(k)$, essential use is made of $W(\chi, \psi)$ as a normalizing factor for the locally integrable functions which represent the characters associated to representations induced from non-split Cartan subgroups. (See Langlands' [15], Chap.7, p. 130, where $\lambda(K/k, \psi)$ is used to denote $W(\chi, \psi)$.) For similar reasons, the quadratic Gaussian sums appear in the Labesse-Langlands' study of L-indistinguishability for representations of $SL_2(k)$ in the normalization of transfer factors arising from endoscopy. (See Shelstad [44], p. 196).

Another equally important application of the quadratic Gaussian sums $W(\chi, \psi)$ appears in the determination of the character formulas for the representations of $SL_2(k)$, a work which has been carried out in great depth by Sally and Shallika. Our interest in these matters has been greatly motivated by the desire to extract arithmetic information about quadratic extensions from their work on the Plancherel formula (see Sally-Shalika [21], p. 1235, where $H(\Phi, V)$ is used to denote $W(\chi, \psi)$; see also Sally [20], p. 396, where $\kappa(\Phi, V)$ is used to denote the quadratic Gaussian sum). This is quite natural since after all the family of Cartan subgroups of $SL_2(k)$ is parametrized by the factor group $k^\times/(k^\times)^2$ whose harmonic analysis is controled by the quadratic Gaussian sums. There are two aspects of this work which we want to mention here briefly. The first is related to the shape of the Plancherel formula for $SL_2(k)$ and is connected to distinguished classes of Cartan subgroups. For example, if L/k is a biquadratic extension and K_1, K_2, K_3 are the three quadratic extensions of k with corresponding quadratic characters $\chi_i : k^\times \longrightarrow \{\pm 1\}$, $i = 1, 2, 3$, then Witt's well known criterion for embeddability of L/k into a quaternion extension

[9] No complete published proof of Dwork's Second Identity exists in the case of a finite extension $k/\mathbf{Q}_2$.

of k is equivalent to the identity

$$W(\chi_1,\psi)W(\chi_2,\psi)W(\chi_3,\psi) = 1,$$

and this in turn suggests that the terms in the Plancherel formula parametrized by the Cartan subgroups can be grouped into chunks which posses a richer arithmetic structure which is often more evident in the classification of monomial and non-monomial two dimensional representations of the absolute Weil group W_k of k.

The second deals directly with the harmonic analysis on the abelian group

$$\mathcal{W} = k^\times/(k^\times)^2.$$

More explicitly, when $\mathcal{W}$ is viewed as a vector space over the binary field $\mathbf{F}_2$, its dimension is $r = 2 + ef$, where e is the absolute ramification degree and f is the residue class field degree of k. If we write the Hilbert norm residue symbol of k in the form $(a,b)_k = (-1)^{[a,b]}$ with $[a,b] \in \mathbf{F}_2$, then $[a,b]$ is a symmetric bilinear form on $\mathcal{W}$ with values in $\mathbf{F}_2$ and its matrix with respect to a basis can be computed in those cases where explicit local reciprocity laws are known. For example Serre (see *A Course in Arithmetic* [41], p.23) has given the following values for the local completions of $\mathbf{Q}$:

- For $k = \mathbf{R}$ the matrix is simply (1),
- For $k = \mathbf{Q}_p$, $p \neq 2$, taking as generators for $\mathcal{W}$, p, u, where u is a quadratic non-residue, the matrix is $\begin{pmatrix} 0 & 1 \\ 1 & 0 \end{pmatrix}$ if $p \equiv 1 \pmod 4$, and $\begin{pmatrix} 1 & 1 \\ 1 & 0 \end{pmatrix}$ if $p \equiv 3 \pmod 4$.
- For $k = \mathbf{Q}_2$ with basis $\{2, -1, 5\}$, the matrix is $\begin{pmatrix} 0 & 0 & 1 \\ 0 & 1 & 0 \\ 1 & 0 & 0 \end{pmatrix}$.

If we choose a basis consisting of r elements in $k^\times$, say $g_1, g_2, \ldots, g_r$, the study of the spectral and arithmetic properties of the matrix $\mathcal{M} = ([g_i, g_j])$, $i, j = 1, 2, \ldots, r$, for arbitrary local fields is of some interest. A related problem, which is of fundamental importance for the representation theory of reductive algebraic groups over local fields arises from Kazhdan's study of character identities under lifting and their relation to the local character expansions of admissible irreducible representations near semi-simple elements. The question can be simply stated in the following terms:

Problem 3. *Study the spectral and arithmetic properties of the character table of the group $\mathcal{W}$ for a local field k of residue class field characteristic 2.*

The problem is equivalent to the study of the $r \times r$ matrix whose entries are Hilbert residue symbols $\mathcal{M} = \Big((g_i, g_j)_k\Big)$. In the tame case a great deal of knowlege is available concerning $\mathcal{M}$. For a very explicit description of the relevant results concerning *local character expansions* **in the tame case** $p \neq 2$ and their relation to the theory of the matrix $\mathcal{M}$, the reader would do well to consult the very interesting paper of Magdy Assem (see [47], §3.2) where a partial description of the work of Sally for $SL_2(k)$ is given.

With the above remarks, incomplete and disconnected as they appear, we end our discussion of the unusual applications of quadratic Gaussian sums.

PART II: Local Theory of Quadratic Gaussian Sums

The Sign of the Gauss Sum

Let k be a finite (local) field extension of the field of dyadic numbers $\mathbf{Q}_2$ (the completion of the field of rational numbers $\mathbf{Q}$ with respect to the dyadic valuation $ord_2(r) = s$, where $r = r'2^s$ and r' is the quotient of two odd integers. Let $\pi = \pi_k$ be a local uniformizing parameter, $\mathcal{O}_k$ be the local ring of integers in k, $P = P_k$ the maximal ideal of $\mathcal{O}_k$. We denote by $\mathcal{T}$ the Teichmuller set of k, a lifting to k of the multiplicative group of the residue class field with the zero element of k adjoined. We denote by $NP_k = q$ the cardinality of the residue class field of k and by e the absolute ramification index of k over $\mathbf{Q}_2$. We have chosen to concentrate on the case $p = 2$ because for quadratic reciprocity law this is the most difficult case. We also hope this case will lead to interesting applications in coding theory and combinatorics. For all the relevant information concerning local fields we refer the reader to Andre Weil's well known monograph "*Basic Number Theory*". The principal object of concern will be the finite ring

$$R_k = \mathcal{O}_k / P_k^m,$$

where the exponent m will be refered as the **conductor**. Note that $m = 1$ refers to the residue class field $\mathbf{F}_k$. The quadratic Gauss sums we are interested in were first introduced and studied in this generality by Hecke (see [8], chap. VIII,p. 197), in his analytic study of relative quadratic extensions of number fields. To define them we consider an arbitrary continuous (quasi-) character $\chi : k^\times \longrightarrow \mathbf{C}^\times$ of conductor $m = 2d + \epsilon$, where ϵ is 0 if m is even and 1 otherwise. Recall that the differential exponent δ of k is defined so that the inverse of the different ideal $\mathcal{D} = (\pi^\delta)$ coincides with the ideal $\mathcal{D}^{-1} = \{x \in k : Tr_{k/\mathbf{Q}_2}(x \cdot y) \in \mathbf{Z}, \text{ for all } y \in \mathcal{O}_k\}$, the orthogonal complement of $\mathcal{O}_k$ with respect to the trace bilinear form. Associated with $\mathcal{D}$ we have the standard additive character $\psi : k^+ \longrightarrow \mathbf{C}^\times$ whose conductor is $\mathcal{D}$, i.e. $\psi(x) = 1$ if $x \in \mathcal{D}^{-1}$ and $\psi(x) \neq 1$ for some $x \notin \mathcal{D}^{-1}$.

DEFINITION. *The root number associated to χ and ψ is the sum*

$$W(\chi, \psi) = q^{-m/2} \sum_u \chi(u/\gamma)\psi(u/\gamma), \tag{13.}$$

where u runs through a complete set of coset representatives for the quotient of unit groups $\mathcal{O}_k^\times/(1 + P_k^m)$ and $\gamma = \pi^{\delta+m}$.

The study of these sums requires some knowledge of the structure of the group of units of the ring R_k. When k is an unramified extension of $\mathbf{Q}_2$, the necessary results can be found in the monograph of Macdonald (see [33], chap. XVI). This is often sufficient for applications to combinatorics and error correcting codes. In general the structure theory of finite rings like R_k follows from the work of Krull (see Lanprecht [17], Theorem 5). In all of these approaches the essential idea is the existence of a Jordan-Holder series

$$\{0\} \subset I^m \subset I^{m-1} \subset \cdots \subset I^1 \subset R_k,$$

which is symmetric with respect to the involution induced by the map of ideal $I^j \mapsto Ann_R(I^j)$. In our particular situation, in which we do not assume that the field k is unramified, the required information follows from the fact that a principal unit u in U^1, i.e. a unit which satisfies $u \equiv 1$ mod. P_k, can be written in the form

$$u \equiv \sum_{0 \leq i < d} t_i \pi^i + \delta_{1,\epsilon} t_d \pi^d + \sum_{d+\epsilon \leq i < m} t_i \pi^i \text{ mod. } P^m,$$

where $\delta_{1,\epsilon} = 1$ if $\epsilon = 1$ and 0 otherwise. The coefficients t_i are taken to lie in the Teichmuller set $\mathcal{T}$. This provides a decomposition of the quotient of unit groups:

$$U^1/U^m \simeq U^1/U^d \times P^d/P^{d+\epsilon} \times P^{d+\epsilon}/P^m,$$

again with the understanding that the middle factor is empty if $\epsilon = 0$.

Using the fact that χ is of conductor m, i.e. is trivial on U^m, we obtain that the map

$$z \mapsto \chi(1+z), \qquad z \in P^{d+\epsilon},$$

defines a non-trivial character on the ideal $P^{d+\epsilon}$. If $\gamma = \pi^{\delta+m}$, then both additive characters $\chi(1+z)$ and $\psi(z/\gamma)$ defined on $P^{d+\epsilon}$ have the same conductor m, and the duality theory of local fields implies there exists a unit $\alpha \in U_k$, unique modulo P^d, such that

$$\chi(1+z) = \psi(\frac{\alpha z}{\gamma}), \qquad z \in P^{d+\epsilon}.$$

In our subsequent discussion we shall refer to the constant α as the **Dwork parameter of** χ. From the product decomposition (3) and the orthogonality relation for character sums in finite abelian groups, a direct calculation leads to the following expression for $W(\chi, \psi)$ which is the central result of the Dwork-Lamprecht theory:

$$\text{(14.)} \qquad q^{\epsilon/2} W(\chi, \psi) = \chi(\alpha/\gamma)^{-1} \psi(\alpha/\gamma) \sum_{x \in \mathcal{O}_k/P_k^\epsilon} \theta(x),$$

where θ is the map defined by $x \mapsto \psi(\alpha x \pi^d/\gamma)\chi(1 + x\pi^d)^{-1}$. Of course, if $\epsilon = 0$, we take the value of the sum to be 1, and in this case the equality (4) is simply $W(\chi, \psi) = \chi(\frac{\alpha}{\gamma})^{-1}\psi(\frac{\alpha}{\gamma})$. This simple fact generalizes the well known result that the Gauss sum (1) is easy to evaluate when n is a square. The interesting case is then, as Hecke, and before him Kummer, had realized, when the conductor m is odd: $\epsilon = 1$. In this situation the function θ defines what A. Weil calls a "**character of second degree**" on $\mathcal{O}_k/P_k = \mathbf{F}_k$:

$$\theta(x+y) = \theta(x)\theta(y) \cdot \psi(\frac{\alpha}{\gamma}\pi^{2d} xy).$$

(See Weil [28], p.145). The sum

$$\Theta(\theta) = \sum_{x \in \mathbf{F}_k} \theta(x),$$

which appears in the expression for $W(\chi, \psi)$ in (4) satisfies a Hasse-Davenport relation which reduces its evaluation to the case of the binary field $\mathbf{F}_2 = \mathbf{z}/2\mathbf{z}$.

Let us now specialize the above considerations to the quadratic case. Let $K = k(\sqrt{\varepsilon})$ be a quadratic extension of k with Galois group $Gal(K/k) = \{1, \tau\}$, and let $\chi_\varepsilon : k^\times \longrightarrow \mathbf{C}^\times$ be the character of order 2 which corresponds by local class field theory to the extension K/k. As is well known, χ_ε coresponds to the Hilbert norm residue symbol of k

$$\chi_\varepsilon(a) = (\varepsilon, a)_k,$$

whose value is 1 if the diophantine equation $\varepsilon x^2 + a y^2 = 1$ has a solution in k and is -1 otherwise.

Remark. The Hilbert norm residue symbol $(\varepsilon, a)_k$ can also be defined as having the value 1 if the quaternion algebra

$$\mathcal{A} := \left\{ \begin{pmatrix} x & ya \\ -y^\tau & x^\tau \end{pmatrix} : x, y \in K = k(\sqrt{\varepsilon}) \right\}$$

is split, and equal -1 otherwise.

If $\varepsilon = \pi$, a local uniformizing parameter for k, then the quadratic character χ_ε has conductor $m(\chi_\varepsilon) = 2e + 1$, and this is the only situation when and odd conductor can occur. Otherwise the conductor of χ_ε, for ε a unit in k, is even and equal to $m(\chi_\varepsilon) = 2(e - s)$, where $s = ord_k(\varepsilon - 1)$.

By an easy modification of the counting arguments given by Serre ([42]), Tunnell ([43], p.79), together with the exercises 3., 4., 5. in (Serre [40], p. 191) we can prove the following result. Let k be a dyadic local field, that is a finite extension of $\mathbf{Q}_2$ of absolute ramification index e. As ususal let i be a non-negative integer and put

$$U_k^i := \{u \in \mathcal{O}_k^\times : u \equiv 1 \text{ mod. } P_k^i\}.$$

We let $\mathbf{F}_q$ be the residue class field of k. Let π be a fixed local uniformizing parameter for k.

Proposition. *If t is an odd integer and $1 \leq t < 2e$ and $\varepsilon \in U_k^{2e-t}$, then the Artin Schreier equation*

$$x^2 - x = a_\varepsilon, \qquad (a_\varepsilon = \frac{\varepsilon - 1}{4}),$$

defines a totaly ramified quadratic extension $K = k(\sqrt{\varepsilon})$ of k and the ramification filtration of $Gal(K/k) = \{\pm 1\}$ is given by

$$Gal(K/k) = G_{-1} = G_0 = \cdots = G_t \not\supseteq G_{t+1} = \{1\},$$

so that the conductor of the associated quadratic character

$$\chi_\varepsilon : k^\times / N(K^\times) \longrightarrow \{\pm\}$$

is given by $m(\chi_\varepsilon) = 1 + t = 2d \leq 2e$. The number of such quadratic extensions is q^{d-1}, and counting the trivial extension and the unramified, there is a total of $2q^e$ extensions of even conductor. The character χ_ε is given explicitely by the Hilbert norm residue symbol $\chi_\varepsilon(x) = (\varepsilon, x)_k$. If ε is a unit, the quadratic extension $K = k(\sqrt{\varepsilon\pi})$ has ramification filtration

$$Gal(K/k) = G_{-1} = G_0 = \cdots = G_{2e} \not\supseteq G_{2e+1} = \{1\}$$

and the associated quadratic character

$$\chi_{\varepsilon\pi} : k^\times / N(K^\times) \longrightarrow \{\pm 1\}$$

has conductor $m(\chi_{\varepsilon\pi}) = 2e + 1$. The number of such extensions is $2q^e$.

Since, as is well known from local class field theory, the total number of quadratic extensions of k, including the trivial one, is equal to the order of the group $k^\times/(k^\times)^2$ which is $4q^e$, about half of the quadratic extensions are of Artin-Schreier type and half are of Kummer type. More precisely, in the notation of the proposition, if $\chi_\pi(x) = (\pi, x)_k$ is the quadratic character of the Kummer extension $K = k(\sqrt{\pi})$, all quadratic characters of k of conductor $2e+1$ are obtained from χ_π by twisting by characters χ' of even conductor, that is to say they are of the form

$\chi_\pi \cdot \chi'$. This has the important consequence that all the quadratic characters of k of odd conductor have the same character of the second degree Θ_π, namely the character of second degree associated to χ_π. This in turn implies that the theta sums associated to all the quadratic characters of odd conductor of k are all the same. It is not difficult to show[10] that in the quadratic case

$$W(\chi_\varepsilon, \psi)^2 = (\varepsilon, \varepsilon)_k = \chi_\varepsilon(-1).$$

Thus the latter value is 1 if $\sqrt{-1}$ belongs to k and -1 otherwise. This gives a very precise value to the square root $\sqrt{(\varepsilon, \varepsilon)_k}$: it is 1 if $(\varepsilon, \varepsilon)_k = 1$ and $i = \sqrt{-1}$ if $(\varepsilon, \varepsilon)_k = -1$. We can thus formulate :

The Problem of the Sign of the Gauss Sum: *To determine the sign of the expression*

$$W(\chi_\varepsilon, \psi) = \pm\sqrt{(\varepsilon, \varepsilon)_k}$$

in terms of arithmetic quantities associated to k and ε.

This is by far a more difficult problem than that solved in Gauss evaluation of (1). As Hecke appears to have realized, the determination of the sign lies deeper than the law of quadratic reciprocity for number fields. Hecke had in fact noted (see [8], p. 218) that for the proof of the appropriate generalization, what is needed is a relation of the form

$$\frac{W(\chi_\varepsilon, \psi)W(\chi_{\varepsilon'}, \psi)}{W(\chi_{\varepsilon\varepsilon'}, \psi)} = (\varepsilon, \varepsilon')_k, \tag{15.}$$

as well as its global analogues (see Tate [34], p.127 for a thorough discussion of (5)).

Status of the problem. The root numbers $W(\chi, \psi)$ play a central role in the local theory of local constants developped by Dwork, Langlands, and Deligne. The study of root numbers for characters of order two for extensions k of $\mathbf{Q}_p$ with p odd, a case which is relatively tame in any sense of the word, was solved completely by Dwork in his thesis in which he gave an exact formula (see [3], p.95 formula 2(a)). This formula will be recalled below. In the general case, Dwork and Langlands did not determine explicitly the value of the Dwork parameter α, but carried out a detailed and rather lengthy determination of the functorial properties of $\alpha = \alpha(\chi)$ under twisting and composition with the norm map. This was sufficient for the needs of the theory of local constants, but does not provide in any obvious sense a solution to the problem of "**the sign of the Gauss sum**". In joint work with Aaron Wan[11], the present authors have found a determination of the Dwork parameter $\alpha(\chi)$ for all quadratic characters of a local field k. Our approach is based on the version of the Shafarevitch-Hasse-Kneser explicit quadratic reciprocity laws developped by Shiratani in [36]. It is expected that the explicit formula found for α will lead to a simple practical determination of the actual sign. These researches will be described else where.

As an illustration of the results that are possible, we describe below Dwork's formula for the tame case which occurs when p is odd. We also describe the simplest

[10]Simply square the sum defining the root number and use the orthogonality relations to cancel all the terms except those on the main diagonal.

[11]Aaron Wan, PhD. Thesis, CUNY

cases: the Gauss sums associated to the seven quadratic extensions of $\mathbf{Q}_2$, and to the fifteen quadratic extensions of $k = \mathbf{Q}_2(\sqrt{-1})$.

Dwork's Root Number Formula (p odd). Let k be a finite estension of the p-adic field $\mathbf{Q}_p$, with p an odd prime number. Let d be the exponent of the absolute different of k, and e the absolute ramification degree of k. We fix once and for all a local uniformizer $\pi = \pi_k$ for k. Let $\mathbf{F}_q$ be the residue class field of k, and let T be the inertia subfield of k. We fix a choice $\mathcal{T}$ of Teichmuller representatives for $\mathbf{F}_q$ in T and set

$$p = \pi^e \sum_{i=1}^{\infty} \eta_i \pi^i,$$

with $\eta_i \in \mathcal{T}$. Let $\chi(x) = (\pi, x)_k$ be the character corresponding to the extension $K = k(\sqrt{\pi})$. As in earlier sections, we denote by ψ the standard additive character of k. Let the valued of the quadratic Gaussian sum in (1.) for n=p be denoted by

$$g_p = \frac{1}{\sqrt{p}} \sum_{s \in \mathbf{Z}/p\mathbf{Z}} e^{-2\pi i s^2/p} = \frac{1+i^p}{1+i}.$$

Proposition. *With notations and assumptions as above we have*

$$-W(\chi, \psi) = \chi(\eta)(-g_p)^{[T:\mathbf{Q}_p]},$$

where

$$\eta = \sum_{i \geq 0} \eta_i Tr_{K/k}(\pi^{e-d-1+i})$$

Remark. The proof of the above result is given in Dwork's Columbia University Thesis (1954). It is also found in J. Gechter's M.I.T. Thesis (1974) as reported in ([48], p. 36), where it is also stated that Stark has suggested an analytic proof based on the form of the functional equation for ring class characters. The formula is also implicitly derived in some of the work of Frohlich and Martinet. See also recent work of Mbodj [49] and of Szechtman [50]

First Example: Quadratic Gauss Sums over $\mathbf{Q}_2$

In addition to the trivial extension $F = \mathbf{Q}_2$, there are seven distinct quadratic extensions whose associated root numbers we want to describe. We shall denote these by K_i, i=1,2,...,7. Since the ramification index is e=1, aside from the trivial character of $\mathbf{Q}_2^\times$, there is a unique unramified character (of conductor 0), two characters of conductor 2 and four of conductor 3. We shall use the notation of Artin-Tate for the values of $(\varepsilon, a)_{\mathbf{Q}_2}$ ([15], p. 156). We fix the canonical additive character of $\mathbf{Q}_2$ by setting $\psi(x) = e^{2\pi i \lambda(x)}$, where $\lambda(x)$ denotes the rational number obtained by removing from x the integral part.

1. We take the unit $\varepsilon_1 = 1 + 2^2 = 5$ and define $K_1 = \mathbf{Q}_2(\sqrt{5})$. The unramified character χ_1 is given by $\chi_1(2) = -1$ and $\chi_1(u) = 1$ if u is a unit. In this case we clearly have

$$W(\chi_1, \psi) = 1.$$

2. Let $\varepsilon_2 = -1 = 1 + 2 + 2^2 + \cdots$ and put $K_2 = \mathbf{Q}_2(\sqrt{-1})$. The corresponding character χ_2 is of conductor 2 and is defined by $\chi_2(2) = 1$, and $\chi_2(u) = (-1)^{\frac{u-1}{2}}$.

Since the conductor is even and clearly $\chi_2(1+z) = \psi(\frac{z}{2^2})$, for $z \equiv 0(\text{mod. } 2)$, we can take the Dwork parameter to be $\alpha = 1$ and

$$W(\chi_2, \psi) = \chi_2(\frac{1}{2^2})\psi(\frac{1}{2^2}) = i.$$

3. The product $\chi_3 = \chi_1\chi_2$ is the character of conductor 2 associated to the unit $\varepsilon_3 = -5 = 1 + 2 + 2^3 + 2^4 + 2^5 + \cdots$. We take $K_3 = \mathbf{Q}_2(\sqrt{-5})$. As in the previous case we have $\chi_3(1+z) = \psi(\frac{z}{2^2})$, for $z \equiv 0$ (mod. 2) with Dwork parameter $\alpha = 1$ and root number

$$W(\chi_3, \psi) = \chi(\frac{1}{2^2})\psi(\frac{1}{2^2}) = i.$$

4. Let $K_4 = \mathbf{Q}_2(\sqrt{2})$ and note that for u a unit the corresponding character is simply the Legengre symbol $\left(\frac{2}{u}\right)$: that is $\chi_4(u) = (-1)^{\frac{u^2-1}{8}}$. The conductor of this character is 3. We note that $\chi_4(1+z) = \psi(\frac{z}{2^3})$ for $z \equiv 0(mod\, 2^2)$ and the Dwork parameter is again $\alpha = 1$. The character of second degree on $\mathbf{F}_2 = \mathbf{Z}/2\mathbf{Z}$ is given by $\theta : x \mapsto \chi_4(1+2x)\psi(\frac{2x}{2^3}) = i^{-x}$. The root number is given by

$$W(\chi_4, \psi) = \chi_4(\frac{1}{2^3})\psi(\frac{1}{2^3})\Theta(\theta)$$

$$= \chi_4(2)e^{\pi i/4}\frac{1}{\sqrt{2}}(1 + i^{-1}) = \chi_4(2) = 1.$$

This is another way of viewing the well known classical evaluation of the Gauss sum

$$\sum_{\substack{1 \le n < 8 \\ (n,2)=1}} \chi_4(n)e^{2\pi i/8} = 2\sqrt{2}.$$

(See Rademacher [35], p.80).

5. Let $K_5 = \mathbf{Q}_2(\sqrt{10})$. Since $10 = 2 \cdot (1 + 2^2)$, the character $\chi_5(x) = \chi_1(x)\chi_4(x)$ has conductor 3 and Dwork constant $\alpha = 1$. The nature of the character of second degree associated to χ_5 is clearly the same as for χ_4. In fact the root number is given by

$$W(\chi_5, \psi) = W(\chi_1\chi_4, \psi) = \chi_1(\frac{1}{2^3})W(\chi_4, \psi) = -1.$$

6. The product of the unramified character χ_1, the character χ_2 of conductor 2, and the character χ_4 yields a character $\chi_6 = \chi_1\chi_2\chi_4$ of conductor 3 associated to the quadratic extension $K_6 = \mathbf{Q}_2(\sqrt{-10})$. Since $\chi_6(1+z) = \chi_4(1+z) = \psi(\frac{z}{2^3})$ for $z \equiv 0(mod.\, 2^2)$, we can take $\alpha = 1$. The character of second degree is given in this case by $\theta : x \mapsto \chi_6(1+2z)\psi(\frac{2x}{2^3}) = i^x$, and we have

$$W(\chi_6, \psi) = \chi_6(\frac{1}{2^3})\psi(\frac{1}{2^3})(\frac{1+i}{\sqrt{2}}) = -i.$$

7. The character $\chi_7 = \chi_2\chi_4$ is of conductor 3 and has Dwork parameter $\alpha = 1$: $\chi_7(1+z) = \chi_4(1+z) = \psi(\frac{z}{2^3})$ for $z \equiv 0(mod.\, 2^2)$. The character of second degree is $\theta : x \mapsto \chi_7(1+2x)\psi(\frac{2x}{2^3}) = i^x$, and the root number is

$$W(\chi_7, \psi) = \chi_7(\frac{1}{2^3})\psi(\frac{1}{2^3})(\frac{1+i}{\sqrt{2}} = i.$$

There is a more succint and suggestive way of writing the above numerical values which may serve as an indication of what may be happenning in the general situation. If $a, b \in \mathbf{Q}_2^\times$ are written in the form

$$a = 2^\alpha u, \qquad b = 2^\beta v,$$

where u and v are 2-adic units in $U = \mathbf{Z}_2^\times$, then the Hilbert norm residue symbol has a very simple expression

$$(a,b)_2 = (-1)^{\varepsilon(u)\varepsilon(v)+\alpha\omega(v)+\beta\omega(u)}.$$

(See Serre, *Corps Locaux*, Chap. XIV, §4.) Here $\varepsilon, \omega : U/U^3 \longrightarrow \mathbf{Z}/2\mathbf{Z}$ are the homomorphisms defined by

$$\varepsilon(z) \equiv \frac{z-1}{2} \text{ mod. } 2 = \begin{cases} 0 & z \equiv 1 \text{ mod. } 4 \\ 1 & z \equiv 3 \text{ mod. } 4 \end{cases}$$

$$\omega(z) \equiv \frac{z^2-1}{8} \text{ mod. } 2 = \begin{cases} 0 & z \equiv \pm 1 \text{ mod. } 8 \\ 1 & z \equiv \pm 5 \text{ mod. } 8 \end{cases}$$

Let

$$\underbrace{\chi_1, \chi_{-1}, \chi_3, \chi_{-3},}_{\chi_j} \quad \underbrace{\chi_2, \chi_{-2}, \chi_6, \chi_{-6}}_{\chi_{2j}}$$

be the characters corresponding to the elements in $\mathbf{Q}_2^\times/(\mathbf{Q}_2^\times)^2$. We then have

$$W(\chi_j, \psi) = i^{\varepsilon(j)}$$

$$W(\chi_{2j}, \psi) = (-1)^{\omega(j)} i^{\varepsilon(j)}.$$

Here $i = \sqrt{-1}$. These two formulas can be combined to yield the elegant expression

$$\frac{W(\chi_a, \psi)}{\sqrt{(a,a)_2}} = (-1)^{\omega(a')m(\chi_a)},$$

where $m(\chi_a)$ denotes the conductor of χ_a, and $a' = a$ if a is odd and $= \frac{a}{2}$ if a is even.

Remark. The determination of the root numbers for quadratic characters has applications to the study of the root number of two dimensional Galois representations. Of particular interest is the twisting of L-functions of elliptic curves by quadratic extensions of $\mathbf{Q}$. We shall pursue this applications elsewhere.

Second Example: Quadratic Gaussian sums over $\mathbf{Q}_2(i)$ Let $k = \mathbf{Q}_2(i)$ and choose as local uniformizing parameter $\pi = 1 - \sqrt{-1}$. The quadratic character of $k/\mathbf{Q}_2$ has conductor 2 and the different of k is the ideal $\mathcal{D}_k = (\pi^2)$. The quotient group $U_k^1/(U_k^1)^2$ is of order 8, and is generated by the images of the three units

$$\varepsilon_1 = 1 + \pi^4, \quad \varepsilon_2 = 1 + \pi, \quad \varepsilon_3 = 1 + \pi^3.$$

The Artin-Schreier theory shows that the conductors of the characters $\chi_{\varepsilon_1}, \chi_{\varepsilon_2}, \chi_{\varepsilon_3}$ corresponding to the three quadratic extensions $K_1 = k(\sqrt{\varepsilon_1})$, $K_2 = k(\sqrt{\varepsilon_2})$, and $K_3 = k(\sqrt{\varepsilon_3})$ are respectively

$$m(\chi_{\varepsilon_1}) = 0, \quad m(\chi_{\varepsilon_2}) = 4, \quad m(\chi_{\varepsilon_3}) = 2.$$

The conductors of the remaining quadratic characters for the Artin-Schreier extensions are calculated easily from the above values. If $\chi_\pi(x) = (\pi, x)_k$ denotes the character of the Kummer extension $K = k(\sqrt{\pi})$, which is of course of conductor 5, then all the other characters of conductor 5 are obtained from the relation

$\chi_{\varepsilon\pi} = \chi_\pi \cdot \chi_\varepsilon$. As usual, we take for ψ the standard additive character of k. The following is the table of values for the quadratic Gaussian sums for the field k.

character χ	$W(\chi, \psi)$	*conductor* $m(\chi)$
Id	1	
χ_{ε_1}	1	0
χ_{ε_2}	1	4
χ_{ε_3}	-1	2
$\chi_{\varepsilon_2\varepsilon_3}$	1	4
$\chi_{\varepsilon_1\varepsilon_2}$	1	4
$\chi_{\varepsilon_1\varepsilon_3}$	-1	2
$\chi_{\varepsilon_2\varepsilon_2\varepsilon_3}$	1	4
χ_π	1	5
$\chi_{\varepsilon_2\pi}$	1	5
$\chi_{\varepsilon_3\pi}$	-1	5
$\chi_{\varepsilon_2\varepsilon_3\pi}$	1	5
$\chi_{\varepsilon_1\pi}$	-1	5
$\chi_{\varepsilon_1\varepsilon_2\pi}$	-1	5
$\chi_{\varepsilon_1\varepsilon_3\pi}$	1	5
$\chi_{\varepsilon_1\varepsilon_2\varepsilon_3\pi}$	-1	5

This example is special in the sense that the complex unit i already lies in k, and hence $(\varepsilon, \varepsilon)_k = \chi_\varepsilon(-1) = 1$. The details of the calculation will be given elsewhere.

Determination of the Dwork Parameter α. We consider here only the case of characters of odd conductor, the case of even conductor being similar. We already know that there are $2q^e$ characters of conductor $2e+1$ and all of these are twists $\chi \cdot \chi_\pi$ of the basic character $\chi_\pi(x) = (\pi, x)_k$ associated to the extension $K = k(\sqrt{\pi})$ with characters of even conductor $m(\chi) \leq 2e$. In this section we shall derive an explicit value for the Dwork parameter of the root number. A brief indication of the evalution of the theta sum will also be given. The derivation of this section depends on the explicit results of Shiratani, a reformulation of which is given in the Appendix.

Let ψ_k be the standard additive character of k and denote its conductor by $n(\psi_k)$; recall also that $m(\chi_\pi) = 2e+1$. The Dwork parameter $\alpha = \alpha_{\chi_\pi}$ is the number

$$\alpha = \alpha_0 + \alpha_1\pi + \cdots + \alpha_{e-1}\pi^{e-1},$$

such that

$$\chi_\pi(1+z) = \psi_k(\frac{\alpha z}{\gamma}) \qquad \text{for } z \in P_k^{e+1},$$

where $\gamma = \pi^{m(\chi_\pi)+n(\psi_k)}$.

We note that any unit in U_k^{e+1} has a representation of the form

$$1 - z \equiv 1 - a_{e+1}\pi^{e+1} - \cdots a_{2e}\pi^{2e} \text{ mod. } \pi^{2e+1}$$

$$\equiv \prod_{e<j\leq 2e} (1 - a_j\pi^j) \text{ mod. } \pi^{2e+1},$$

where the $a_j's$ are Teichmuller representatives. Thus we have

$$\chi_\pi(1-z) = \prod_{e<j<2e} \chi_\pi(1-a_j\pi^j)\cdot\chi_\pi(1-a_{2e}\pi^{2e}).$$

If $\Pi_K = \sqrt{\pi_k}$ is taken as alocal uniformizing parameter in $K = k(\sqrt{\pi_k})$, we see easily that for j odd and with $a_j = b_j^2$, $(1-a_j\pi^j) = N_{K/k}(1+b_j\Pi_K)$ and therefore $\chi_\pi(1-a_j\pi^j) = 1$. Thus in the expression for $\chi_\pi(1-z)$ we need only keep the factors with j even. The problem is then reduced to finding an α such that

$$\prod_{e<2j<2e} \chi_\pi(1-a_{2j}\pi^{2j})\cdot\chi_\pi(1-a_{2e}\pi^{2e}) = \psi_k(\frac{\alpha z}{\gamma}).$$

We now proceed to find constants b_j such that

$$\chi_\pi(1-a_j\pi^j) = (-1)^{Tr_{T/\mathbf{Q}_2}(b_j a_j}, \quad (e<j<2e).$$

We shall use a simple lemma from dyadic analysis whose proof is reminescent of Lucas' derivation of binomial congruences. First we note that with the dyadic expansion

$$2 = \pi^e\sum_{i=1}^{\infty}\eta_i\pi^i,$$

we can introduce a power series in $\mathcal{O}_K[x]$:

$$w(x) = x^e\sum_{i=1}^{\infty}\eta_i x^i.$$

Lemma. *With notation as above we have*

$$\frac{1}{w(x)^2} \equiv x^{-2e}\sum_{J=0}^{\infty} b_J x^J \ mod.\ (2\mathcal{O}_k[x]),$$

where

$$b_J = \sum_{\substack{J=2^1i_1+\cdots+2^ki_k\\ i_1,\ldots,i_k\geq 0}} \Big(\prod_{j=1}^{k}\eta_{i_j}^{2^j}\Big),$$

and the sum is taken over all solutions in positive integers $i_1, i_2, \ldots$ of the equation $J = 2^1i_1+\cdots+2^ki_k$ with the restriction that $i_1,\ldots,i_k\geq 0$.

Proof. Recalling the dyadic expansion

$$-1 = 1+2+2^2+\cdots,$$

we obtain

$$\begin{aligned} w(x)^{-2} &= x^{-2e}\Big(\sum_{i\geq 0}\eta_i x^i\Big)^{-2} \\ &= x^{-2e}\prod_{j\geq 1}\Big(\sum_{i\geq 0}\eta_i x^i\Big)^{2^j} \\ &\equiv x^{-2e}\prod_{j\geq 1}\Big(\sum_{i\geq 0}\eta_i^{2^j}x^{2^j i}\Big) \text{ mod. } (2\mathcal{O}_k[x]) \\ &\equiv x^{-2e}\sum_{J\geq 0} b_J x^J \text{ mod. } (2\mathcal{O}_k[x]). \end{aligned}$$

The most essential step being the fact that the multinomial coefficients satisfy

$$\binom{2^j}{k_1, k_2, \ldots, k_m} \equiv 0(\text{ mod. } 2), \quad (j \geq 1),$$

provided that all $k_i > 0$, a result that can be proved easily by induction.

The above lemma together with Shiratani's Theorem B (see the Appendix) give for t satisfying $e+1 \leq 2e-2t \leq 2e$,

$$\chi_\pi(1+a\pi^{2e-2t}) = (-1)^{Tr_{T/\mathbf{Q}_2}(b_{2e-2t}a)},$$

where

$$b_{2e-2t} = \sum_{\substack{0 \leq i_1, i_2, \ldots, i_k \\ 2^1 i_1 + 2^2 i_2 + \cdots + 2^k i_k = 2t}} \eta_{i_1}^2 \eta_{i_2}^{2^2} \cdots \eta_{i_k}^{2^k},$$

and

$$2 = (\sum_{j=0}^{\infty} \eta_j \pi^j) \pi^e.$$

This is a straight forward calculation of the constant term in the Laurent expansion of the power series

$$\frac{-ax^{2t}}{w(x)^2(1+ax^{2t})^2},$$

a task that is greatly simplified by the dyadic lemma proved above.

We now proceed to the calculation of Dwork's parameter α for the character χ_π. Since $m(\chi_\pi) = 2e+1$, letting

$$1+z = 1 + z_{e+1}\pi^{e+1} + \cdots + z_{2e}\pi^{2e},$$

we get

$$\chi_\pi(1+z) = \prod_{e+1 \leq s \leq 2e} \chi_\pi(1+z_s\pi^s).$$

However, as noted in the lemma in the Appendix, when s is odd and $b_s^2 = z_s$, we have

$$1 + z_s\pi^s \equiv N_{k(\sqrt{\pi})/k}(1 + b_s\sqrt{\pi}^s)$$

modulo π^{2e+1}. Therefore,

$$\chi_\pi(1+z) = \prod_{e+1 \leq 2j \leq 2e} \chi_\pi(1+z_{2j}\pi^{2j}).$$

Moreover, we have, for odd s, $2e > s \geq e+1$,

$$\psi(\frac{\alpha\pi^s}{\gamma}) = \chi_\pi(1+\pi^s) = 1.$$

Assume $z_s = 0$ when s is odd. Then

$$\psi(\frac{\alpha z}{\gamma}) = \chi_\pi(1+z) = \prod_{e+1 \leq 2j \leq 2e} \chi_\pi(1+z_{2j}\pi^{2j}) = (-1)^{\sum_j Tr(b_{2j}z_{2j})}.$$

Here we have put $Tr = Tr_{T/\mathbf{Q}_2}$. On the other hand we have

$$\psi(\frac{\alpha z}{\gamma}) = \psi_0(Tr_{k/\mathbf{Q}_2}(\frac{\alpha z}{\gamma})) = \pm 1,$$

so that

$$Tr_{k/\mathbf{Q}_2}(\frac{\alpha z}{\gamma}) = \frac{a_{-1}}{2} + a_0 + a_1 2 + \cdots,$$

where $a_i = 0$ or 1. Thus

$$\psi(\frac{\alpha z}{\gamma}) = (-1)^{2\Lambda(Tr_{k/\mathbf{Q}_2}(\frac{\alpha z}{\gamma}))}$$

and this in turn implies that

$$Tr_{k/\mathbf{Q}_2}(\frac{2\alpha z}{\gamma}) \equiv \sum Tr(b_{2j} z_{2j}) \text{ mod. } 2.$$

Since $z = z_{e+1}\pi^{e+1} + \cdots + z_{2e}\pi^{2e}$, we have

$$Tr_{k/\mathbf{Q}_2}(\frac{2\alpha z}{\gamma}) = \sum_{s=e+1}^{2e} Tr_{k/\mathbf{Q}_2}(\frac{2\alpha z_s \pi^s}{\gamma}).$$

Let

$$\alpha = \alpha_0 + \alpha_1 \pi + \cdots + \alpha_{e-1}\pi^{e-1}.$$

Thus we have

$$\begin{aligned}\frac{2\alpha_s \pi^s}{\gamma} &= \frac{\pi^e(\eta_0 + \eta_1\pi + \cdots)(\alpha_0 + \alpha_1\pi + \cdots + \alpha_{e-1}\pi^{e-1}) z_s \pi^s}{\pi^{\delta+2e+1}} \\ &= \frac{\{\sum_{i=0}^{\infty}(\sum_{j=0}^{i} \eta_j \alpha_{i-j})\pi^i\} z_s \pi^{s+e}}{\pi^{\delta+2e+1}} \\ &= \sum_{i=0}^{\infty} z_s (\sum_{j=0}^{i} \eta_j \alpha_{i-j})\pi^{i+s-\delta-e-1}.\end{aligned}$$

Recall that

$$Tr_{k/\mathbf{Q}_2}(\mathcal{D}^{-1}) = \mathbf{Z}_2,$$

so that

$$Tr_{k/\mathbf{Q}_2}(2\mathcal{D}^{-1}) = (2),$$

and therefore

$$Tr_{k/\mathbf{Q}_2}(\pi^{-\delta+e}) \equiv 0 \text{ mod. } 2.$$

Hence, when taking the trace, we only consider

$$i + s - \delta - e - 1 < \delta + e$$

which implies that $i < 2e+1-s$. In other words, modulo 2, the following congruence holds

$$Tr_{k/\mathbf{Q}_2}(\frac{2\alpha z_s \pi^s}{\gamma}) \equiv Tr_{k/\mathbf{Q}_2}(\sum_{i=0}^{2e-s} z_s (\sum_{j=0}^{i} \eta_j \alpha_{i-j}) \frac{\pi^{i+(s-2e)}}{\pi^{\delta-e+1}}.$$

Since we already have

$$Tr_{k/\mathbf{Q}_2}(\frac{2\alpha z}{\gamma}) \equiv \sum_{e+1 \le 2j \le 2e} Tr(b_{2j} z_{2j}) \text{ mod. } 2,$$

it follows that

$$\sum_s Tr(\sum_{i=0}^{2e-s} z_s (\sum_{j=0}^{i} \eta_j \alpha_{i-j}) Tr_{k/T}(\frac{\pi^{i+(\delta-2e)}}{\pi^{\delta-e+1}})) \equiv \sum_s Tr(b_s z_s)$$

modulo 2. Here $e+1 \leq s \leq 2e$ and s is even. (Recall that we assume $z_s = 0$ for odd s). Rewrite $2e - s = \ell$. Then we have

$$\sum_{\ell} Tr(\sum_{i=0}^{\ell} z_{2e-\ell}(\sum_{j=0}^{i} \eta_j \alpha_{i-j}) Tr_{k/T}(\frac{\pi^{-\delta+i}}{\pi^{\delta-e+1}})) \equiv \sum_{\ell} Tr(b_{2e-\ell} z_{2e-\ell})$$

modulo 2. Here $0 \leq \ell \leq e-1$ and ℓ is even. This congruence is equivalent to $\psi(\frac{\alpha z}{\gamma}) = \chi_\pi(1+z)$ when we assume $z_s = 0$ for odd s.

The following is then our main result:

Main Theorem. *With notations and assumptions as above, the coefficients* $\alpha_0, \alpha_1, \ldots, \alpha_{e-1}$ *which define the* **Dwork Parameter** *for* χ_π

$$\alpha := \alpha_0 + \alpha_1 \pi + \cdots + \alpha_{e-1} \pi^{e-1}$$

are determined by solving the triangular system:

$$b_{2e-\ell} = \eta_0 \alpha_0 Tr_{k/T}(\frac{\pi^{-\ell}}{\pi^{\delta-e+1}}) + (\eta_0\alpha_1 + \eta_1\alpha_0) Tr_{k/T}(\frac{\pi^{-\ell+1}}{\pi^{\delta-e+1}}) + \cdots$$

$$+(\eta_0\alpha_\ell + \eta_1\alpha_{\ell-1} + \cdots + \eta_\ell\alpha_0) Tr_{k/T}(\frac{1}{\pi^{\delta-e+1}})$$

for $0 \leq \ell \leq e-1$. *We take* $b_{2e-\ell} = 0$ *when* ℓ *is odd.*

One must of course show that

(1.) The triangular system of linear equations for the α_i is consistent, and
(2.) The solutions $\alpha_0, \alpha_1, \ldots, \alpha_{e-1}$ satisfy

$$\psi(\frac{\alpha z}{\gamma}) - \chi_\pi(1+z)$$

with no particular assumption on z except that it lies in U_k^{e+1}.

We recall that once a local uniformizing parameter π for k has been chosen, all quadratic extensions K of k of odd conductor have

$$\chi_{K/k}(x) = \chi_\pi(x)\chi'(x),$$

with $\chi_\pi(x) = (\pi, x)_k$ and χ' a quadratic character of $k^\times$ of even conductor $m(\chi') \leq 2e$. Therefore all quadratic extensions of k of odd conductor have associted to them a single **character of second degree**, namely the one corresponding to χ_{pi} and which we denote by θ_π.

Denote by θ_k the **standard character of second degree** of k (see Dwork's Thesis, chapter II) and define the associated theta sum by

$$\Theta_\pi = \sum_{x \in \mathcal{O}_k/P_k} \theta_\pi(x).$$

If α is the Dwork parameter of χ_π, and $\psi_k(\alpha x/\pi) = (-1)^{Tr_{T/\mathbf{Q}_2}(\eta x)}$ is satisfied for all integers $x \in \mathcal{O}_T$ (T the inertia field), and β is a unit in U_T such that modulo π we have $\eta \equiv \beta^{-2}$, and

$$\theta_\pi(\beta x) = \theta_k(x)\psi_k(cx),$$

Dwork has shown (see Thesis, chap. II) that

$$\Theta_\pi = \bar{\theta}_k(c)\big(-\frac{1+i}{\sqrt{2}}\big)^f.$$

Since the Dwork parameter for $\chi_{K/k} = \chi_\pi \cdot \chi'$ can be found easily from the Dwork parameter α_π of χ_π and α' of χ', the above results together with formula (14.) give the complete determination of the quadratic Gaussian sum $W(\chi_{K/k}, \psi_k)$ as a function of the parameter c. The details of the proof will be given elsewhere.

Appendix: The Quadratic Hilbert Norm Residue Symbol

Explicit formulas for the norm residue symbolare well known. We collect here a number of results that apply particularly to the quadratic case and are due to Shiratani [36].

Let k be a local field which is a finite extension of $\mathbf{Q}_2$. Let $\mathcal{O}_k$ be its ring of integers, P_k its maximal ideal, $\pi = \pi_k$ a local uniformizing parameter, $\mathbf{F}_k = \mathcal{O}_k/P_k$ the residue class field, $\mathcal{T}_k$ a set of Teichmuller representatives for $\mathbf{F}_k^\times$ in k and $T = \mathbf{Q}_2(\{t : t \in \mathcal{T}\})$ the inertia subfield of k. Let f be the residue class field degree and e the absolute ramification index. We denote by δ_k the differental exponent of k over $\mathbf{Q}_2$.

The quadratic extensions of k are grouped naturaly into two classes: those of even conductor and theose of odd conductor (or equivalently into extensions of Artin-Schreier type or of Kummer type). If K/k is one such quadratic extension and its relative different is denoted by $\mathcal{D}_{K/k} = (\Pi_K^{\delta_{K/k}})$, then we know that $\delta_{K/k} = 1+t$ is even if and only if $\delta_{K/k} \leq 2e$ and $\delta_{K/k} = 1+t$ is odd if and only if $\delta_{K/k} = 2e+1$. If $\delta_{K/k} = 1+t$ is even, then t is odd and $K = k(\sqrt{\varepsilon})$, where ε is a principal unit satisfying

$$\varepsilon \equiv 1 \text{ mod. } \pi^{2e-t}.$$

In this case K/k is also an Artin-Schreier extension generated by the root of the equation $x^2 - x = a_t$, $(a_t = \frac{\varepsilon-1}{4})$, and $ord_k(a_t) = -t$. The Hilbert norm residue symbol of K/k is

$$\chi_\varepsilon : k^\times \longrightarrow \mathbf{C}^\times, \quad \chi_\varepsilon(\alpha) = (\varepsilon, \alpha)_k.$$

Its value is 1 if $\alpha = N_{K/k}(\beta)$ and -1 otherwise. The conductor of χ_ε is

$$m(\chi_\varepsilon) = t + 1 = 2d$$

and $\chi_\varepsilon(\varepsilon') = 1$ for any unit $\varepsilon' \in U_k^{2d}$. If we write $t = 2t' - 1$ and put

$$\Pi_K = (1 - \sqrt{\varepsilon})\pi^{-(e-t')},$$

one verifies readily that Π_K is a local uniformizing parameter for K. We denote its norm by

$$\tilde{\pi}_k = N_{K/k}(\Pi_K),$$

and observe that $ord_k(\tilde{\pi}) = 1$, i.e., $\tilde{\pi}_k$ is a local uniformizing parameter for k. From the representation of 2 as a power series in terms of the local uniformizing parameter with coefficients in $\mathcal{O}_T$

$$2 = \pi_k^e \sum_{i=0}^{\infty} \eta_i \pi^i,$$

we consider the power series in $\mathcal{O}_T[x]$ given by

$$w(x) = x^e \sum_{i=0}^{\infty} \eta_i x^i.$$

Similarly, we associate with the unit $\varepsilon = 1 - a\pi^{2e-t}$, $(a \in \mathcal{T})$ the power series

$$\varepsilon(x) = 1 - ax^{2e-t},$$

and with the unit $\varepsilon' = 1 - a'\pi^{2e-u}$, the series

$$\varepsilon'(x) = 1 - a'\pi^{2e-u}.$$

We now suppose that both t and u are odd integers.

Theorem A. *With notations as above, we have*

$$\chi_\varepsilon(\varepsilon') = (\varepsilon, \varepsilon')_k = (-1)^{Tr_{T/\mathbf{Q}_2}(b)},$$

where

$$b = D_0\Big\{\frac{1-\varepsilon(x)}{\varepsilon(x)} \cdot \frac{1-\varepsilon'(x)}{\varepsilon'(x)} \Big/ w(x)^2\Big\},$$

where $D_0\{f\}$ denotes the coefficient of x^0 in the Laurent series of f in powers of x.

Remark. An anlogous expansion is also available in the case when u is not odd (see Shiratani [36], p. 425, Eqn. (25)).

We next consider the quadratic extension $K = k(\sqrt{\pi})$, which has Hilbert norm residue symbol

$$\chi_\pi : k^\times \longrightarrow \mathbf{C}^\times, \quad \chi_\pi(a) = (\pi, a)_k,$$

of conductor $m(\chi_\pi) = 2e + 1$. Here $\Pi_K = \sqrt{\pi_k}$ serves as a local uniformizing parameter for k and $N_{K/k}(\Pi_K) = -\pi_k$. We observe the following simple lemma.

Lemma. *For $a \in \mathcal{O}_T$, we have*

$$N_{K/k}(1 - a\Pi_K^t) = 1 + a^2\pi^t, \quad t \equiv 1(2).$$

Let $\varepsilon' = 1 - a\pi^{2t}$ and define $\varepsilon(x) = 1 + ax^{2t}$, for $a \in \mathcal{T}$.

Theorem B. *With notations as above we have*

$$\chi_\pi(\varepsilon') = (\pi, 1 - a\pi^{2t})_k = (-1)^{Tr_{T/\mathbf{Q}_2}(b)},$$

where

$$b = D_0\Big\{\frac{1-\varepsilon(x)}{\varepsilon(x)} \Big/ w(x)^2\Big\}.$$

(For a proof see Shiratani [36], p. 429).

Remark. The interested reader will also find in Shiratani's paper a derivation of the Shafarevitch-Hasse-Knesser formula for the Hilbert norm residue symbol in terms of the Artin-Hasse exponential function. (See [36], p. 429 and p. 430).

References:

1. **G. E. Andrews**, *The Theory of Partitions*, Encyclopedia of Math. and its Applications, Addison-Wesley Publ. Company, London, 1976

2. **E. Landau**, *Elementary Number Theory*, Chelsea Publishing Co., New York, 1966.

3. **L.K.Hua**, *Additive Theory of Prime Numbers*, American Mathematical Society, R. Island, 1965.

4. **R. C. Vaughan**, *The Hardy-Litlewood Method*, Cambridge University Press, Cambridge, England, 1981.

5. **V. Guillemin** and **S. Sternberg**, *Geometric Asymptotics*, American Mathematical Society, Providence, R. Island, 1977.

6. **L. Hormander**, *The Analysis of Linear Partial Differential Operators I*, Springer-Verlag, New York, 1983.

7. **J. J. Duistermat**, *The light in the neighborhood of a caustic*, Seminair Bourbaki, 29e annee 1976/77, no. 490, pp. 1-11.

8. **J.-I Igusa**, *A stationary phase formula for p-adic integrals and its applications*, Algebraic Geometry and its Applications, Springer-Verlag, 1994, pp. 175-194.

9. **J.-I Igusa**, *Forms of Higher Degree*, Lecture Notes, Tata Institute of Fundamental Research, 1978, Bombay.

10. **J. Denef**, *Report on Igusa's Zeta Function*, Seminaire Bourbaki, n° 741, Asterisque 201-203(1991) pp. 359-386.

11. **P. Deligne**, *Les Constantes des Equations Fonctionnelles des Fonctions L*, In Modular Functions of One Variable, Lecture Notes in Math., no. 349, Springer-Verlag, 1973, New York.

12. **B. Dwork**, *On the Root Number in the Functional Equation of the Artin-Weil L-function*, (Thesis, Columbia University, July 19, 1954).

13. **B. Dwork**, *On the Artin Root Number*, Amer. Jour. of Math., vol. 78(1956), pp. 444-472.

14. **H. Jacquet and R.P. Langlands**, *Automorphic Forms on GL(2)*, Lecture Notes in Math. no. 114, Springer-Verlag, New York, 1970.

15. **R. P. Langlands**, *Base Change for GL(2)*, Annals of Math. Studies, no. 96, Princeton Univ. Press, 1980, Princeton.

16. **R. P. Langlands**, *On the Functional Equation of the Artin L-functions*, Notes, Yale Univ. (preprint).

17. **E. Lamprecht**, *Allgemeine Theorie der Gaussschien Summen in endlichen kommutativen Ringen*, Math. Nachr., vol. 9(1953), pp. 149-196.

18. **E. Kummer**, *Note sur une expression analogue a la resolvante de Lagrange pour la equation* $z^p = 1$, Atti dell'Accademia Pontifica de Nuovi Lincei VI, (1852-1853), pp. 237-241.

19. **Carlos Julio Moreno**, *Algebraic Curves Over Finite Fields*, Cambridge Tract in Mathematics, No. 97, Cambridge, England 1991.

20. **P. J. Sally**, *Character formulas for* SL_2, Proc. of Symposia in Pure Math., vol. 26(1973), pp. 395-400.

21. **P. J. Sally and J. Shalika**, *The Plancherel Formula for SL(2) over a local filed*, Proc. Nat. Acad. Sci. U.S.A., vol. 63(1969), pp. 661-667.

22. **P. J. Sally and M. H. Tableson**, *Special Functions on Locally Compact Fields*, Acta Math., vol. 116(1966), pp. 279-309.

23. **J. A. Shalika**, *A theorem on semi-simple P-adic groups*, Ann. of Math., vol. (19), pp.226-242.

24. **W. Casselman**, *On the representations of* $SL_2(k)$ *related to binary quadratic forms*, Amer. Jour. of Math., vol. 94 (1972), pp. 810-834.

25. **M. Deuring**, *Asymptotische Enwicklungen der Dirichletschen Reihen*, Math. Annalen, vol. 168 (1967), pp. 1-30.

26. **N. Katz**, *An Estimate for Character Sums*, Jour. A.M.S., vol. 2 (1989), 197-200.

27. **E. Hecke**, *Lectures on The Theory of Algebraic Numbers*, Graduate Texts in Math., Springer-Verlag, New York, 1981.

28. **A. Weil**, *Sur Certain Groups D'Operateurs Unitaires*, Acta Mathematica, vol. 111 (1964),143-211.

29. **C. L. Siegel**, *Uber das quadratische Reciprozitatsgesetz in algebraischen Zahlkorpern*, Nachrichten der Akademie der Wissenschaften in Gottingen, Math-phys. Klasse (1960), pp. 1-16.

30. **M. C. Berg**, *The Fourier-Analytic Proof of Quadratic Reciprocity*, John Wiley and Sons, New York, 2000.

31. **T. Kubota**, *A generalized Weil type representation and a function analogous to* e^{-x^2}, Bull. A.M.S., vol. 81 (1975), pp. 902-903.

32. **M. Gromov**, *Soft and Hard Symplectic Geometry*, Proc. International Congress of Math., vol. 2 (1986), pp. 81-98, Berkely.

33. **B. McDonald**, *Finite Rings with Identity*, Pure and Applied Mathematics, Marcel Dekker, Inc, New York, 1974.

34. **J. Tate**, *Local Constants*, in Algebraic Number Fields (L-functions and Galois Properties), Editor A. Frohlich, Academic Press, London, 1977.

35. **H. Rademacher**, *Lectures on Elementary Number Theory*, Blaisdell Publishing Co., New York, 1964.

36. **K. Shiratani**, *On the quadratic norm symbol in local number fields*, J. Math. Soc. Japan, vol. 13, No.4 (1961), 416-430.

37. **E. Artin and J. Tate**, *Class field Theory*, W.A Benjamin, Inc., New York, 1967.

38. **E. F. Assmus and J.D. Key** ,*Designs and their Codes*, Cambridge tracts in Mathematics, vol. 103, Cambridge Univ. Press, Cambridge, England, 1992.

39. **A.R. Hammons, P.V. Kumar, A.R. Calderbank, N.J. A. Sloane, and P. Sole**, *The* $\mathbf{Z}_4$*-linearity of Kedock, Preparata, Goethals, and related codes*, IEEE Trans. Inform. Theory, vol. 40 (1994), 301-319.

40. **J.-P. Serre**, *Corps locaux*, Actualites scientifiques et industrielles 1296, Herman, 1968, Paris.

41. **J.-P. Serre**, *A Course in Arithmetic*, Graduate Texts in Math., Springer-Verlag, 1970, New York.

42. **J.-P. Serre**, *Une "formule de masse" pour les extensions totalment ramifee de degre donne d'un corps local*, C.R. Acad. Sc. Paris, t. 286 (5 juin 1978), Serie A, pp. 1031-1036.

43. **J. Tunnell**, *On the local Langlands' conjecture for GL(2)*, Invent. Math., vol. 46(1978), pp. 179-198.

44. **D. Shelstad**, *Notes on L-indistinguishability*, Proc. of Symposia in Pure Math.,vol. 33(1979), part 2, pp. 193-203.

45. **Wilson Zuniga**, *Igusa's local zeta functions of semiquasihomogeneous polynomials*, (preprint).

46. **R. Livne**, *The average distribution of cubic exponential sums*, Crelle, vol.375/376 (1987), pp. 362-379.

47. **Magdy Assem**, *The Fourier Transform and Some Character Formulae for p-adic* SL_ℓ, ℓ *a Prime*, Amer. Jour. of Math., vol. 116(1994), pp. 1433-1467.

48. **J. Gechter**, *Artin Root Numbers for Real Characters*, Proc. A.M.S., vol. 57, no. 1 (1976), pp.35-38.

49. **O. Mbodj**, *Quadratic Gauss Sums,*, Finite fields Appl. vol. 4(1998), pp. 347-361.

50. **F. Szechtman**, *Quadratic Gauss Sums over Finite Commutative Rings*, Jour. Number Theory, vol. 95 (2002), pp. 1-13.

DEPARTMENT OF MATHEMATICS, CUNY, NEW YORK, NY, 10016-4309
E-mail address: cmoreno@gc.cuny.edu

DEPARTMENT OF MATHEMATICS, CUNY, NEW YORK, NY, 10016-4309
E-mail address: nt1a@yahoo.com